DATE DUE

| | | | |
|---|---|---|---|
| | | | |
| | | | |
| | | | |
| | | | |
| | | | |
| | | | |
| | | | |
| | | | |
| | | | |
| | | | |
| | | | |
| | | | |
| | | | |
| | | | |
| | | | |
| | | | |
| | | | |

# EXPLORATION OF THE DEEP CONTINENTAL CRUST

Edited by
H.-J. Behr, Göttingen
C. B. Raleigh, Palisades

# Observation of the Continental Crust through Drilling I

Proceedings of the International Symposium held in Tarrytown, May 20–25, 1984

Edited by

C. Barry Raleigh

With 211 Figures

Springer-Verlag
Berlin Heidelberg New York Tokyo

Dr. C. BARRY RALEIGH
Lamont-Doherty Geological Observatory
of Columbia University
Palisades, NY 10964, USA

ISBN 3-540-15873-1 Springer-Verlag Berlin Heidelberg New York Tokyo
ISBN 0-387-15873-1 Springer-Verlag New York Heidelberg Berlin Tokyo

Library of Congress Cataloging in Publication Data. Main entry under title: Observation of the continental crust through drilling, I. (Exploration of the deep continental crust) 1. Earth–Crust–Congresses. 2. Boring–Congresses. I. Raleigh, C. Barry. II. Series. QE511.028 1985 551.1'3 85-22175



Offsetprinting: Beltz Offsetdruck, Hemsbach/Bergstraße.
Bookbinding: J. Schäffer OHG, Grünstadt.
2131/3130-543210

# PREFACE

Drilling deep into the earth holds a fascination for earth scientists derived in part from the fact that the drill hole is the ultimate test of a hypothesis. When surface exploration methods have been fully utilized and all the geological inferences drawn about the structure beneath the surface, we must finally drill to sample directly the third dimension of the crust of the earth.

The drill is thus the tool of choice of the energy and minerals resources industry. Because of high cost, drilling has been only sparingly used for solving fundamental problems in the earth sciences. But now, having used the quite sophisticated methodology of exploration geophysics, the exciting structural detail emerging from seismic reflection profiling in particular has led several nations to begin a major program of scientific drilling to solve some of the major problems in the earth sciences.

What is described in this volume are the blueprints for national research programs in France, the Federal Republic of Germany, Japan and the United States. The Soviet Union has already embarked on a major drilling effort, the results of which are soon to be published. Results, of course, are still few, and this first volume is more concerned with the problems to be solved.

Within the U.S.A., some of the principal scientific issues to be addressed by drilling are described here: active major transform faults, vast hydrothermal circulation systems such as Yellowstone, magmatic intrusion into the crust at Long Valley, and sites where drilling would aid greatly in unraveling the history of continental breakup and collision.

Scientific drilling will place demands on the drilling and logging technologies designed primarily for sedimentary basins. A part of this volume is devoted to description of the measurement systems available and to some of the problems surrounding high-temperature drilling. In future volumes, we hope to see published rapid development of techniques for drilling in hostile environments.

This is the first of a series of international symposia to be held on scientific drilling. The programs described here are, at least in three nations, already under way. The next decade will undoubtedly see some of the most exciting discoveries in the exploration of the earth as drilling proceeds.

C. BARRY RALEIGH

CONTENTS

*Review of National Drilling Programs*



*Thermal Regimes*



*Geochemistry and Mineral Resources*



*In Situ Borehole Measurements*

*Advances in Drilling and Logging Technology*



*Recent Results from Scientific Drilling*



*Deep Structures of Continents*

## LIST OF CONTRIBUTORS

You will find the addresses at the beginning of the respective contributions

# REVIEW OF NATIONAL DRILLING PROGRAMS

# A U.S. NATIONAL CONTINENTAL SCIENTIFIC DRILLING PROGRAM

Francis G. Stehli[1]

Geophysical measurements have provided a wealth of indirect information about properties of the crust and have led to exciting hypotheses regarding crustal structure, materials, and deep environments. Direct observation of the crust below its surface has been limited. Industrial drilling has resulted in great numbers of shallow wells, but few have penetrated more than 5 km, and they are localized in sedimentary basins productive of oil or gas. No well in North America has penetrated as deeply as 10 km. Since continental crust is generally thicker than 30 km, it is clear that a vast volume of rock awaits direct observation.

It is clear that many experiments in crustal geology of great scientific interest have been sufficiently defined to indicate that exploration by the drill is an essential next step in achieving better understanding. When man has explored unknown terrains, there have always been surprises, and there will be more in a Program of Continental Scientific Drilling. If our knowledge of the earth's crust – the most important portion of the earth to us who live upon and draw resources from it – is to develop in an orderly way, the time to initiate a carefully considered National Program of Continental Scientific Drilling is at hand.

The need for a national program was foreseen by a number of people and discussed in conferences and reports (Shoemaker 1975; U.S. Geodynamics Committee 1979; Board on Earth Sciences 1983). In 1978, a Continental Scientific Drilling Committee (CSDC) was set up under the leadership of Eugene Shoemaker within the framework of the National Research Council, and it outlined the scope of a possible program. The Committee continues to operate with about 15 members representing many of the disciplines and perspectives found in industry, federal and state geological surveys, government laboratories, and in universities. The Committee itself cannot cover all pertinent fields and concerns in adequate depth, so it has established Panels and ad hoc Task Groups to provide greater breadth. The following Panels and Task Groups are now active in considering potential sites for drilling experiments:

- Panel on Basement Structures and Deep Continental Basins – Jack E. Oliver, Cornell University, Chairman
- Panel on Mineral Resources – J. James Eidel, Coastal Mining Company, Chairman
- Task Group on Tonopah – Roger P. Ashley, U.S.G.S., Menlo Park, Chairman
- Task Group on the Creede Mining District – Phillip M. Bethke, U.S.G.S., Reston, Chairman

[1]Chairman, Continental Scientific Drilling Committee; Dean, College of Geosciences, University of Oklahoma, Norman, OK 73019, USA

Observation of the Continental Crust through Drilling I (ed. by C.B. Raleigh)

Panel on Thermal Regimes – John F. Hermance, Brown University, Chairman

Task Group on Yellowstone – Robert O. Fournier, U.S.G.S., Menlo Park, Chairman

Task Group on the Caldera – Richard F. Taschek, Los Alamos, Chairman

Panel on Downhole Measurements and Fault Zone Drilling – Mark D. Zoback, U.S.G.S., Menlo Park, Chairman

Each Panel has selected one or more sites which it believes should have high priority in a Continental Scientific Drilling Program. At a meeting in Albuquerque in November, 1983, the sites proposed by each Panel were considered by the Continental Scientific Drilling Committee on the basis of the scientific interest of the experiments proposed and feasibility with existing drilling, logging, and measurement techniques and instruments. Highest priority was accorded to a site in the Southern Appalachians close to the COCORP line through the region and designed to test the thin-skinned tectonic hypothesis for construction of the Appalachians. Drilling to the desired depth of about 10 km is feasible with existing technology, and this site offers not only the test of an important hypothesis but also the opportunity to do several secondary experiments as well.

Looking ahead to future, deeper drilling or drilling into hostile thermal or chemical environments, the Committee established a Panel on Drilling, Logging, and Instrumentation Technology, under the chairmanship of Richard Traeger of Sandia National Labs. The work of this Panel indicates that drilling to some of the most exciting future targets, for instance in thermal regimes sites, will require a program of instrument development. Existing logging, coring, and other equipment dependent upon silicon chips, elastomeric seals, and/or electric cables cannot be relied upon at temperatures above about 275°C unless the hole is continuously cooled and few, if any, instruments can withstand waters rich in hostile chemicals, such as hydrogen sulfide. An instrument development program probably should be separate from the Continental Scientific Drilling Program. An additional Panel on Sample Curation and Data Management, chaired by Matt Walton of the Minnesota Geological Survey, is considering how best to conserve, as a national resource, materials and data from wells in a Continental Scientific Drilling Program and how to manage data and materials efficiently.

Once a dedicated scientific hole in the Southern Appalachians had been assigned first priority, a special Task Group on the Southern Appalachians was set up under the chairmanship of Charles Mankin of the Oklahoma Geological Survey, to begin evaluating the drilling and logging program for such a well and the costs associated with it. Though the Task Group is still at work, preliminary information suggests the hole will take about three years to drill and will cost $40-$45 million.

A successful National Continental Scientific Drilling Program must be just that – a national program – in which all the concerned government agencies play a role. Agency missions will, of course, control the roles to a considerable extent. The Department of Energy, for instance, because of its concern with geothermal energy resources, would be expected to play a major role in drilling into thermal regimes targets. The U.S. Geological Survey, with a mission in mineral resource evaluations and a concern about earthquake hazards, would play a role in these areas. The National Science Foundation, with a basic science mission, would be deeply involved in much of the program. To a lesser degree, the Nuclear Regulatory Commission and the Department of Defense may have interests in the program. A promising step has been the development of an Interagency Coordination Committee to draw together the concerned groups and to secure funding for the program. Another fundamental necessity for a successful program is the incorporation into it of the expertise which resides in the industrial community, for it, after all, does almost all the drilling in this country. Ways are now being sought to involve industry into the planning structure of a program. Eventually, a drilling program may find it necessary to operate through a JOIDES-like structure that can integrate planning and develop mechanisms to contract pre-drilling site surveys, drilling, logging, and the like.

Several recent events have focused attention on a National Continental Scientific Drilling Program and have brought closer the hope that the program can be funded. The first of these is the heroic efforts of Wilfred Elders of the University of California at Riverside, to obtain support for deepening a commercial well in the Salton Sea geothermal field, where very high temperatures and metamorphism in progress may be encountered. Support now seems assured through DOE and a drilling contractor and project manager are being selected.

The second step resulted from a briefing to the Office of Science and Technology Policy by Don L. Anderson and Charles L. Drake, in which they presented several possible research initiatives in the earth sciences. As a result of this meeting, OSTP has strongly supported a Continental Scientific Drilling Program as a promising, basic science initiative. In response to advice from OSTP to prioritize the potential drilling sites, CSDC selected the Southern Appalachian hole to recommend for a first attempt.

A third significant event has been the incorporation into the 1985 National Science Foundation budget of funds for preliminary site surveys in connection with siting the Southern Appalachian hole. The next step will be to obtain the funding needed to commence drilling during fiscal year 1986. At the moment, there is much reason to be optimistic about the program and the potential for funding.

With the initiation of a national program likely, it is well to consider some of the problems that must be overcome if it is to be successful. The first is that progress toward an integrated national program involving all interested groups must be continued. The establishment of the Interagency Coordinating Group is a good first step. Secondly, communication of news about the program and research opportunities it represents must be broadened. In part, this can be achieved through symposia, talks at meetings, and through the Drilling Early Warning (DEW) Newsletter, distributed by CSDC, but additional mechanisms, especially for reaching industrial interests, are needed. Thirdly, ways must be devised to fund programs so that preliminary site surveys and selection procedures on several sites can progress concurrently with drilling, sampling, logging, measurements, and establishment of permanently instrumented stations at other sites, while still allowing adequate money for scientific work on the materials and data and for curation and data management. Fourth, independent programs of instrument development must be carried on so that the most exciting and difficult environments can eventually be penetrated. Fifth, we must keep in mind that drilling is a tool of last resort in conducting experiments and that, because of its cost, it should be used only after other methods have been exhausted. Finally, many of the most fascinating problems – that is, those that deal with crustal processes – are of international scope and probably should be considered as international experiments and should be jointly funded and undertaken. Unlike the holes drilled in the Ocean Drilling Program, those on land are always on territory claimed by someone, and integrating an international program will require diplomatic, as well as scientific, innovation for success.

It is important to think where on the entire earth each experiment can be best performed and then to attempt to secure agreements to allow drilling at that place. Continued drilling solely within the confines of the United States may be necessary, but, in the long run, this approach should be followed only if international agreements cannot be negotiated with reasonable efforts.

## References

Board on Earth Sciences (1983) Opportunities for research in the geological sciences. Natl Acad Washington DC, p 95

Shoemaker EM (ed) (1975) Continental drilling. Carnegie Institution of Washington, Washington DC, p 56

US Geodynamics Committee (1979) Continental scientific drilling program. Natl Acad Sci, Washington DC, p 192

# GERMAN APPROACHES TO CONTINENTAL DEEP DRILLING

Eugen Seibold[1] and Hansjuergen Behr[2]

For most earth scientists outside the USSR, plans for deep continental crustal drilling seem unrealistic. It will be even harder to convince money-spending organizations or personalities of their feasibility. Therefore, we ourselves should continually have in mind an introduction to lectures that Patrick Winston from MIT uses to sell Artificial Intelligence: "If you are a skeptic, I want to make you a believer – and if you are a believer, I want to make you a skeptic."

## Aims

As skeptics, we all know that the different catalogues of aims for deep continental drilling that we would like to achieve have some Utopian aspects. This may even be true for two general ideas some of us promote in Germany.

1. Deep continental drilling, as a project for basic science, should be launched as a demonstration that earth sciences need to become as important sciences as some others, which spend much more money, but have a longer tradition of cooperation, such as nuclear physics or space research.

In Germany, we are stimulated by the historical success of the Deep Sea Drilling Project and thus some members of the Senate Commission for Geoscientific Cooperative Research (Geo-Commission) of the Deutsche Forschungsgemeinschaft (DFG), an organization similar to the National Science Foundation of the U.S., began relevant discussions in 1977. A memorandum was presented to our Minister for Science and Technology (BMFT) in July 1980 and was published as Report XI of the Geo-Commission in 1981. It was a very timely memorandum because a Peer Committee of the Ministry was just setting priorities for big projects in basic sciences such as neutron research, heavy ions, elementary particles, and synchroton radiation, and they added a research vessel and the drilling proposal to their discussion.

2. A large project is normally a good tool to focus scientific potential from different specialities and to expose them to critical comments. Whoever was lucky enough to spend some time aboard *Glomar Challenger* will certainly agree.

[1]Deutsche Forschungsgemeinschaft, Kennedyallee 40, 5300 Bonn 2, FRG

[2]Institut und Museum für Geologie und Paläontologie, Goldschmidtstr. 3, 3400 Göttingen, FRG

Observation of the Continental Crust through Drilling I (ed. by C.B. Raleigh)

3. Specific scientific aims and tasks for deep continental drillers are listed in many catalogues. Some topics in *our* list are:

- to explain geophysical crust heterogeneities (as illustrated by seismic velocities, electric resistance, magnetic properties) by petrological studies, in situ measurements and laboratory experiments to better calibrate existing or planned geophysical surveys and to better understand materials and structures of the upper and lower continental crust, whatever these units may mean.
- to investigate gas/fluid/rock interactions in situ and in the laboratory by analyzing crustal permeabilities, constituents, transport of the mobile phases to better understand crustal deformation, metamorphism, magma generation and ore formation, to pay special attention to the thermal regime in general and around 300°C and to try to learn for Hot Dry Rock Projects.
- to measure physical parameters in order to expand our knowledge of stresses in different crust depths, of earthquake mechanisms, and of geotechnical behaviour of the crust.
- to prepare methods to use the bore-hole as deep crust lab as long as possible, measuring suitable geophysical or geochemical parameters in situ repeatedly. Final depths should be more than 10,000 m.

## Site Selection

To meet all these requirements, it would be ideal to drill at different sites with different aims, as it is planned in the USA. We try to concentrate first on a single hole and to select a site to reach the lower continental crust where we can hope to get some results for all of the aims mentioned above. After detailed discussions of some 40 site proposals by several working parties and a reduction of four sites (Figs. 1 and 2), an international workshop in Neustadt/Weinstraße (November 1983) recommended to continue geoscientific investigations in two areas, and to promote further methodological and technological preparations. Up to now, the DFG has spent 4 million DM from special funds of the Ministry (BMFT) on some of these scientific preparations (1981-1983), and will continue to do so. Final decision for the first drill site is planned in late 1985.

Our central European crystalline basement, as part of our roughly 30 km thick crust, is as variegated as political Europe and has the same complicated history.

1. Different elements were consolidated during several Precambrian and during Caledonian and Hercynian cycles (Fig. 3). Parts were repeatedly remobilized as illustrated by sometimes very long ranges in absolute age datings.

2. Most of these rocks are buried in Central Europe or are additionally influenced by Alpine cycles. Bigger outcrops of crystalline rocks are only to be found in areas uplifted and eroded essentially in Cenozoic time, as, for example, in the Black Forest and in the forests bordering the Bohemian Massif, besides the smaller Harz, Spessart, and Odenwald mountains (Fig. 4).

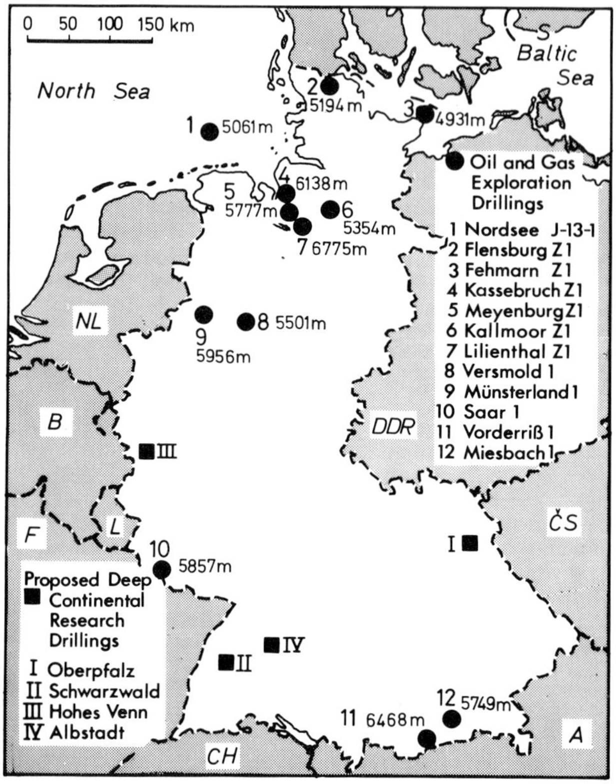


Fig. 1. Location of oil and gas exploration drillings deeper than 5000 m *(dots)*, and locations of discussed deep continental research drill sites *(squares)* in the Federal Republik of Germany. (After Vidal et al. 1981)

Direct information from covered crystalline rocks from oil industry bore-holes penetrating the sedimentary sequences or from xenoliths is rare.

Therefore, in November 1983, the Geocommission and the International Consulting Committee recommended both an eastern (Oberpfalz = Upper Palatinate) and a western site area (Schwarzwald = Black Forest) (Figs. 4 and 5) to be further investigated by continuous seismic reflection and further geological, geophysical, geothermal, geochemical, and petrological exploration. Additionally, shallower holes in both areas should give better estimates of temperature and, eventually, geochemical gradients. The pre-site surveys are in a final stage now.

Up to now, both areas have their pros and cons depending on the main aims to be achieved and we have now a neutral situation because, considering our present state of knowledge, it is easy to predict that both sites will produce surprises. In both areas mineralizations, thermal and $CO_2$ springs are known. In the Black Forest, granulites in the gneisses signal deeper crustal rocks. The central part is a young uplift and very active present circulation of fluids is obvious. But perhaps interesting messages from the mantle were therefore lost.

The Oberpfalz situation is a more sealed one but perhaps also tectonically a more complicated one, situated near the boundary (suture line) of two classic units of the Variscan Central European Cordillera (see Fig. 4).

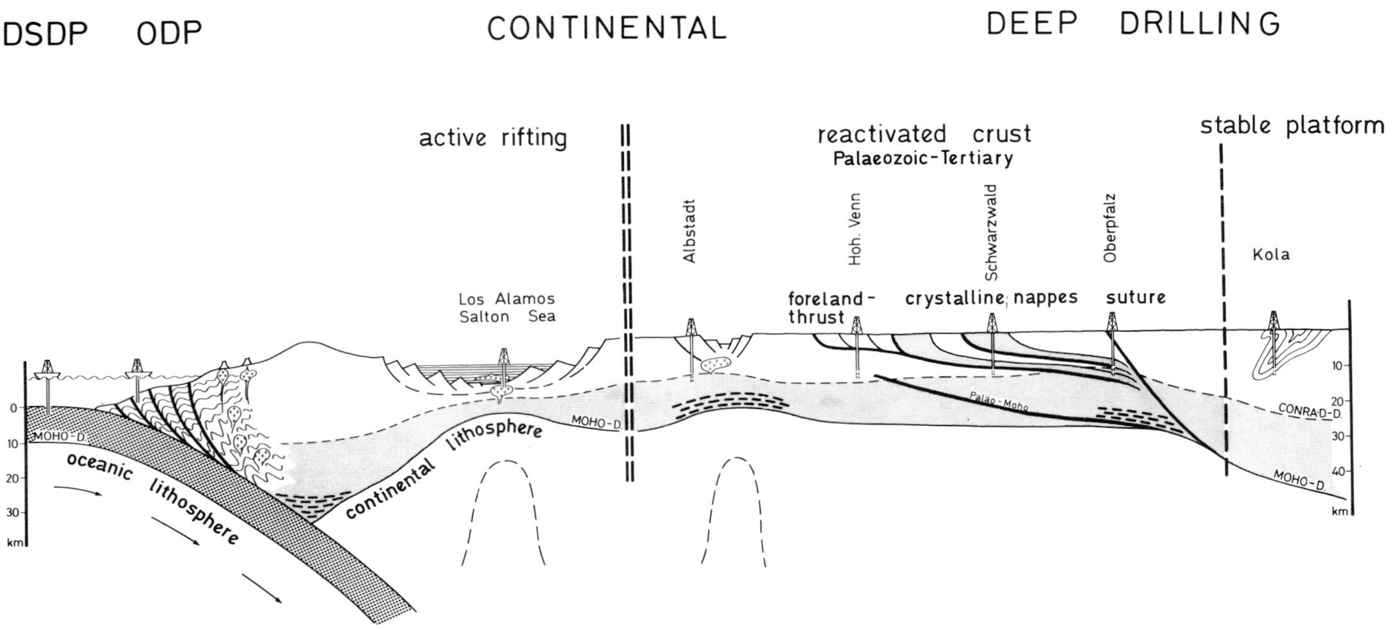


Fig. 2. Schematic overview of the position of deep drilling projects in the oceanic and continental crust. The present and planned borings for scientific purposes are situated in various types of crust. The Ocean Drilling Program (ODP) as a successor of the Deep Sea Drilling Program (DSDP) is aimed preferably at the transitional zone from oceanic to continental crust. In addition to geothermal borings (Los Alamos), continental deep drillings mainly serve to the exploration of crustal structures (Althaus et al. 1984b)

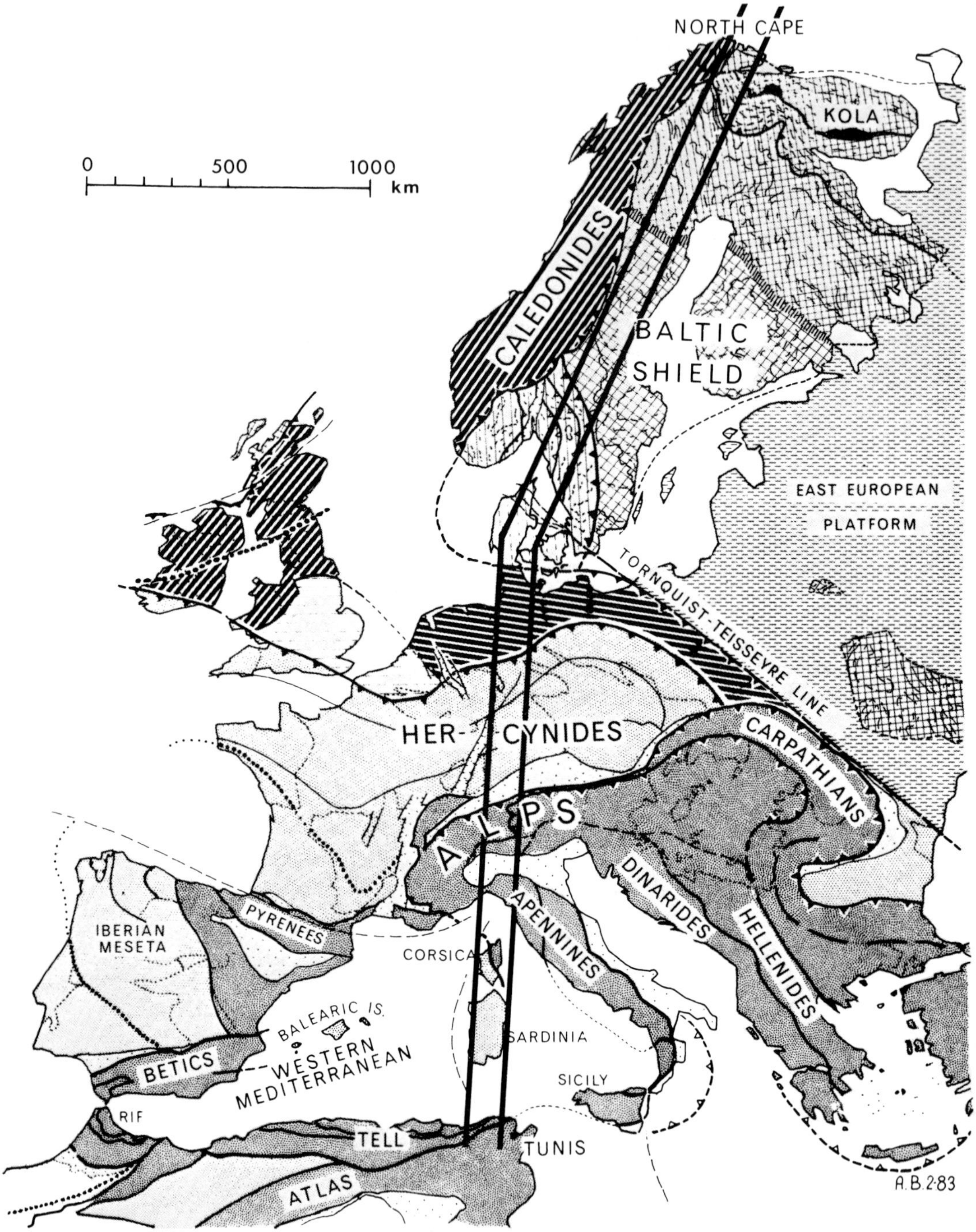


Fig. 3. Main Tectonic units of Europe (after Berthelsen 1983) and the area of the European Geotraverse EGT. (Giese 1983)

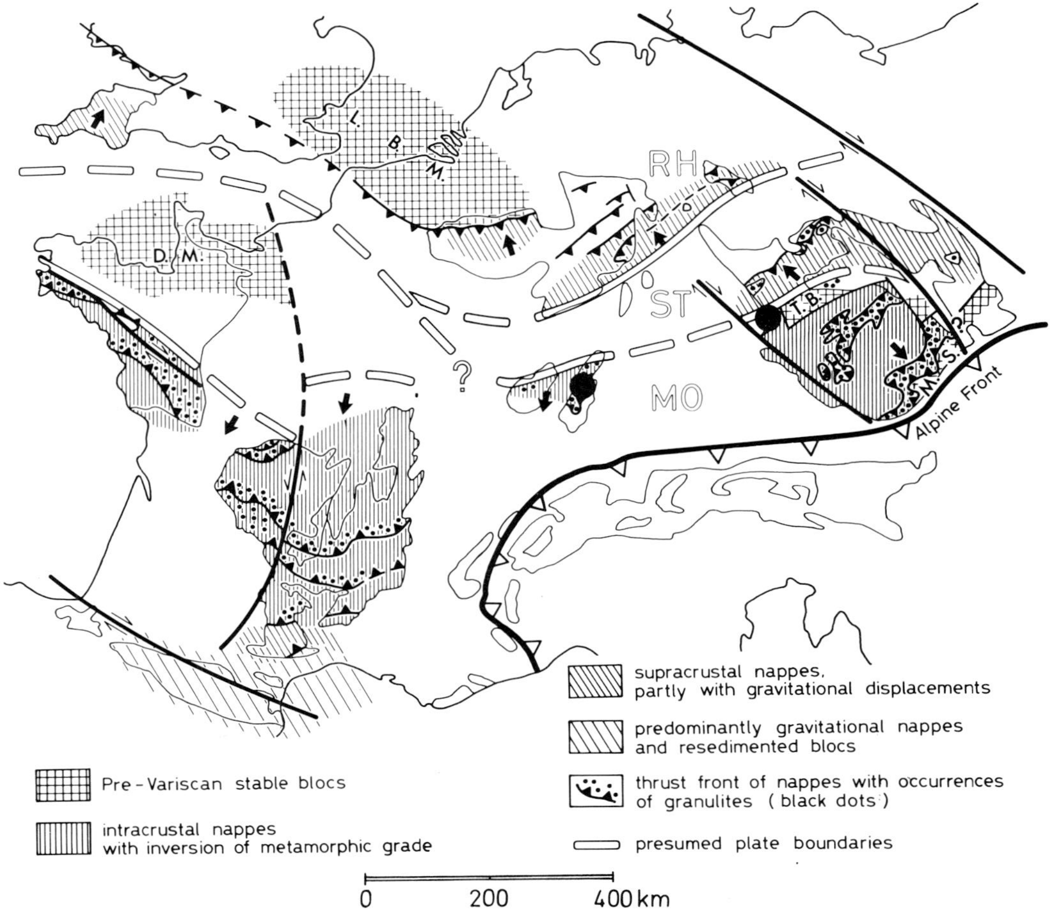


Fig. 4. Structural subdivisions of the Variscan Belt of Central Europe, showing the distribution of several types of thrust complexes and the possible sites *(black dots)* for a crystalline deep drilling in the Oberpfalz and Schwarzwald. *Arrows* main transport direction; *RH* Rhenohercynian Zone; *ST* Saxothuringian Zone; *MO* Moldanubian Region; *L.B.M.* London/Brabant Massif (Caledonian); *D.M.* Domnoneo-Mancellian (Cadomian) and Icartien (Lower Proterozoic); *T.P.* Tepla/Barrandean (Cadomian); *M.S.* Moravo-Silesian (Cadomian). (After Behr et al. 1984)

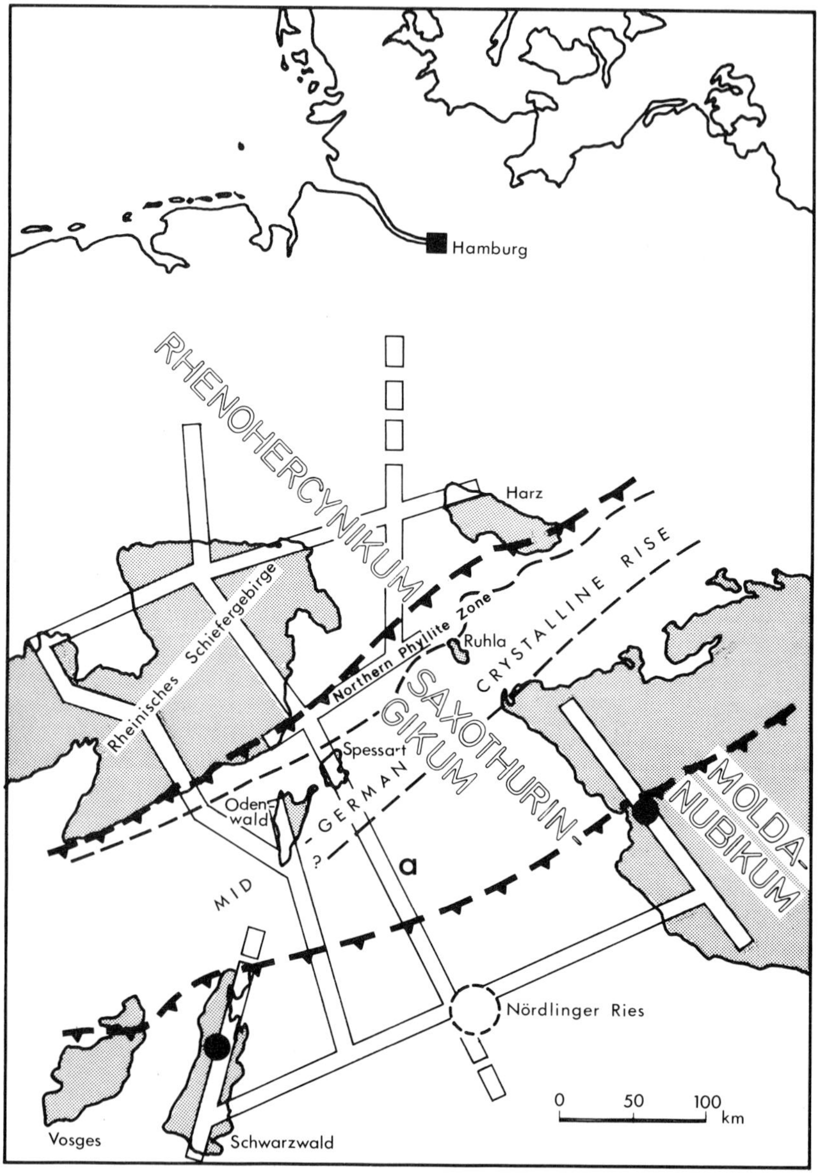


Fig. 5. The German Continental Reflection-Seismic Program traverses (DEKORP) with the possible deep drilling sites Oberpfalz and Schwarzwald. (After Althaus et al. 1984a)

Research during the late 1970's has revealed, in the Variscan basement, a pronounced horizontal-tectonic layering, produced by important intracrustal thrusts as well as thin-skinned decollement structures. These results were an important stimulus for the Continental Deep Drilling concept.

If we look, however, at the heated and ongoing discussion about the evolution of Western and Central Europe's crystalline basement, it is pretty sure that one hole and even a detailed geophysical survey around it will not solve all problems. Perhaps we should learn that many parts were imported from elsewhere and accreted to the North Atlantic Continent as terranes. Even if we are able to put together a mosaic of information, we are not sure whether a system of late Hercynian wrench faults destroyed the mosaic of the believers.

Hopefully, some ongoing geophysical projects will improve this complex situation. Large-scale seismic reflection programs such as COCORP in the US, ECORS in France, BIRPS or MOIST in UK, and DEKORP in Germany, or the European Geotraverse Project (EGT), since November 1982 an Additional Activity of the European Science Foundation (Fig. 3), are all important contributors producing a meaningful mosaic and some of them will directly gain by drilling somewhere in Central Europe. Naturally, we are glad to learn from and to contribute to the International Lithosphere Program.

The German Continental Reflection Program DEKORP (Fig. 5) is especially connected to the EGT and to the Deep Drilling Project KTB. This program is financed (about 6 million DM/year) by our Ministry (BMFT), too. It began in 1983 and is planned to produce annually approximately 200 km long profiles for the next decade. Numerous university departments and geological surveys participate. The first 200 km long profile was just finished (Fig. 5, a) with reflectors down to about 60 km. Perhaps some of them separate thrust units. This looks impressive for Germany, but modest by US standards where COCORP is said to have accomplished 5200 km up to now.

Additionally, plans have been discussed to cooperate directly with Swiss and French geophysicists, e.g., within a Reflection Survey of the Rhinegraben Rift System, especially stimulated by the exciting COCORP results in the Rio Grande Rift Area. In the southern Vosges, as in the Black Forest, 60-km-long base lines for undershooting the Graben are planned perpendicular to its axis. A 32-km Split Dip Spread in the Graben proper and, parallel to the Graben, 80-km Expanding Spread Profiles for control of velocity in the Vosges and the Black Forest will complete this experiment.

Even if everything runs smoothly, the results will solve only some old problems and produce many new ones. But enough skepticism. As believers, we easily can predict progress from both drill sites in structural and dynamic geology in general and especially for the bordering area of a continent welded together from Laurentia, now called the US, Canada and Greenland and Fennosarmatia. Therefore, both parts now separated by the Atlantic should be interested.

All the other aims such as research in fluids, heat, stresses, deep crustal labs, are valid for all deep holes planned or drilled in Laurentia and Fennosarmatia. Hopefully, all of them will add exciting and surprising results to the Kola-Drill-Hole results that we have heard about up to now.

## Problems

Up to now, we were optimists and we have to continue to be optimistic when we mention some more open problems.

1. Is the German scientific potential sufficient for such a large experiment?

In a general way, we hope so. The DFG, for example, has brought together during the last years scientists with similar problems in special programs like the Priority Program *Vertical Movements and Their Causes as Exemplified in the Rheinische Schild Area* (1976-1982), or the Göttingen Special Collaborative Program *Earth Crust* (1969-1981). Recently, books like *Plateau Uplift* or *Intracontinental Fold Belts* have summarized the results. An active Priority Program *Kinematics of Rock-Forming Processes* will be terminated 1986. Furthermore, we shall take care to develop and to finance accompanying scientific programs to strengthen the potential for evaluating the drilling as the Priority Program *Lower Continental Crust* or the Special Collaborative Program 108 *Stress in the Lithosphere* in Karlsruhe.

Nevertheless, we hope that many colleagues from abroad can be encouraged to contribute with their own ideas, to participate in measurements and in the evaluation of geophysical, geochemical, petrological data and materials.

2. What is the technological state of the art?

Although previous experience with drilling and measuring at such depths is limited, the consulting committee is optimistic to be able to cope with electronics and measurement techniques under possible aggressive fluids, high pressures and temperatures above about 250°C. To get better ideas about the temperature gradient to be expected, a ca. 2500 m deep preliminary drill hole is planned at the final drill site before the actual deep drilling begins. It allows a geological study of the uppermost profile sections in which afterwards a casing must be set.

Naturally, many technical problems remain. For example, up to now, even in German sediment deep drill holes, the drill bit touched the rocks only 36.8% of the time.

3. What are the ideas for managing such an experiment?

Several models have been discussed. Some of them are very much related to the Deep Sea Drilling Project. Ideally, the organization should:

- guarantee close cooperation of universities, Geological Surveys, and industry,
- enable foreign participation,
- guarantee for the next decade effective continued preparation, drilling, evaluation, and publication of the results,
- take care of follow-up like deep crustal lab studies or further research with the core material and other data.

4. What are the estimated costs?

Again, we are both skeptics and believers. Skeptic towards all existent figures between 100 and 500 million marks, and believers that earth scientists could use much more and spend the money well.

*Acknowledgments.* Planning of research of this magnitude and breadth could not have been carried out without the help and goodwill of many persons and institutions.

We are particularly grateful to the following colleagues who supported our German plans of Continental Drilling and helped us to complete this brief report: E. Althaus, F.W. Eder, K. Fuchs, F. Goerlich, D. Maronde, K.H. Wedepohl, and W. Ziegler.

Our work could not have succeeded without the help of our Bundesanstalt für Geowissenschaften und Rohstoffe (BGR) and the Bundesministerium für Forschung und Technologie (BMFT).

## References

Althaus E, Behr HJ, Eder FW, Goerlich F, Maronde D, Ziegler W (1984a) Memorandum: Kontinentales Tiefbohrprogramm (KTB) der Bundesrepublik Deutschland – Fortschritte und Stand 1984. DFG-Senatskommission für Geowissenschaftliche Gemeinschaftsforschung, Report

Althaus E, Behr HJ, Eder FW, Goerlich F, Maronde D, Ziegler W (1984b; in press) Kontinentales Tiefbohrprogramm (KTB) (Continental Deep Drilling Program) of the Federal Republik of Germany – advances and status 1984. Report commissioned by the Senate Commission for Geoscientific Cooperative Research (Geo-Commission) of the Deutsche Forschungsgemeinschaft (DFG, German Research Society); Terra cognita

Behr HJ, Engel W, Franke W, Giese P, Weber K (1984; in press) The Variscan Belt in Central Europe: main structures, geodynamic implications, open questions. Tectonophysics

Fuchs K, von Gehlen K, Mälzer H, Murawski H, Semmel A (eds) (1983) Plateau uplift – the Rhenish Shield – a case history. Springer, Berlin Heidelberg New York, p 411

Giese P (1983) Die Europäische Geotraverse (EGT). Mitt XIII Komm Geowiss Gemeinschaftsforsch. Chemie, Weinheim, pp 29-50

Martin H, Eder FW (eds) (1983) Intracontinental Fold Belts – case studies in the Variscan Belt of Europe and the Damara Belt in Namibia. Springer, Berlin Heidelberg New York, p 945

Vidal H, Eder FW, Fuchs K, Goerlich F, Illies H, Neumann J, Ziegler W (1981) Kontinentales Tiefbohrprogramm der Bundesrepublik Deutschland. Mitt XI Komm Geowiss Gemeinschaftsforsch. Boldt, Boppard, p 70

# PROGRAMME ON THE DEEP GEOLOGY OF FRANCE FIRST YEAR'S RESULTS

Claude Megnien and Christian Weber[1]

A programme of geological exploration of the continental lithosphere was established in 1982 by the French scientific community. As in various other countries, this programme is oriented towards deep drilling, despite the fact that this does not constitute the principal objective. The main aim is to understand the deep geology of France. In this country an important part of the Paleozoic basement structures, under the sedimentary basins, is inaccessible to direct observation. Over recent folded ranges, geophysical data available leads to the assumption that the large crystalline massives on the outer border of Alpine Orogen are lying on a sedimentary cover. Progress in our understanding also implies the backing up of geological models with a better comprehension of physical and chemical mechanisms which control the evolution of the earth crust properties. These fundamental mechanisms concern the transfer of energy and mass which affect the crust, but which originate partly from the mantle.

The French programme of deep geology falls within the framework of the general objectives defined by the International Commission on the Lithosphere. It was initiated by the Bureau de Recherches Géologiques et Minières (BRGM) with the collaboration of the Centre National de la Recherche Scientifique (CNRS) and the Direction de la Recherche of the Ministry of Education. It was discussed with the national geological community at a colloquium in June 1982, under the sponsorship of the Ministry of Industry and Research. At this colloquium more than 100 sites for drilling were proposed, and finally 11 principal targets were retained which will be described later.

This project complements two other major French geological programmes. One of these, called ECORS, is concerned with the study of the continental crust by seismic profiles, in which petroleum societies are involved; the other concerns generalized deep geothermal research. The orientations of the deep geology programme are chosen by a large scientific committee, presided over by J. Aubouin. The programme management is assured by C. Megnien.

## Planned Targets

The 11 study areas chosen can be regrouped into three categories: (1) sedimentary basins and basements hidden under basins, (2) basements and related physicochemical phenomena, and (3) recent folded ranges. Figure 1 situates the different targets.

[1]Bureau de Recherches Géologiques et Minières, Orléans, France

Observation of the Continental Crust through Drilling I (ed. by C.B. Raleigh)

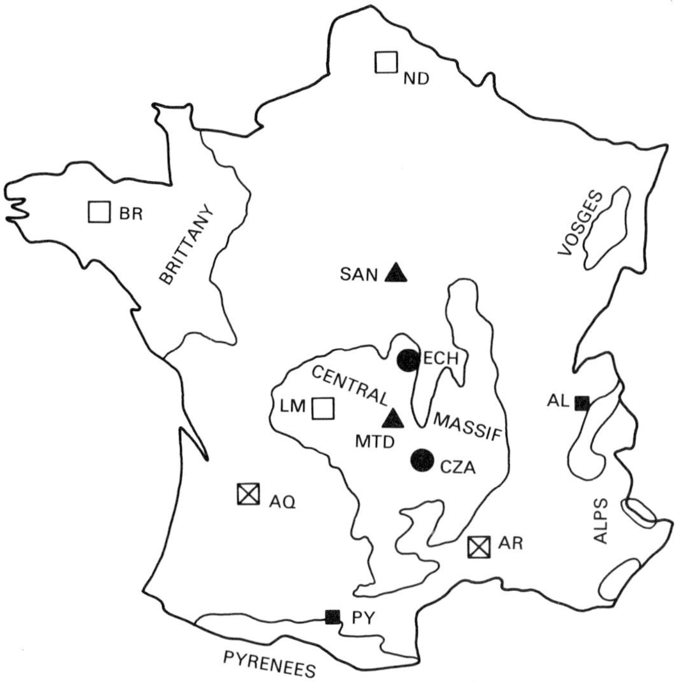


Fig. 1. Approximate location of the 11 targets subjected to preliminary studies. Abbreviations see text

Basins and Basements Under Basins

In spite of numerous petroleum drillings in the three principal French sedimentary areas (Paris Basin, Aquitaine Basin, and Rhodanian Basin), their deep structure is not well understood. In Aquitaine, for example, the presence of basins filled with Cambrian to Carboniferous sediments under the Mesozoic-Cenozoic cover is implied. Their investigation has obvious economic relevance (AQ in Fig. 1).

A geological map of the Paris Basin substratum, the depth of which exceeds 3 km, has been compiled (Fig. 2) on the basis of drilling and geophysical surveys. Two major questionmarks remain, however: the structure of the Variscan front in the North, and the nature and significance of the magnetic body at the origin of the large magnetic anomaly of the Paris Basin. The first target (ND) relates to the northern limit of the Hercynian range, known as the Faille du Midi: what is the extent of the European foreland under the frontal thrust of the Hercynian chain? An ECORS seismic profile, carried out in 1983-1984, should contribute vital information on this subject in the near future. The second subject (SAN) is of importance in understanding the early history of the Paris Basin: does the magnetic body, provoking one of the most important anomalies of all Western Europe, represent evidence of an ante-Permian rift which could have induced the original subsidence of the Paris Basin?

In the south-east of France (AR) it is planned to study the subsidence and the diagenesis of the Ardèche border. It concerns a continental margin superimposed on an important crustal thinning of 50%. Early studies here have shown a very mobile basin during the Mesozoic, with the deposit of 10 km of sediments and a high thermal activity on the slope.

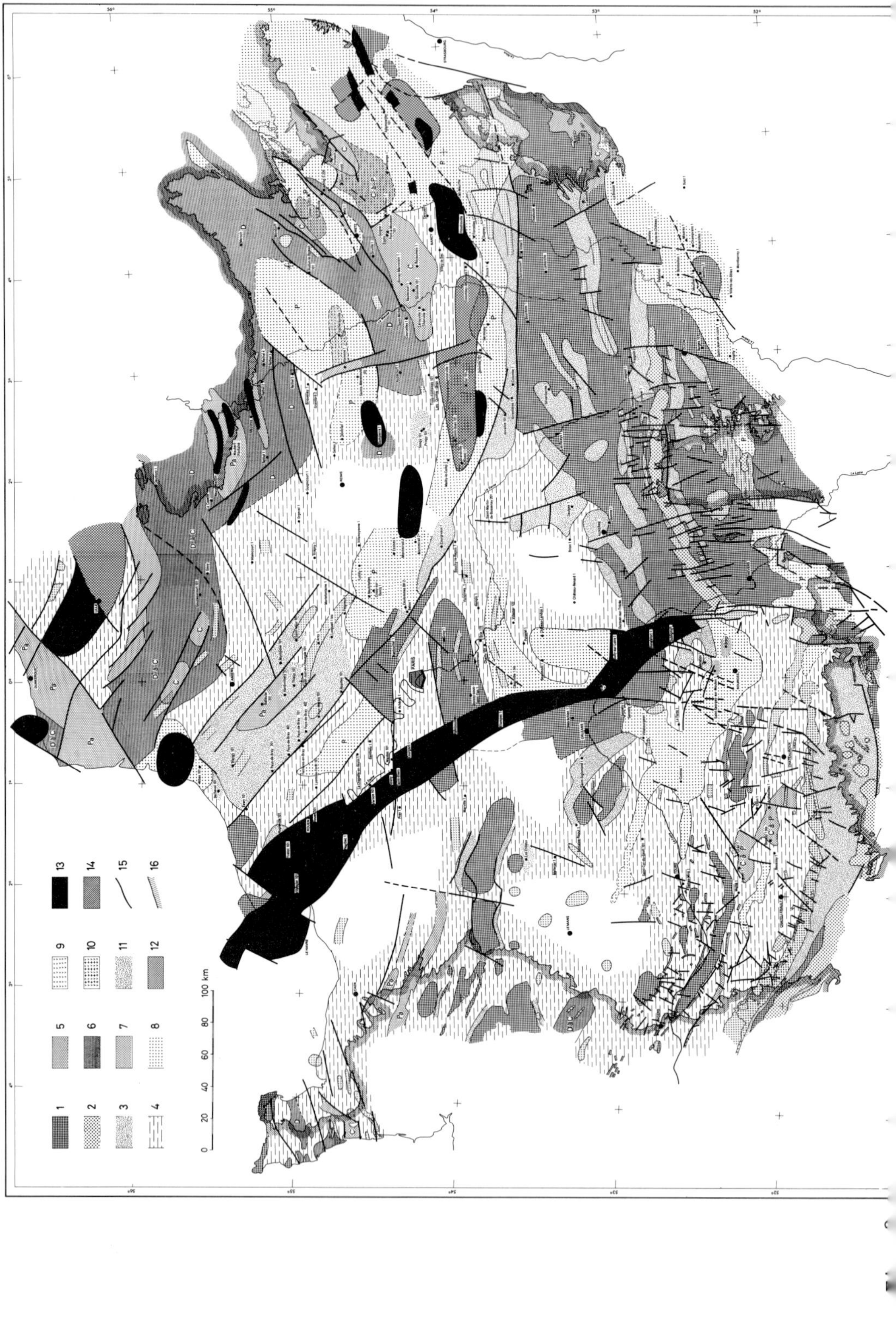
1
2
3
4
5
6
7
8
9
10
11
12
13
14
15
16
0
20
40
60
80
100 km
PARIS
LE MANS
La Loire

Drillings to be done in the basins will be able to exceed a depth of 5000 m. On the other hand, those situated on the basement targets could obtain results of similar interest at shallower depths.

## Basements and Related Physicochemical Phenomena

One of the targets concerns the study in Brittany (BR) of the Precambrian basement of the Variscan domain with its consequences on the structural formation of the chain. The West European continental crust is considered to be "young" compared to the blocks which surround it to the North and South. The goal of the proposed drilling would be to search, under the Upper Proterozoic cover (650 m.y.) in Central Brittany, for the nature and the age of the basement. Geophysical data suggest that it is dense and magnetic. Investigation of the basement is essential to understand the geodynamics of the Cadomian orogen.

Metamorphic phenomena will be better understood from synmetamorph deep seated crustal nappes in the Limousin (LM), Massif Central. Large thrusts bring into contact unities of very different origins. The thrusting upper unit shows a higher metamorphic grade (anatexites), and generally a very high pressure type (eclogites) than the lower autochtonous unit (micaschists). Their study by drilling will allow, over a small distance, the recognition of physical conditions as well as of the nature of phenomena which control the convergent evolution of crustal units of very different crustal origin.

Another target concerns the process of metallogenesis: the study of the deep structure of the leuco-granitic Echassières cupola (ECH), Massif Central. The target provides the best example in France of granitic differentiation with enrichment of W, S, F, Li Ta and Nb up to exploitable grades. The problem is to determine how the magma evolves at depth and how the fluid transfer of metals came about. This model could then be extended to about ten cupolas of the Variscan range which have been discovered in France.

Deep volcanic phenomena are the object of a particular target, the Mont Dore (MTD) in the Massif Central. In this case geophysical and geochemical studies have already been carried out for geothermal purposes. Hypotheses on caldera filling, on the possible nature of the reservoir and on the composition of the magmatic chamber remain to be verified. The top of the chamber is thought to be located at a depth of less than 5 km.

Geothermal exploration in France has also enabled the location of a particular target for the study of hydrothermalism. Such a target is located in a crystalline area on the Cézallier volcano (CZA), above an anomalous zone of the upper mantle beneath the Massif Central.

Fig. 2. Geological map of the Paris Basin deduced from geological data and boreholes (from Debeglia and Weber 1984). *1* granite (sensu lato); *2* leucogranite; *3* metamorphic rocks; *4* undifferentiated Proterozoic or Paleozoic; *5* lower Paleozoic; *6* Devonian; *7* Carboniferous; *8* Permian; *9* mafic rocks; *10* granodiorite; *11* metamorphic rocks and contact metamorphism; *12* iron-bearing beds; *13* deep magnetic body; *14* deep heavy body; *15* fault; *16* limit of the outcropping basement

## Allochtony of Recent Folded Ranges

In the Alps (AL) recent geophysical interpretations in the outer zone of the French part of the belt suggest tangential phenomena much greater than estimated, including crystalline massives. Moreover, the external Savoie sub-Alpine massives are considered nowadays as a cover detached from the crystalline external massives, these massives being themselves thrusted over the autochtonous cover. It is to be wondered if large horizontal thrusts, as in the Appalachians, do not exist. In this case the wide, high resolution seismic profiles are evidently a preliminary to deep drilling. Such is also the case for the Pyrenees (PY), where a part of the large North-Pyrenean crystalline massives is at present thought to be thrusted onto the Mesozoic cover. Indeed, gravity surveys indicate that certain granitic plutons are probably allochtonous. Seismic reflectors have also been detected under the Arize massive, which would certainly represent one of the best targets for deep drilling.

## 1983 Results and Choice of Priority Targets

The year 1983 was devoted to the preliminary study of the 11 targets by many laboratories with which 48 research contracts were made. For its part the BRGM had its teams support these objectives. The first results were presented to the Scientific Committee in March 1984. The preliminary data indicate that the topics related to the basement studies come first in importance. Two priority areas were chosen by the programme committee: the Echassières granite cupola and the geothermalism of Cézallier.

At Echassières the planned drilling is definitive and should reach a depth of around 2000 m. At Cézallier a preliminary test drilling is planned which will reach only 1000 m. The definitive deep drilling should reach a depth of 3000 to 4000 m. At a second level an alternative is possible; it concerns either a drilling at Mont Dore or the recognition of the magnetic anomaly of the Paris Basin. As for the Mont Dore, an initial drilling could be carried out, whereas, for the magnetic anomaly of the Paris Basin, studies will soon be advanced enough to determine the site of a deep drilling.

The results obtained in 1983 at Echassières, Cézallier, and the Magnetic Anomaly of the Paris Basin are described here briefly.

## Echassières Target (Fig. 3)

The theme which corresponds to this target aims to understand metallic concentration processes during the different phases in the evolution of a metallogenic granite.

The Echassières granite was formed in the Upper Carboniferous (300 m.y.) after an epoch of extensive tectonic sheet deformation. Granite has intruded the micaschists and gneiss of the Sioule series. It outcrops over 6 $km^2$, but extends at depth over 45 $km^2$ as shown by gravity data (Gable 1974). A large N.S. seismic refraction profile (Perrier and Rueg 1973) indicates a remarkable velocity stratification in the upper crust (thrusted sheet of Devonian age at 6 km $s^{-1}$ velocity) and confirms the placing of granite under that nappe in a thick metapelitic cover of 8 km, with a 5.7 km $s^{-1}$ velocity (autochtonous series of the

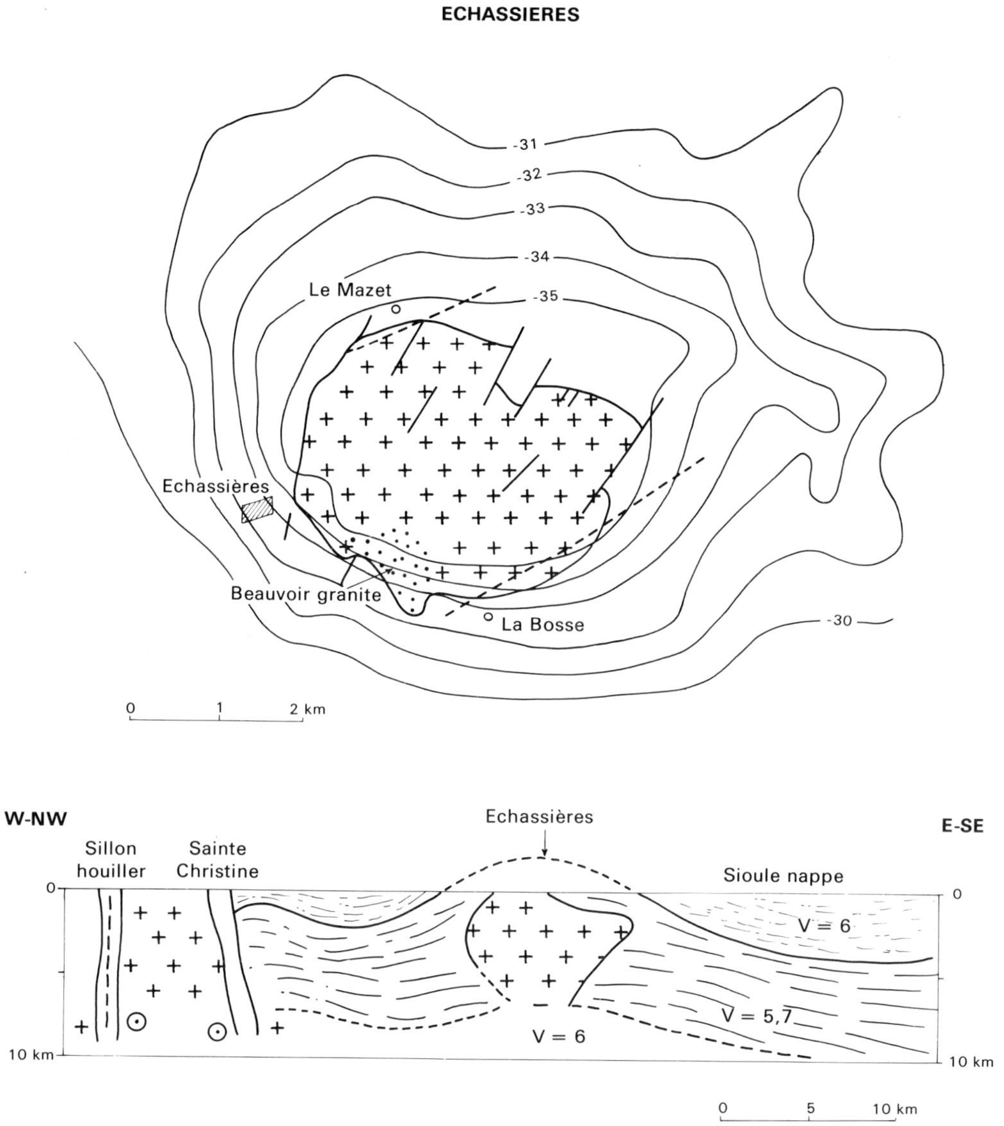


Fig. 3. Interpretative section in the Echassières granitic cupola region. Corresponding geological and gravity map (milligal curves), according to A. Autran (unpublished)

Sioule). Underneath is another crustal layer (V = 6 km $s^{-1}$), similar to that which constitutes the thrusting unit of the Sioule sheet.

Work is continuing this year with complementary geophysical and gravity surveys in order to tighten the measurement grid in view of refining the geometric model of the granite. Complementary electric profiles will be carried out to determine the upper section of the model (granite-micaschists contact). A magneto-telluric survey should enable the

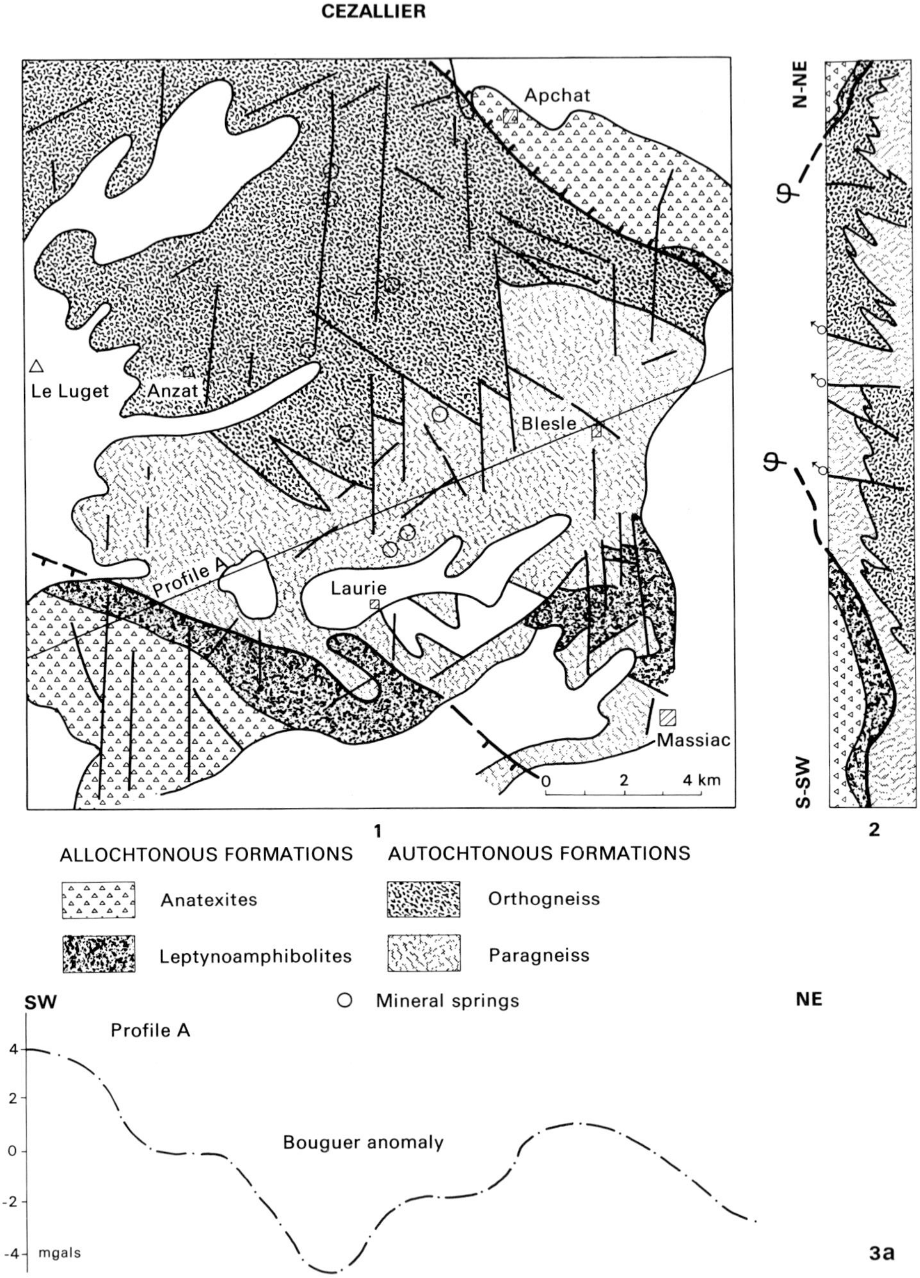


Fig. 4. Geological setting of Cézallier geothermal zone *(1)* with corresponding interpretative curve from Mathonnat (1983) *(2)* and geo-electric and gravity sections. (Gable 1974; Demange and Puvilland 1984; *(3a,b)*)

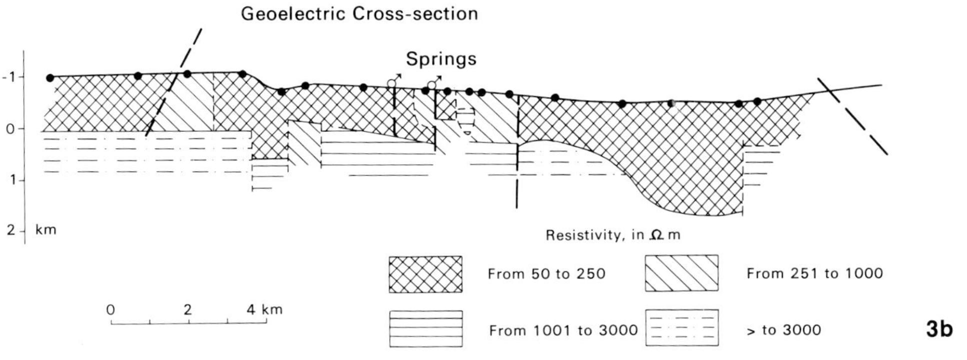


Fig. 4

detection of the schists corresponding to the floor of the granite which is expected to be between 5 and 8 km deep.

At the same time geochemical and structural geological work is developing to choose the best site for the drilling. This should soon be carried out, for a depth in the order of 2000 m is sufficient, with continuous coring, to obtain material necessary for the programmed studies. The results expected are as follows:

- continuous evolution of the various stages of differentiation of the granite,
- geochemical models of interaction between the granitic magmas and fluids,
- concentration and metal fractionation processes by fluid phases,
- tentative quantitative geochemical modelling, taking into account the small volume of the granite (~150 $km^3$).

Moreover, the drilling will provide data on physical parameters concerning the natural fracturing and the stress field at depth. Particular attention will be paid to the measure of thermal flow linked to the geothermal use of the granitic formations.

## Cézallier Target (Fig. 4)

The understanding of geothermal systems in deep basements is one of the important subjects of the programme, since data on geological conditions of indirect evidence of such systems are almost totally lacking, such as the physical and chemical laws which impose on fluid circulations their mobility and composition.

The choice concentrated on the zones of thermomineral carbo-gaseous springs of the Massif Central, in zones of medium geothermal gradient. It was, however, necessary to avoid choosing a drilling site near the warmer springs so as to avoid conflicts between exploiters of mineral water and spas. For this reason the research is oriented towards the Cézallier region, between the Cenozoic volcanic massives of the Mont Dore and of the Cantal, and more precisely to the NE of Massiac (Cantal) where numerous cold or warm (20°C) mineral springs, not yet exploited, are situated. Geochemical studies show that the waters are the result of a mixture between superficial waters and an aquifer of a high temperature. For the latter case the use of classical geother-

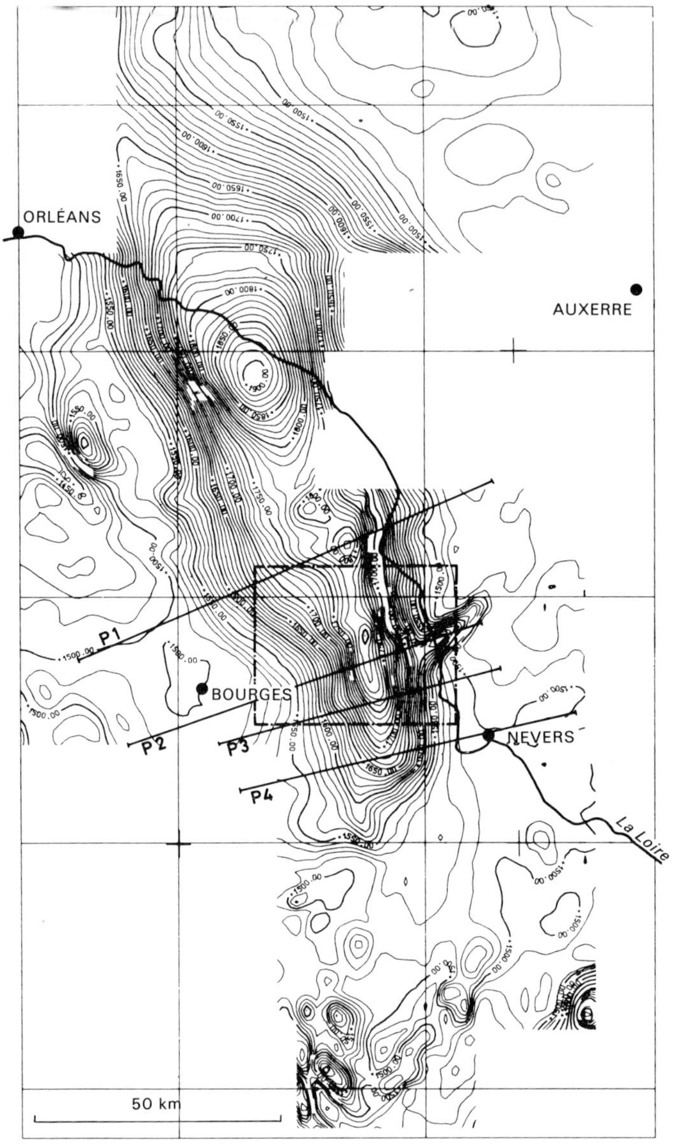


1

Fig. 5. Magnetic anomaly of the Paris Basin. Map of magnetic field reduced to the pole *(1)*. Simultaneous modelling of gravity data *(2)* and evolution of density and magnetization parameters depending on the depth of the abnormal body's upper surface. Guillen and Debeglia 1984; *3*)

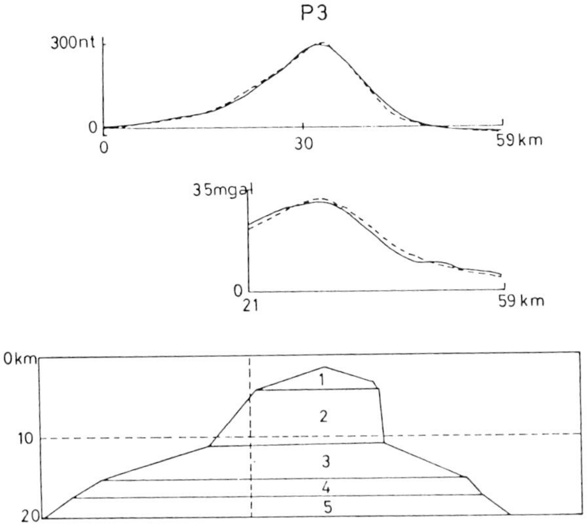


| body | 1 | 2 | 3 | 4 | 5 | |
|---|---|---|---|---|---|---|
| Δρ | 0.12 | 0.09 | 0.06 | 0.03 | 0.00 | cgs |
| χ | 0.0014 | | | | | emu cgs |

Simultaneous inversion of gravity and magnetic data

**2**

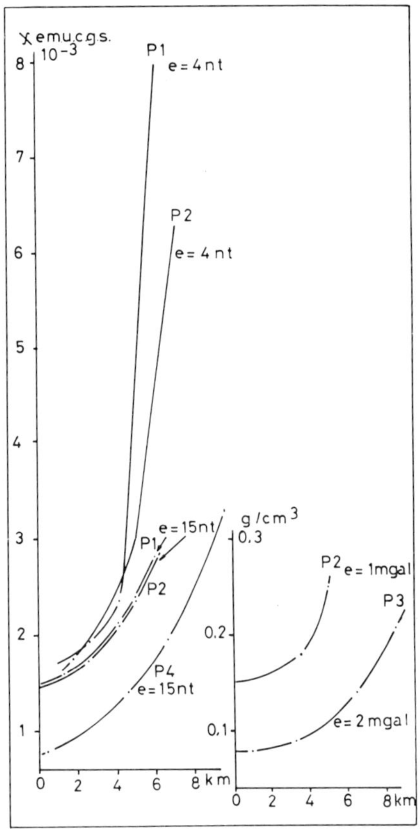


**3**

Fig. 5

mometers NaK, NaKCa, and NaLi show temperatures in the order of 200°C, confirmed moreover by the presence, in a significant quantity, of elements such as Rb, Cs, Sr, and As which become soluble, in the water-rock interaction, only at a high temperature. The springs emerge thanks to a fracture network, oriented N-S and NW-SE, which cuts across the Hercynian basement composed of ortho- and para-gneiss, appearing as a geologic window under a thrust sheet made of leptino-amphibolites and anatexites.

Work in course aims to model gravity and magnetic data as well as heat flow, whereas a magneto-telluric survey is destined to recognize in detail conductive zones (⩽80 Ω m) already shown up between 800 and 2000 m.

In 1985 preliminary drilling is planned at 1000 m in the most favourable zone which, if confirmed, will be followed up by drilling to 3000 or 4000 m.

Expected results concern (1) the understanding of the variation with depth of fluid circulation parameters, in other words the interactions between hydraulic and thermal gradients; (2) the relationship between the permeability and the stress pattern and the natural fracturation; (3) equilibria at different levels of physicochemical exchanges between water and rock. Finally drilling must allow for the possible superposition of fluid networks of different chemical composition by independent thermal loops, either separated or not by hyperthermal clogged-up zones.

## Magnetic Anomaly of the Paris Basin (MAPB) (Fig. 5)

Work achieved in 1983 concentrated principally on a simultaneous inversion of magnetic and gravity data (Guillen and Debeglia 1984). The model retained as the most plausible, in the southern part of the Paris Basin, gives a disymetric body with an eastern vertical or overthrusting flank; its upper surface does not appear to be deeper than 2.5 km, allowing the hypothesis of a negligeable remanent magnetization.

Recent geological work (study of the successive deformation fields, folding, and fracturation, at the top of the anomaly and on the enclosing basement) leads one to suppose that the emplacement of the magmatic rocks at the origin of the MAPB could have taken place at the time of the upper Devonian extension (Castaing and Autran 1984). Subsequently, during the Carboniferous, a system of compression may have created, on the level of the MAPB, a NE thrust accompanied by the positioning of granites on the western flank of the anomaly. This implies a rifting phenomenon, followed by a partial obduction. The faults limiting this rift remained active during the sedimentation of the Paris Basin up until the Lower Quaternary.

The recognition of the MAPB by drilling should thus contribute key information on the structuring of the Hercynian shield of Western Europe and on the origin even of the Paris Basin. Drilling can only be done after more detailed geophysical research, in particular magneto-telluric and seismic studies, which will be carried out in 1984-1985.

## Conclusion

The Deep Geology of France programme is a project which has been widely open to different scientific proposals. Numerous teams collaborate and research themes concern a large variety of fields in the geosciences. During the first two years it was necessary to make progressive choices which were elaborated following detailed studies of each of the proposed targets.

At present the programme is focussing on four objectives: two are well advanced (Echassières granitic cupola and Geothermalism of Cézallier); two others need a few more years of preliminary study (Magnetic Anomaly of Paris Basin and Mont Dore Caldera).

The results of the first phase will be presented at a colloquium planned in Paris on 19 and 20 September 1984, and the communications will be published at the end of the year.

Now that the targets are clearly defined, cooperation with countries having similar projects is desirable. Areas of collaboration could be on themes such as drilling technology, measurements, and in situ testing, and of course of the comparison of scientific results.

References

Autran A et al. (1980) Introduction à la carte tectonique de la France à 1/1 000 000. Mémoire BRGM No 110, p 52

Castaing C, Autran A (1984) Les différents contextes tectoniques et géodynamiques de l'anomalie magnétique du Bassin de Paris à partir de la fin du Proterozoique. BRGM Open file report

Colloque National (1982) Programme Géologie Profonde de la France, Paris, 15 juin 1982, vol 1. Communications, p 555, vol 2. Rapport généraux et comptes rendus, p 124. Doc BRGM No 39 et 40

Debeglia N, Weber C (1984, in press) Geological mapping of the basement of the Paris Basin (France) by gravity and magnetic data interpretation. In: Hinze WJ (ed) Utility of regional gravity and magnetic anomaly map. Publ Soc Exploration Geophysicists

Demange J, Puvilland P (1984) Reconnaissance géothermique haute énergie dans la région du Cézallier (Massif Central). BRGM Open file report

Gable R (1974) Un exemple d'application de la méthode gravimétrique à l'étude structurale d'une coupole granitique: Echassières (Allier). DEA Univ Paris VI

Guillen A, Debeglia N (1984) Contribution à l'étude de l'anomalie magnétique du Bassin de Paris, thème 2, géologie profonde de la France. BRGM Open file report

Mathonnat M (1983) La série métamorphique du Cézallier, Massif Central français. Thèse, Univ Clermont-Ferrand

Perrier G, Rueg JC (1973) Structure profonde du Massif Central français. Ann Géophysique 29 (No 4): 435-502

# DEEP DRILLING PROGRAMS IN BELGIUM

J. Bouckaert[1]

## Introduction

As a result of three important orogenic events (Cadomian, Caledonian, Variscan), the rather small surface of the Belgian country (30,570 km$^2$) can be subdivided into three main geological regions: a northern one, composed mostly by gentle dipping strata to the north, and north-east since devonian times, a southern one intensively affected by the Variscan orogeny and a central part forming the so-called (London)-Brabant Massif (Fig. 1).

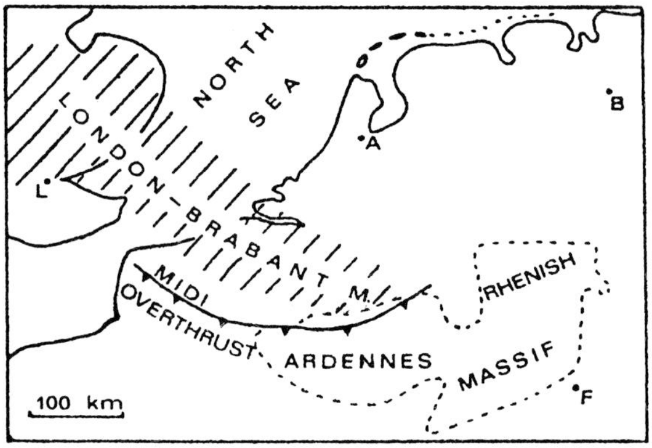


Fig. 1. Location of the London-Brabant Massif in relation to the Midi Overthrust and the outcropping Pre-Permian deposits of the Ardennes and Rhenish Massif. *L* London; *A* Amsterdam; *B* Berlin; *F* Frankfurt

## The Southern Part of the Brabant Massif

The northern front of the Variscan deformation in Belgium and neighbouring parts of France and the Federal Republic of Germany is formed by a major thrust-fault complex, which, for the sake of simplicity, is here called the Midi Overthrust.

The Midi Overthrust and frontal imbrications separate the allochthonous deposits which are outcropping in the Dinant Synclinorium to the south from the (partly underlying) autochthonous rocks which are outcropping in the Namur Synclinorium to the north. These underlying autochthonous rocks may be much younger (e.g., Upper Carboniferous) than the immediately overlying allochthonous deposits (e.g., Lower Devonian) as has been observed in coal mines and boreholes.

[1]Geological Survey, Jennerstraat 13, B-1040 Brussel, Belgien

Observation of the Continental Crust
through Drilling I (ed. by C.B. Raleigh)

As far back as 1913, it was suggested (Fourmarier) that the autochthonous coal-bearing Upper Carboniferous rocks of the Namur Synclinorium might extend southwards below the Dinant Synclinorium into at least the Dinant region (Fig. 2). This would mean that in fact the Dinant Synclinorium might be a large-scale tectonic nappe structure.

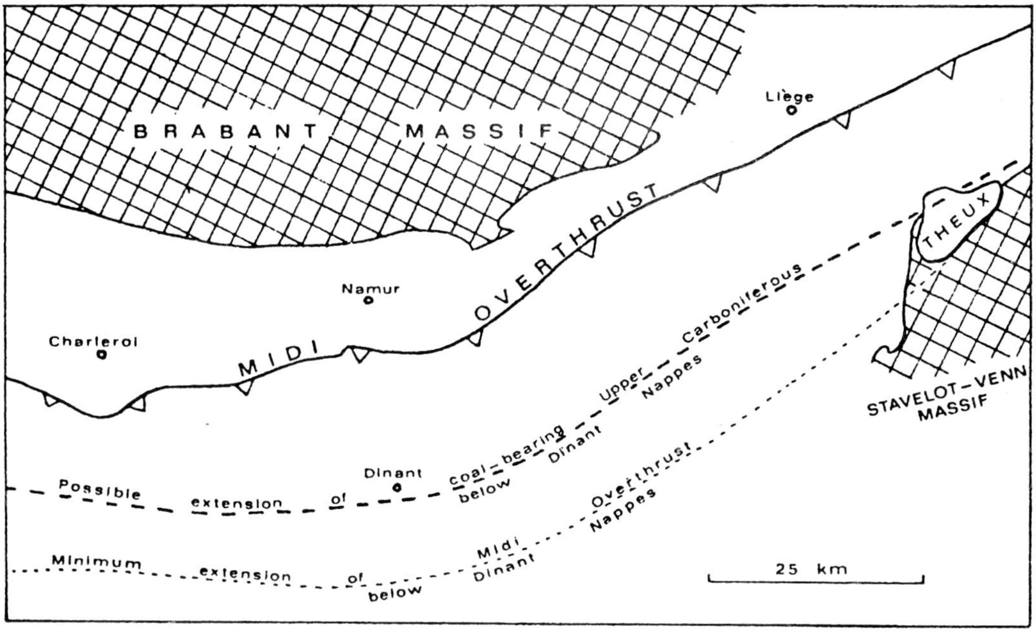


Fig. 2. Possible extension of autochthonous coal-bearing Upper Carboniferous strata and suggested minimum extension of Midi Overthrust (Fourmarier 1913)

Tectonic nappes occur in many places of the external zones of the Variscan belt in Europe, although their size may be less spectactular than that of the Alpine thrust sheets. Well-known examples have been described from northern Spain (Julivert 1971). Recently, Variscan nappe structures have also been discovered in Czechoslovakia (Dvorak, Friakova et al. 1981). Thrust sheets of the same age are also known from the Appalachian belt in eastern North America (Jacobeen and Kanes 1975).

The geological mapping of Belgium (1:40 000 scale) was fulfilled in 1924. In this way Belgium was one of the first countries of the world to have a two-dimensional knowledge of the underground.

The complicated structures as mentioned above induced the Belgian Geological Survey (BGS) in 1953 to start with a vertical investigation program by boreholes and Geophysical Surveys. (A total length of 34 022 m was drilled by the end of June 1984.)

One of the first attempts to explore the southward extension of the Midi Overthrust was made by the Belgian Geological Survey in 1953 at Wêpion, south of the city of Namur (Graulich 1961). But it was not until 1963 that a seismic survey by a consortium of oil companies revealed the continuation of the Dinantian (= Lower Carboniferous) and

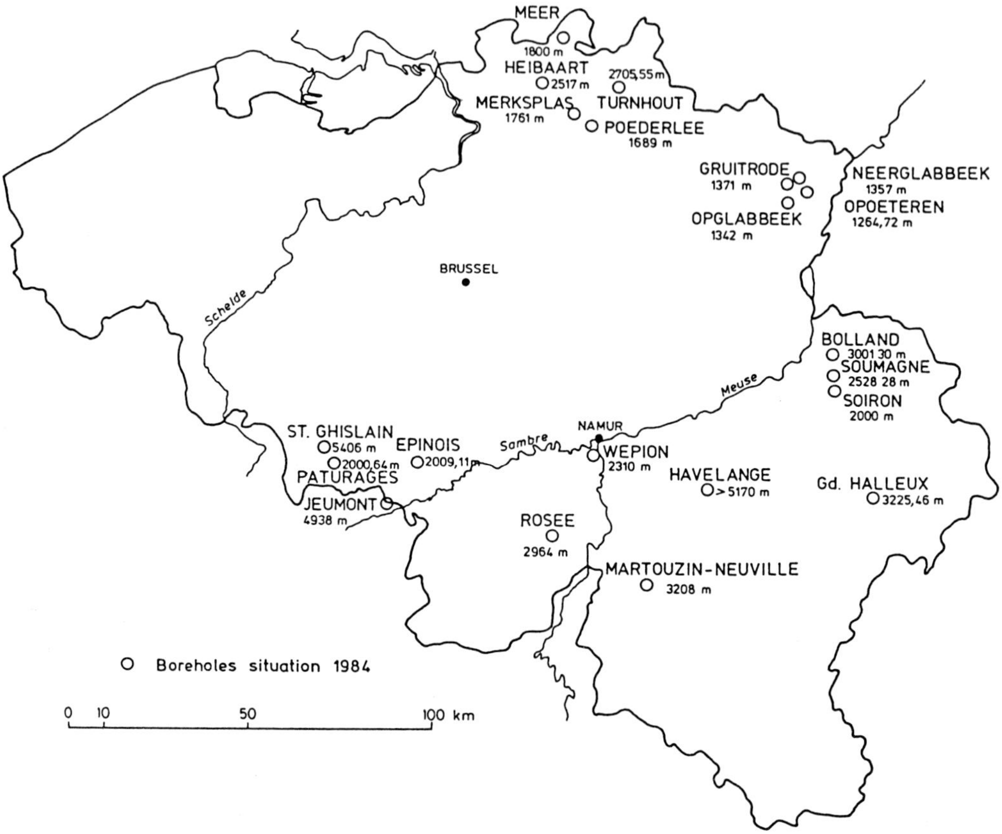


Eifelian-Frasnian carbonates below the Midi Overthrust over a distance of some 40 km to the south in northern France. This distance may be considered as a minimum, since this was determined by the length of the seismic profiles. Subsequently, this seismic interpretation was confirmed by the Jeumont borehole (Clement 1963), where not only coal-bearing Upper Carboniferous strata occur below the overlying Lower Devonian, but also Dinantian and Eifelian-Frasnian carbonates (Fig. 3).

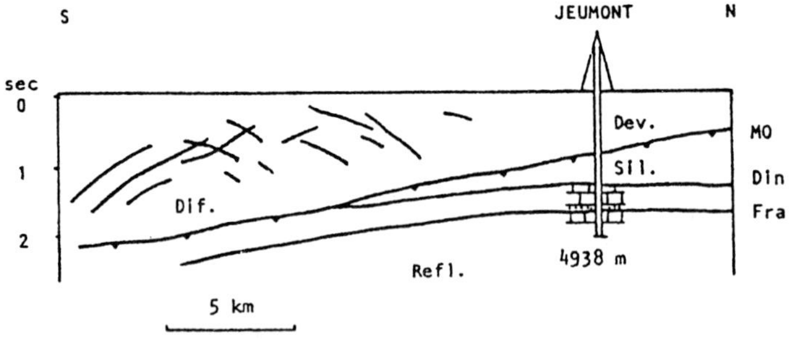


Fig. 3. Seismic profile 415 through Jeumont borehole (northern France), showing diffractions above the Midi Overthrust *(MO)* and strong reflectors representing the top of the Dinantian and the top of the Frasnian carbonates below the thrust plane. (After Clement 1963)

Later, a seismic reflector – presumably representing the Midi Overthrust – has been detected by several seismic surveys (by Shell in 1965, by the Belgian Geological Survey in 1976-1978) in the Condroz-Famenne region of southern Belgium (Fig. 4), (cf. Bless, Bouckaert et al. 1977b).

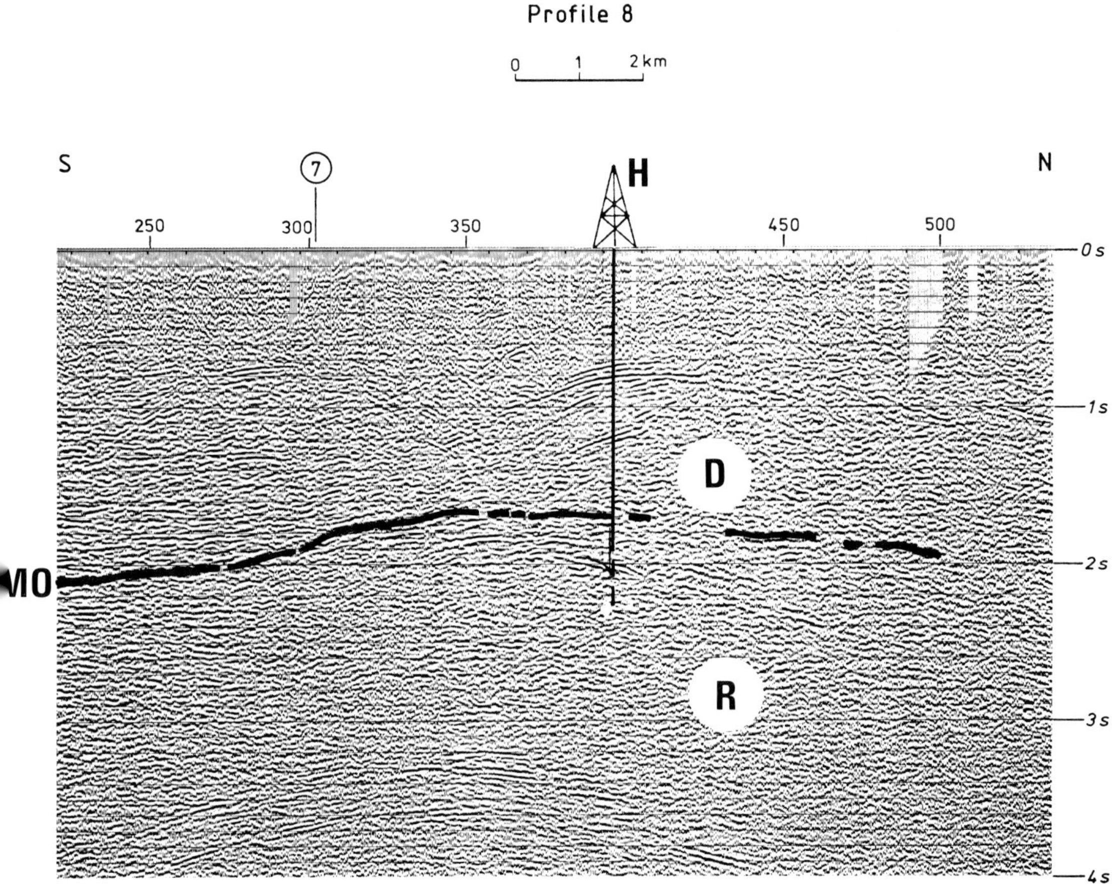


Fig. 4. Seismic profile through Havelange borehole. Showing upper zone with diffractions *(D)* and lower zone with regular reflections *(R)* separated by strong reflector *(MO*, here interpreted as Midi Overthrust)

Finally, a similar reflector below the Cambro-Ordovician rocks of the Stavelot-Venn Massif, southeast of Aachen in the Federal Republic of Germany, was recognized by Meissner et al. (1981; Fig. 5).

These investigations suggest a minimum extention of the Midi Overthrust over an area of 175 × 40 km (Fig. 6). However, it seems reasonable to believe that this overthrust extends southwards far beyond the area studied thus far.

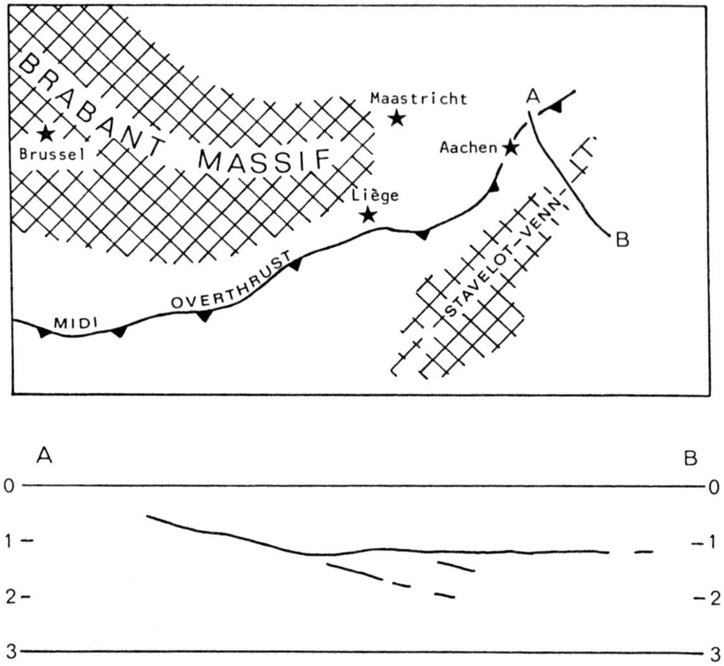


Fig. 5. Seismic reflector at 1.2 s two-way travel time below Stavelot-Venn Massif (after Meissner et al. 1981), suggesting the presence of an important thrust-fault at 3-4 km depth

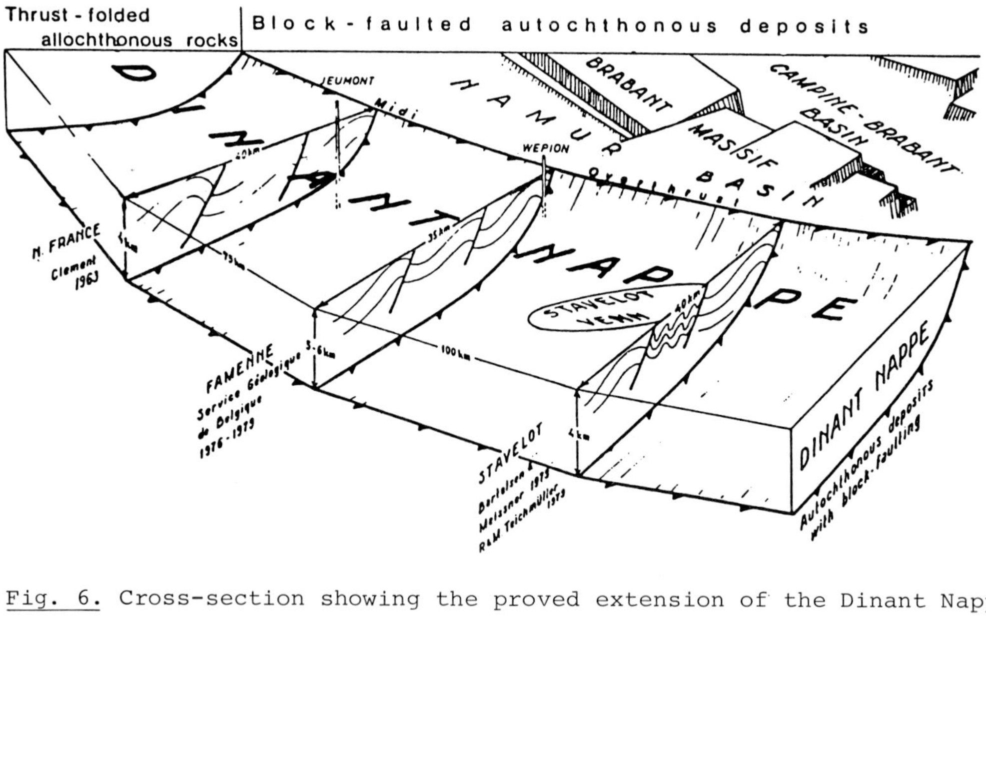


Fig. 6. Cross-section showing the proved extension of the Dinant Nappes

One of the most fascinating questions is that of the nature and age of the rocks immediately below the Midi Overthrust. Those immediately above the thrust plane consist of thrust-faulted and folded Lower Devonian to Cambrian rocks. Those below the thrust plane seem to be more or less flat-lying and affected by block-faulting and large updoming structures of unknown origin. These latter features may have deformed the Midi Overthrust and overlying Dinant Nappe as has been observed in the Famenne region. However, the age of the youngest deposits below the thrust plane is largely dependent on the displacement along the same.

The Jeumont borehole in northern France has proved the southward extension of Upper Carboniferous (coal-bearing) deposits at least 15 km south of the present northern outcrop of the Midi Overthrust. And the seismic section in that area suggests that Dinantian carbonates may extend up to at least 40 km south of that line.

However, it might well be that large parts of the Dinant Nappes rest directly upon a Cambro-Silurian or even Precambrian basement.

This problem may be (partly) resolved by deep boreholes. The Belgian Geological Survey has started a deep boring in the Famenne region at Porcheresse-Havelange (Fig. 7), where the presumed Midi Overthrust has been traced by a seismic survey at a depth of about 5200 m.

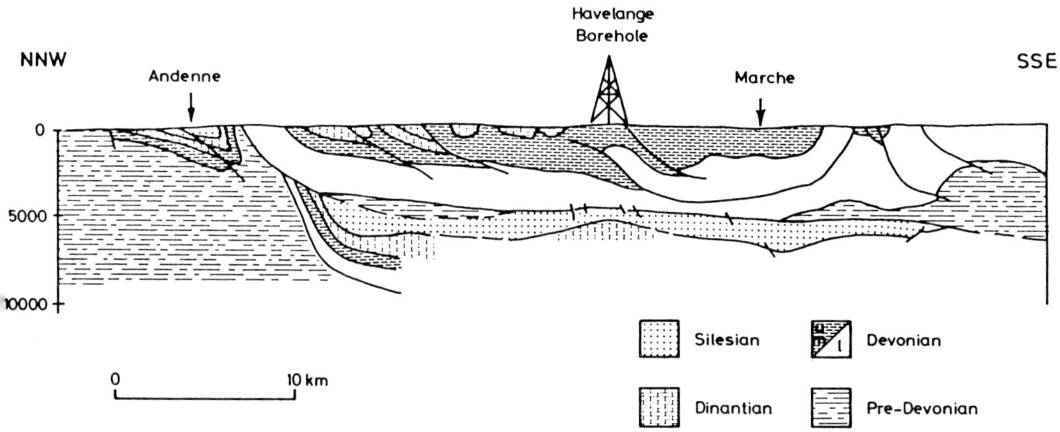


Fig. 7

Whatever the outcome of the above investigations may be, there seems to be little doubt that the Dinant Nappes show all the characteristics of typical thin-skinned fold-and-thrust belts. The thin-skinned nature of such belts is determined by the fact that only beds above a given stratigraphic horizon are involved in the folding and thrusting. Such a limiting horizon may be near the crystalline or metamorphic basement, or it may consist of a relatively weak rock-type (shale, evaporite). Examples of the latter are the Rome Shale in the southern Appalachian Valley and Ridge Province, and the Triassic salt beds in the French Maritime Alps. As pointed out by Chapple (1978), there is not necessarily a mechanical contrast between the relatively undeformed basement rocks and the thrust-folded cover. It is the layer of weakness that determines the thin-skinned nature of the belt.

Exceptions to this rule are also known. In northern Spain, the rather thin sequence of Cambrian Lancara carbonates commonly forms the basal layer of the Hercynian nappe structures (Julivert 1971). Practically always, the nappe structures are broken up into several smaller, but interrelated sheets (cf. Bless et al. 1977a). The term Dinant Nappes is therefore preferred over the term Dinant Nappe.

The second point in common to thin-skinned fold-and-thrust belts is the wedge-shaped form of the deposits before the deformation, that is more or less preserved in the eventual fold-and-thrust belt (Chapple 1978). The thickest part of the wedge is found at the back end from where the thrust comes, whereas the frontal portions are much thinner. These differences in thickness of the original sedimentary wedge are enhanced by the structural shortening during the folding and thrusting process. It is also this strong shortening of the wedge that explains how the front may only have been moved over a few (tens) of kilometers, whereas at the same time the rear end has travelled some 100-200 km.

## The Northern Part of the Brabant Massif

Since 1979 there has been a renewed exploration for coal, geothermal power, and storage possibilities in the Carboniferous basin north and north-east of the Brabant Massif. The initial phase of coal exploration by the B.G.S. was confined to the north-eastern area of the basin.

Maybe even more important as the discovery of a new coalfield is the presence there of an almost 300 m thick sequence of the so-called Neeroeteren.

Sandstones of Lower Westphalian D-age (Fig. 8).

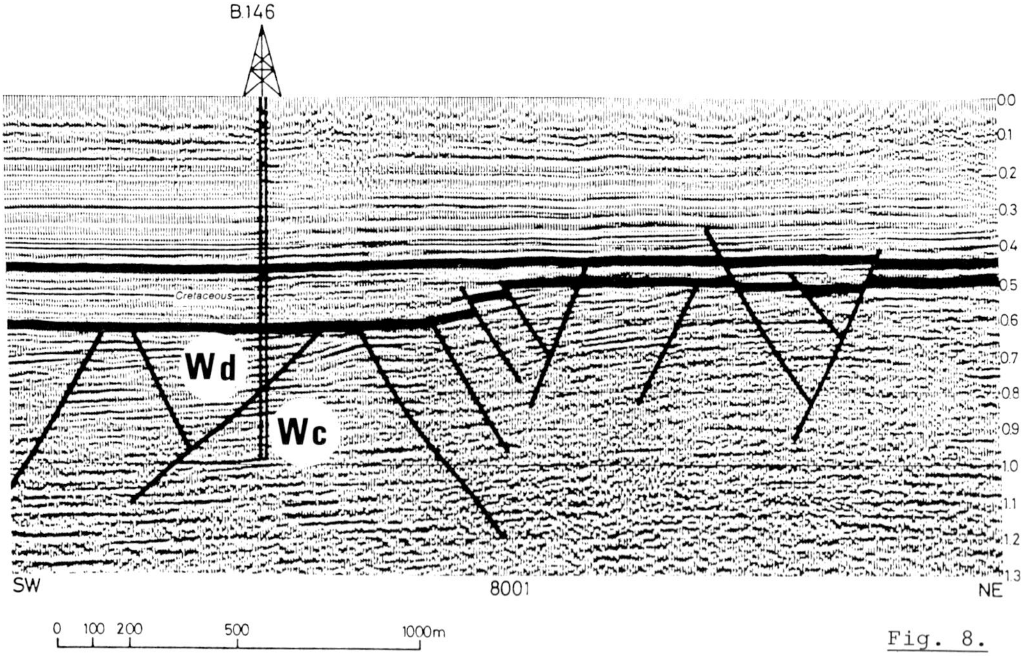


Fig. 8.

The white-grey, feldspar-rich sandstones show a relatively high porosity (average porosity 18%), notwithstanding the frequent occurrence of post-sedimentary vermicular kaolinite crystals. However, the permeability of these sandstones is medium (usually below 50 md). This means that the Neeroeteren Sandstones represent a potential reservoir for gaseous hydrocarbons. It is expected that these extend to the north-east below the Roer Valley Graben or Central Netherlands Graben, whereas a change towards more clay/silt-dominated deposits has been observed in western directions (Fig. 9; Van Wijhe and Bless 1974; Bless, Bouckaert, Paproth 1982).

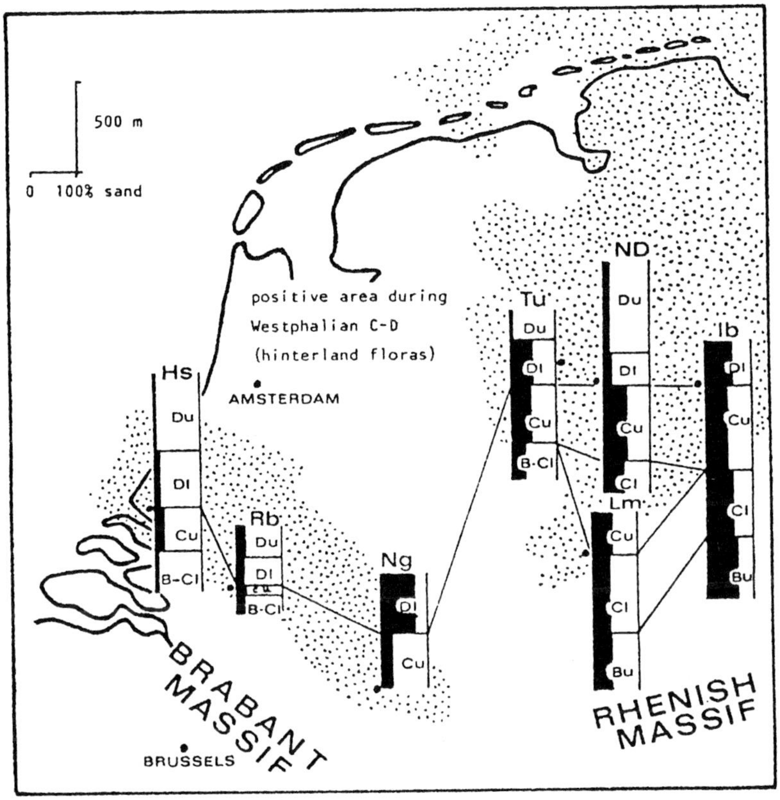


Fig. 9. Average sandstone percentage in Upper Westphalian *B*, Lower and Upper Westphalian *C*, and Lower and Upper Westphalian *D* in boreholes Hellevoetsluis *(Hs)*, Rijsbergen *(Rb)*, Neerglabbeek *(Ng)*, Tubbergen-8 *(Tu)*, Nord-Deutschland-8 *(ND)*, Obbenbüren *(Ib)*, and Lippermulde *(Lm)*. *Stippled areas* indicate the present distribution of Westphalian C-D deposits in northern Belgium, the Netherlands, and north-western Germany

Fig. 8. Seismic profile through Neerglabbeek borehole (B146), showing Uplift in pre-Cretaceous rocks in the north-eastern part of the profile. Top and base of the Cretaceous are indicated. *Wd* Westphalian D; *Wc* Westphalian C. (Dusar 1982)

The Neeroeteren Sandstones might be considered as the southern pendant of the partly slightly older (Westphalian C) gas-productive Tubbergen Sandstones in the eastern Netherlands. But maybe, one might envisage the possibility that some of the shales intercalated in this Neeroeteren Sandstone sequence act as stratigraphic traps somewhere in the not yet explored southeastern portion of the Roer Valley Graben (Fig. 10).

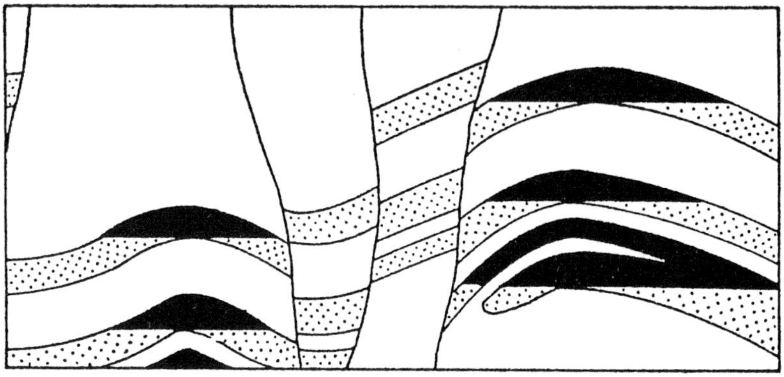

Fig. 10. Working model of possible occurrence of gas-bearing sandstones in the Westphalian D deposits of the Roer Valley Graben (not to scale). The clay/silt-dominated upper portion of the Westphalian D sequence might act as a seal for such hypothetical gas reservoirs. *White* clay/silt-dominated intervals; *stippled* sand-dominated intervals; *black* gaseous hydrocarbons

In the north, unusually thick Namurian Strata are shown to be related to the existence of a deep basin. The Namurian basin seems to be limited by a listric growth fault in regard to the Lower Carboniferous Limestone.

The Upper Carboniferous in the area contained enough organic matter to have acted as a source rock. In the absence of Permian till Lower Cretaceous formations, the listric fault and its associated roll-over antiform supply with a trap structure containing both sandstone beds and sealing shale beds (Fig. 11, Vandenberghe 1984).

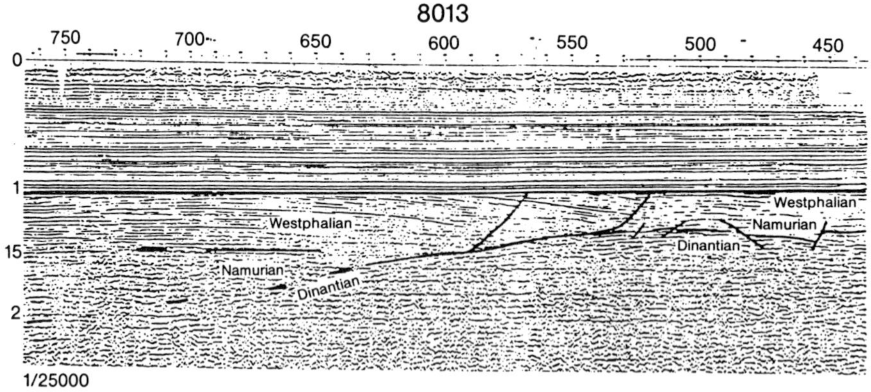


Fig. 11.

## Future Investigations

It is planned for the near future (1985) to survey by seismic methods (Vibroseis) a Geotraverse-profile in a N-S direction over the all of the Brabant-Massif.

On the other hand, we hope to fulfill in collaboration with a West German oil company a joint seismic survey in order to follow the extension of the Midi Overthrust in southern Belgium.

Ultra-deep boreholes are not foreseen at this moment.

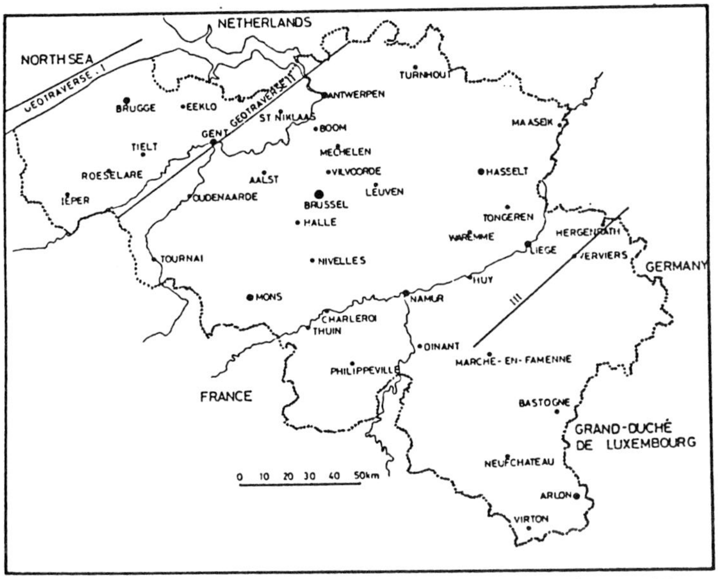


I or II : 1985 III: VIBROSEIS i. coll. with Germany: 1985

## References

Bless MJM, Bouckaert J, et al. (1977b) Y a-t-il des hydrocarbures dans le Pré-Permien de l'Europe occidentale? Serv Géol Belg Prof Pap 148: 1-58

Bless MJM, Bouckaert J, et al. (1980b) Pre-Permian sedimentation in NW Europe: Pre-Permian depositional environments around the Brabant Massif in Belgium, the Netherlands and Germany. Sedimentary Geol 27 (1): 1-81

Bless MJM, Bouckaert J, Paproth EV (1982) Recent exploration in Pre-Permian rocks around the Brabant Massif in Belgium, the Netherlands and the F.R.G. Geofiles, 1-63

Bless MJM, et al. (1981) Paleogeography of Upper Westphalian deposits in NW Europe with reference to the Westphalian C North of the mobile variscan belt. Med. Ryk geol Dienst WS 28 (5): 101-127

---

Fig. 11. Interpretation of the seismic data as a listric fault. Rollover structure is present

Chapple WM (1978) Mechanics of thin-skinned fold-and-thrust belts. Geol Soc Am Bull 89: 1189-1198
Clement J (1963) Résultats préliminaires de campagnes géophysiques de reconnaissance dans les permis de recherches "Arras et Avesnes" de l'Association Shell Francaise – P.C.R.B. – SAFREB. Objectifs du forage profond Jeumont-Marpent No 1. Ann Soc Géol Nord 83: 237-241
Dusar M (1982) Exploration for coal in the Belgian Campine. Publ Nat Hist Genootschap Limburg 32 (1-4): 27-39
Dvorak J, Friakova O, et al. (1981) Paleozoikum ve vrtu Opatovice I na Hranicku (SV Morava). Cas Slez Muz Opava (A) 30: 211-229
Fourmarier P (1913) Les phénomènes de charriage dans le bassin de Sambre-Meuse et le prolongement du terrain houiller sous la faille du Midi dans le Hainaut. Ann Soc Géol Belg 40: B192-B235
Graulich JM (1961) Le sondage de Wépion. Mém Expl Cartes Géol Min Belg 2: 1-102
Jacobeen F, Kanes WH (1975) Structure of Broadtop Synclinorium, Wills Mountain Anticlinorium, and Allegheny Frontal Zone. AAPG Bull V 39 (7): 1136-1150
Julivert M (1971) Décollement tectonics in the Hercynian Cordillera of northwest Spain. Am Jour Sci 270: 1-29
Meissner R, Bartelsen H, Murawski H (1981) Thin-skinned tectonics in the northern Rhenish Massif, Germany. Nature Land 290 (5805): 399-401
Vandenberghe N (1984) The subsurface geology of the Meer area in North Belgium and its significance for the occurrence of hydrocarbons. J Petroleum Geol 7 (1):55-66
Van Wijhe DH, Bless MJM (1974) The Westphalian of the Netherlands with special reference to the miospore assemblages. Geol Mijn bouw (53): 295-328

# SCIENTIFIC CONTINENTAL DRILLING PROJECT IN JAPAN AND SCIENTIFIC GOALS

Nobuaki Niitsuma[1]

## Introduction

The Japanese Island Arcs locate between the Eurasian continent and the Pacific Ocean, and the crustal structure represents both continental and oceanic characteristics. The Island Arc involves continental crust formed by the accretionary process. The continental crust is separated from the Eurasian Continent by the marginal seas with oceanic crust. Studies of the geologic history of the continental crust suggest that one should consider the continental crusts as being linked to the global tectonic events, but as an isolated event. Naturally, the optimum scientific goal of the proposed continental drilling is to focus on the ultimate understanding of the dynamics of the crust.

The prospect of a scientific continental drilling project was considered as a part of the International Lithosphere Project in the later 1970's, based upon dramatically improved information. However, the drilling itself has not been carried out in the following decade. The Japanese earth science community strongly urges a start in scientific continental drilling in the final decade of this century.

## Activity for Scientific Continental Drilling

As a part of combined efforts, our organization has distributed the publication *Scientific Boring News Letter* to directly and indirectly related communities since 1981. The News Letter contains information on the activities of Coordinating Committee 6 *Continental Drilling* of the Lithosphere Project, recent drilling operations and plans in Japan, core sample availability, technical notes, related project of earth science, *Drilling Early Warning Newsletter* of the United States, etc. In 1982, we published a thorough data book on 428 holes drilled in Japan up to that year with 1:5000 geologic columnar sections.

## History of Japanese Scientific Drilling

Japanese scientific drilling was initiated by the project of drilling a coral reef sequence at Kitadaito-jima in 1934 and 1936. The drill hole reached the depth of 431.67 m, however, did not reach the volcanic basement rocks. The thick limestone and calcareous sandstone indicate

[1]Institute of Geosciences, Shizuoka University, 422 Shizuoka, Japan

Observation of the Continental Crust
through Drilling I (ed. by C.B. Raleigh)

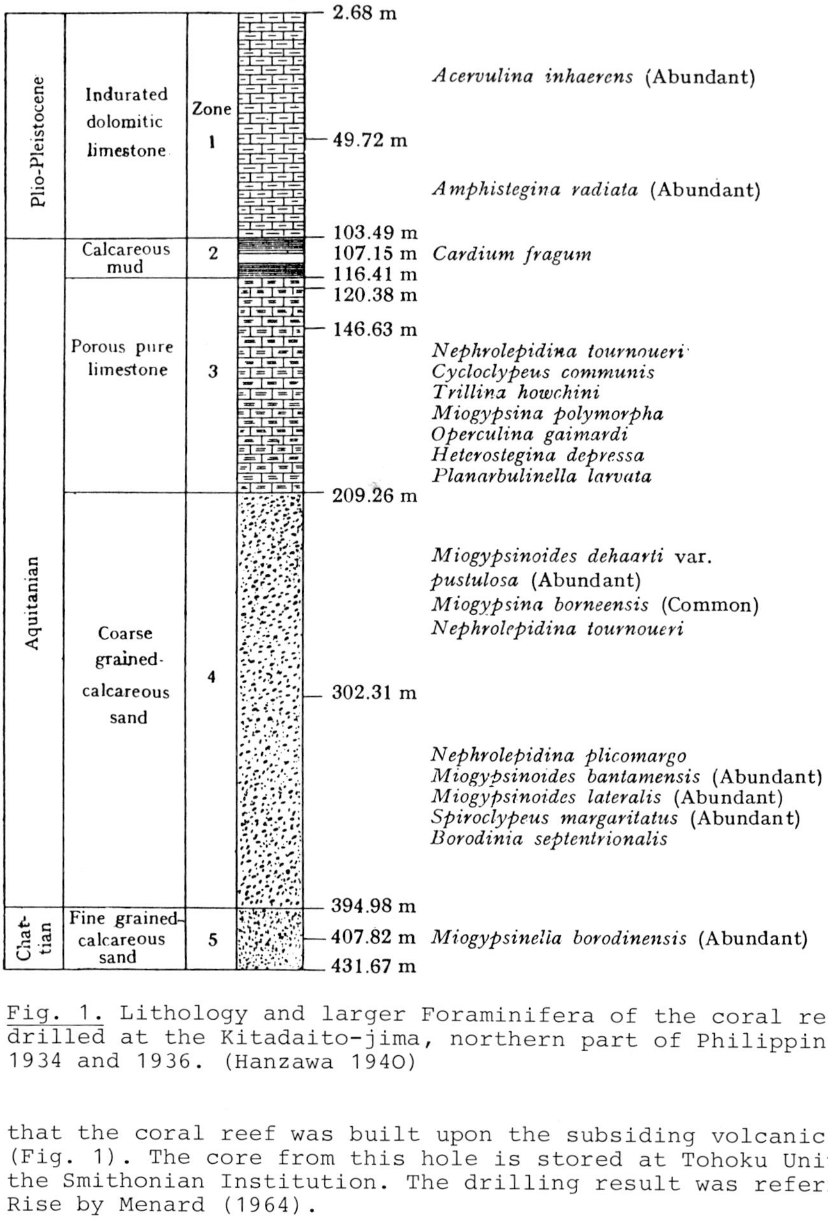


Fig. 1. Lithology and larger Foraminifera of the coral reef sequence, drilled at the Kitadaito-jima, northern part of Philippine Basin, in 1934 and 1936. (Hanzawa 1940)

that the coral reef was built upon the subsiding volcanic basement (Fig. 1). The core from this hole is stored at Tohoku University and the Smithonian Institution. The drilling result was referred as Dawin Rise by Menard (1964).

More systematic continental drilling was planned under the Upper Mantle Project in the 1960's by our community, similar or related to the Russian Super Deep Boring and the Mohole Project in the United States. The three sites were considered for drilling during the site survey (Yoshida et al. 1973), and a site in north-eastern Japan was proposed

Fig. 2. Typical cross-section of North-East Japan Arc, which was considered in the stage of Upper Mantle Project (Kitamura and Onuki 1973). *Arrow* represents the proposed drilling site in 1968

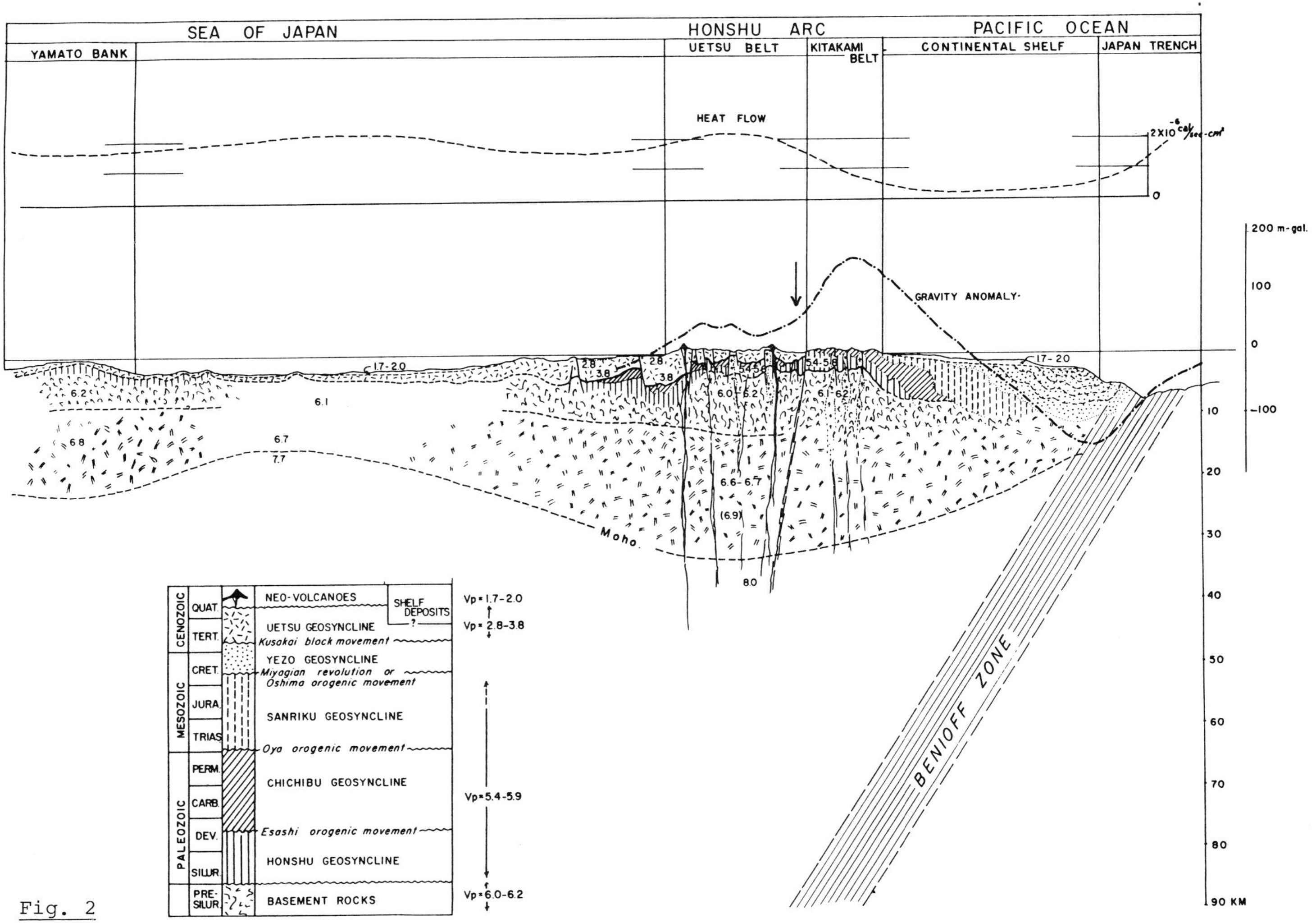
SEA OF JAPAN
HONSHU ARC
PACIFIC OCEAN
YAMATO BANK
UETSU BELT
KITAKAMI BELT
CONTINENTAL SHELF
JAPAN TRENCH
HEAT FLOW
2X10^-6 cal/sec-cm²
0
200 m-gal.
100
0
-100
GRAVITY ANOMALY
17-20
17-20
2.8
3.8
2.8
3.8
5.4-5.8
5.4-5.8
6.2
6.1
6.7
7.7
6.8
6.0
6.2
6.1
6.2
6.6-6.7
(6.9)
8.0
Moho
BENIOFF ZONE
10
20
30
40
50
60
70
80
90 KM
CENOZOIC
MESOZOIC
PALEOZOIC
QUAT.
TERT.
CRET.
JURA.
TRIAS.
PERM.
CARB.
DEV.
SILUR.
PRE-SILUR.
NEO-VOLCANOES
SHELF DEPOSITS
?
UETSU GEOSYNCLINE
Kusakai block movement
YEZO GEOSYNCLINE
Miyagian revolution or Oshima orogenic movement
SANRIKU GEOSYNCLINE
Oya orogenic movement
CHICHIBU GEOSYNCLINE
Esashi orogenic movement
HONSHU GEOSYNCLINE
BASEMENT ROCKS
Vp=1.7-2.0
Vp=2.8-3.8
Vp=5.4-5.9
Vp=6.0-6.2

Fig. 2

as the final one. The Japanese Committee of the Upper Mantle Project recommended drilling down to 5000 m depth in 1968.

Figure 2 shows a typical cross-section of the North-East Japan Arc at the stage of discussion where scientists were concerned with the sedimentary sequences with geosynclinal sedimentary bodies. The major purpose of the drilling was to collect the basement rocks overlaid by Silur-Devonian Geosyncline. The seismic velocity of the turget rock was 6.0-6.2 km $s^{-1}$. The sequence with such velocity was thought to be the pre-Silurian basement rocks in the most part of Japan. The Wadati-Benioff Zone was illustrated as shear zone or fault zone, but not a descending slab. Most site survey efforts were spent on the determination of the seismic velocity. As a result, the 6.0-6.2 km $s^{-1}$ layer was found below the depth of 3 km in the proposed site. The 5000-m of drill hole should reach and drill through the pre-Silurian basement rock.

During the planning of the drilling in late 1960's, the new concept of plate tectonics emerged. The new concept called attention on the importance of the horizontal movements during tectonic developments, especially in the plate boundaries. In the next decade, the 1970's, the Geodynamics Project was organized by a number of Japanese geoscientists in order to clarify the evolution of the plate subduction zone. A great number of age determinations have been made on the sedimentary rocks using planktonic microfossils. Thus we are capable to depict detailed tectonic histories in the Neogene, and now know that the majority of the so-called Paleozoic sediments are, in fact, the Mesozoic subduction complex. Paleomagnetic results suggest a considerable horizontal movement of the crust.

While the plan of the continental drilling for the pre-Silurian basement was suspended, a scientific drilling of 1934 m depth was made for direct observation of the focus of earthquake at Matsushiro, central Japan, in 1969, and 200-m and 1423-m holes were made for reconstruction of Pleistocene paleoenvironments at Lake Biwa in 1971 and 1983. Japanese government and enterprises carried out drillings for oil-gas, metal mining and geothermal prospections and exploitations. These results are incorporated in the *Data Book of Boring*, as mentioned above (Fig. 3). The drillings ranged between 500 m to 5300 m. The drillings for exploitation are still active and 3000-5000 m drillings are being carried out for oil and gas and the number of wells are about 15 on land and 7-8 off shore in a year. Ten to 15 holes with 500-2700 m depth are bored for metal mining in a year. Twenty holes with 1000-3000 m depth are bored for geothermal energy in a year.

## Evolution and Crustal Structure of Japanese Island Arcs

This section will give a general view of Japanese geology, based on the updated age determination, paleomagnetic data with tectonic considerations and the confined drilling results. Figure 4 shows a typical cross-section of the North-East Japanese Arc. We see high gravity anomaly along the fore-arc and the outer-arc belt. The heat flow raises along the volcanic front. The Wadati-Benioff zone is regarded as a descending slab.

Results of drilling of the oil wells show that the thick Neogene sediments accumulated in a limited zone with a few tens of kilometers width along the present Japan Sea coast (Fig. 5). The Neogene sediments are

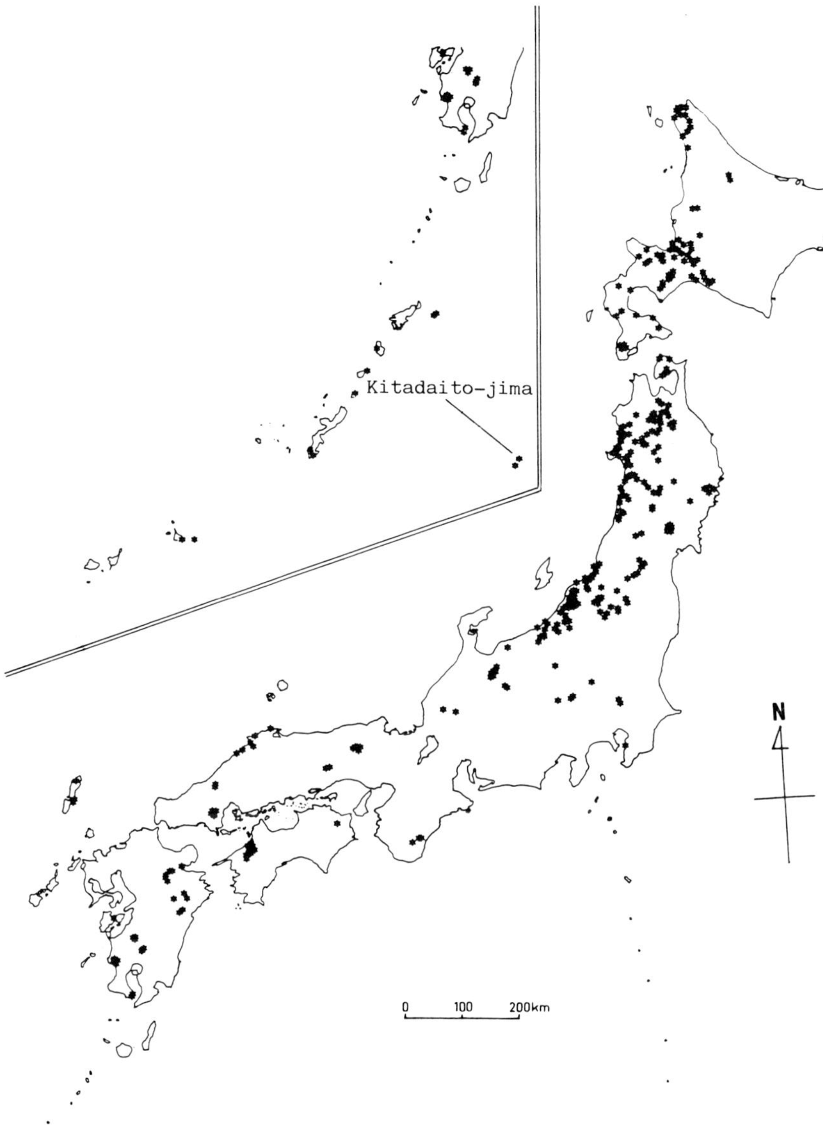


Fig. 3. Locations of the drilling sites which are incorporated in the Data Book of Boring in Japan. The coral reef drilling site of the Kitadaito-jima is also included

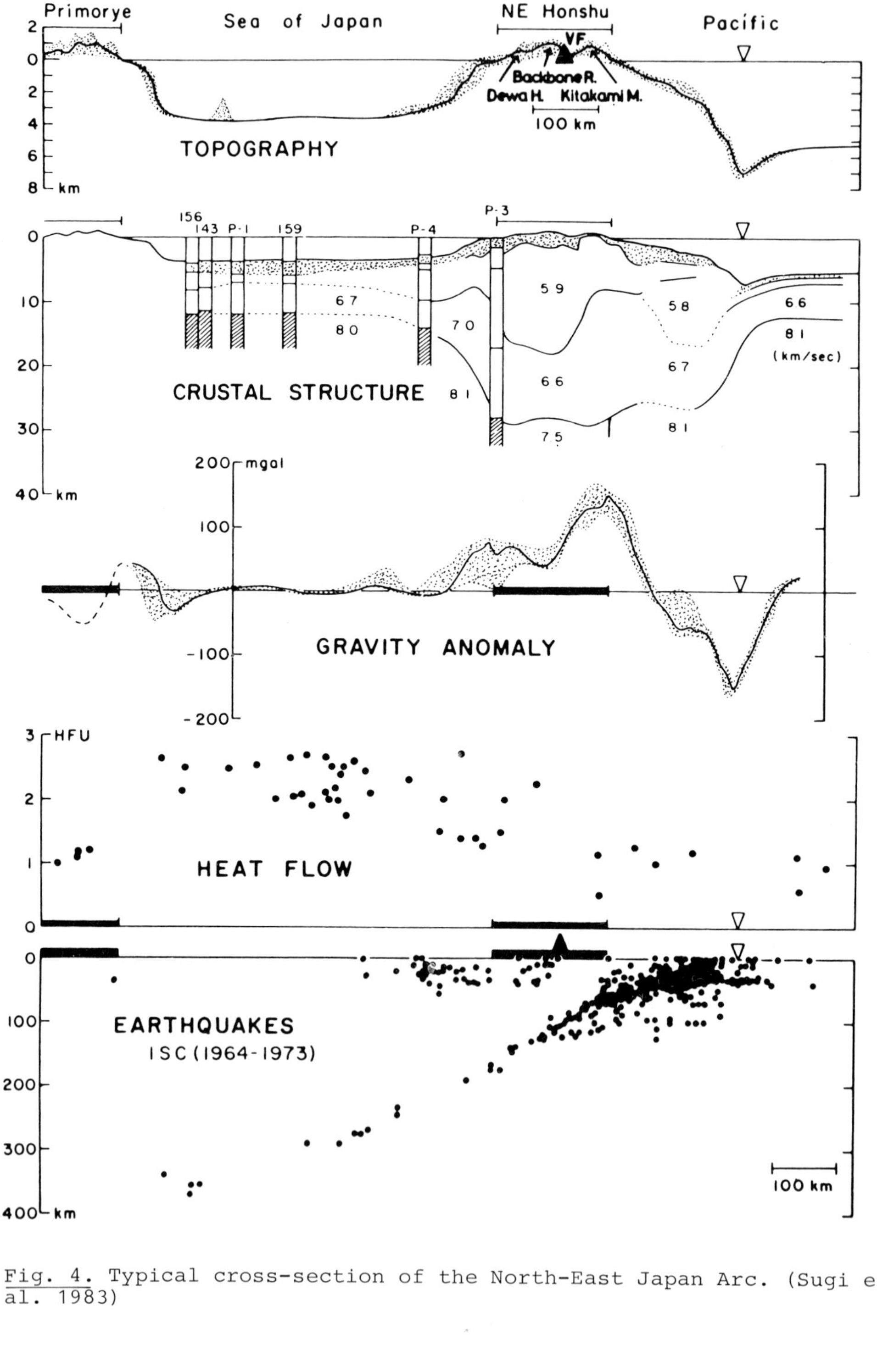


Fig. 4. Typical cross-section of the North-East Japan Arc. (Sugi et al. 1983)

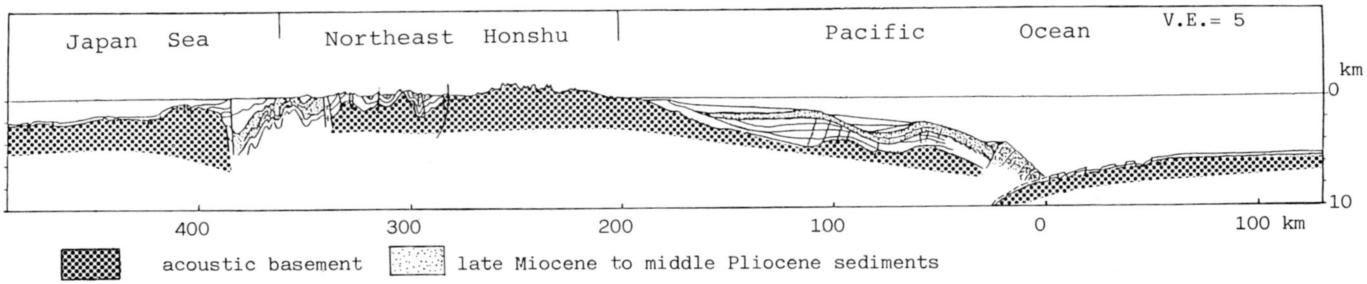


Fig. 5. Geologic cross-section of North-East Japan, based on Ishiwada et al. (1977) and Niitsuma and Akiba (in press)

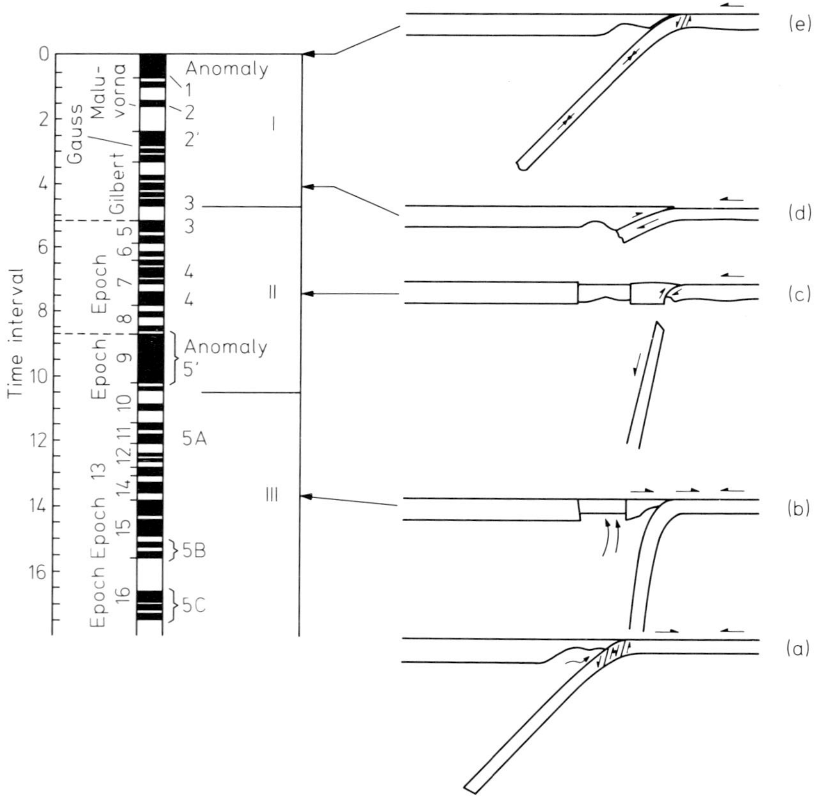


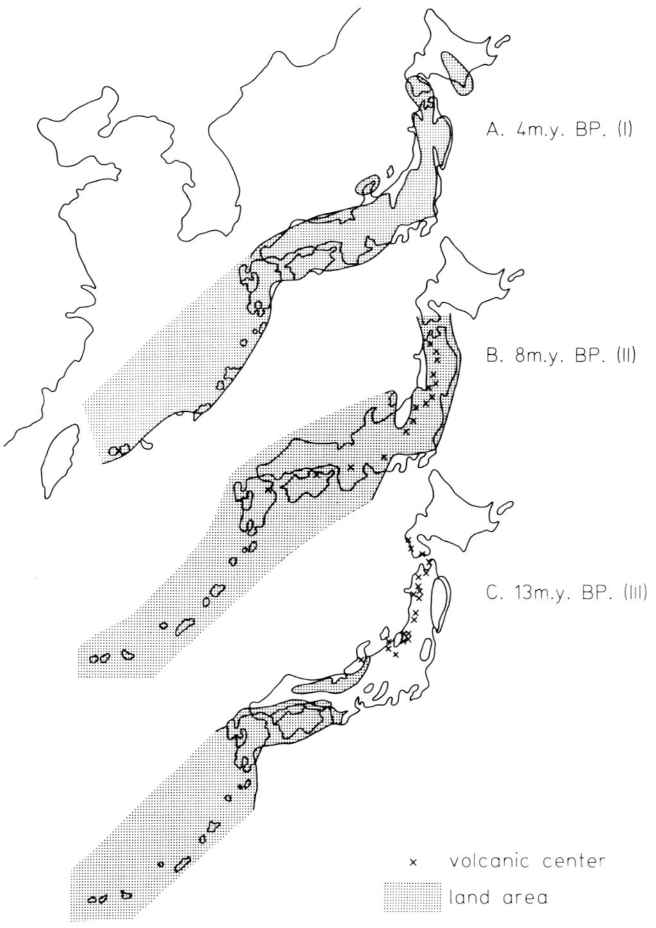


Fig. 6. Schematic "cyclic evolutionary model" for the plate subduction in North-Eastern Japanese island arc system and geochronological classification of tectonic evolution. *Right* paleogeographic maps of the Japanese island arc area in the each tectonic evolutionary time interval. (Niitsuma 1978)

thicker than 5000 m and their basements have not been reached. The depth of the basements is comparable or deeper than the depth of the acoustic basement in the Japan Basin of the Japan Sea with the oceanic crustal structure. The sediments which deposited in the trough are younger than middle Miocene. These facts suggest that the productive oil field might be limited to the narrow and deep trough along the coast. The trough was formed in the middle Miocene. The siliceous mud-stone of the middle Miocene was intruded by dolerite sheets; the in-trusion is related to the formation of the abnormal trough which was formed by rifting of the continental crust and lithosphere.

In the middle Miocene, marine sediments widely covered the Japanese Island Arc area (Fig. 6). The paleogeography during this stage is sim-ilar to that of the present Mariana-Bonin Arc, where a main part of the volcanic arc is situated below the sea level and the island arc con-sists of nonvolcanic basement. In the present Mariana Arc, we know that the back arcs are rifting. The rifting can be correlated to the for-mation of the oil field along the Japan Sea coast and the opening of the Japan Sea. In the next stage of the Mariana-type paleogeography, 10 m.y. the Japanese Island Arc changed abruptly to the features sim-ilar to the present Japanese Island Arc. This paleogeographic evolu-tion suggests that the type of the Island Arc, for example, Mariana-type, can be changed in a time series. "Cyclic evolutionary model for the plate subduction (Niitsuma 1978) is one explanation for the evo-lutionary events of the Japanese Island Arcs (Fig. 6). A descending slab grows in length. The down-pull force increases. Paleogeographic setting of the Mariana-type arc is formed by this down-pull force en-vironment. Finally, the fully extended slab detaches and this triggers abrupt changes in the surface paleogeography of the area.

A detailed reconstruction of paleobathymetry has been made by Kitazato (1983), based on the benthic foraminiferal fauna (Fig. 7). We can see another trough with submarine volcanos in the outer zone of the oil field. The activities of the submarine volcanos were related to the formation of Kuroko Ore Deposits.

Paleomagnetic evidence shows that the opening of the Japan Sea occur-red at the middle Miocene, 15-12 m.y. (Otofuji and Matsuda 1983). This means that all of the events mentioned above were related to or a part of the marginal sea opening and rifting of continental lithosphere (Fig. 8).

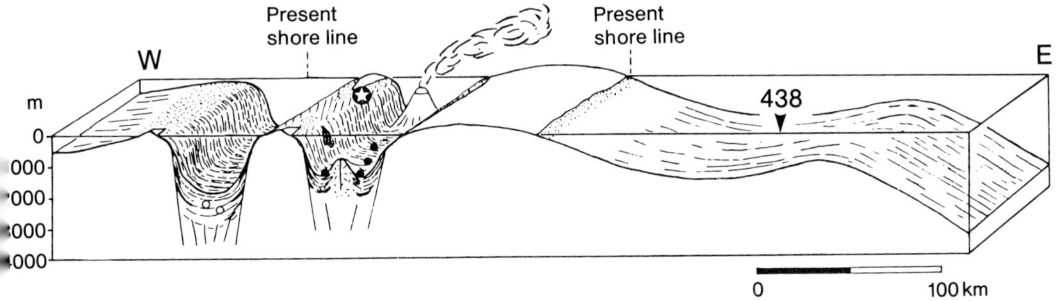


Fig. 7. Schematic bird's-eye view of the paleotopography of the North-east Honshu Arc during the early Middle Miocene. (Kitazato 1983)

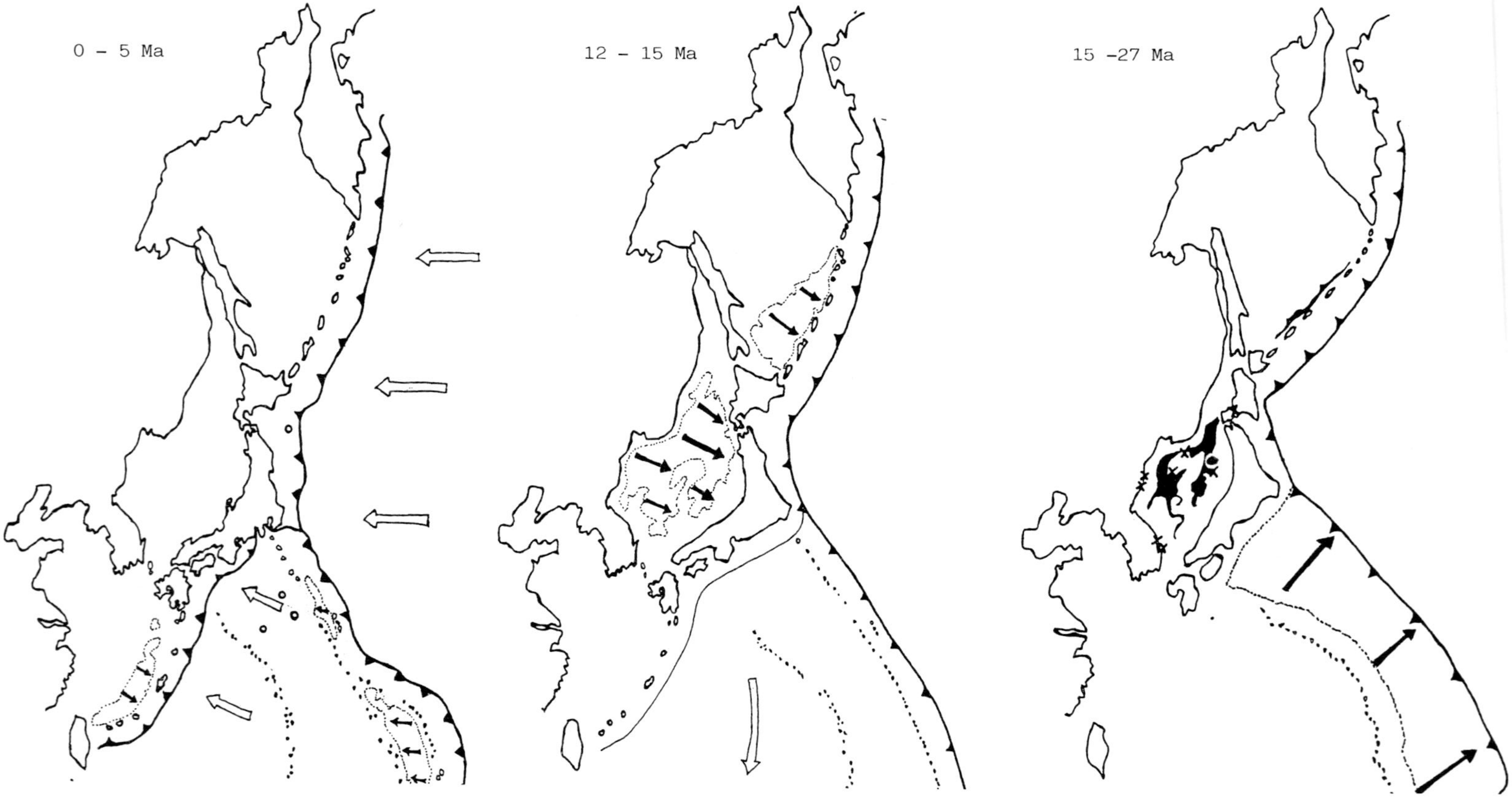


Fig. 8. Opening of the marginal seas *(black arrow)* and plate motion *(open arrow)* around the Japanese Island Arcs

In southwest Japan, noticeable events became apparent; southwest Japan faces to the Shikoku Basin, the east part of Philippine Sea. The Shikoku Basin had opened prior to the opening of the Japan Sea. The opening of the Japan Sea made southwest Japan overlay the Shikoku basin, while off-ridge volcanic activities were still present. During this limited stage of the opening of Japan Sea, specific volcanic activities were observed in the outer zone of the southwest Japan (Takahashi 1980). The volcanic activities were characterized by the great amount of pyroclastic flow, referred to as granitoid in the Kumano-type and Valles-type volcanic activities, and high magnesian andesitic volcanos (Fig. 9). The formation of granitoids in the deeper portion can be classified in the formation of the continental crust along the continental margin (Fig. 10).

One can say that these events in the middle Miocene, which were induced by the descending slab, were directly relating to the rift and formation of continental crust, and such conspicuous crustal evolution, happened in a short duration of 2-3 m.y.

The basement of the Neogene sedimentary sequences can be divided into three divisions (Fig. 11); (1) Cretaceous-Paleogene subduction complex, (2) Jurassic subduction complex, containing exotic blocks of Paleozoic ages, (3) "Older Terrane", probably fragments of continent, exotic blocks. All those sections are cut through by strike slip mobile zones involving ultramafic rocks (Taira et al. 1983). The reconstruction on the Late Cretaceous subduction zones is shown in Fig. 12, based on the analysis of sedimentary facies and igneous activity.

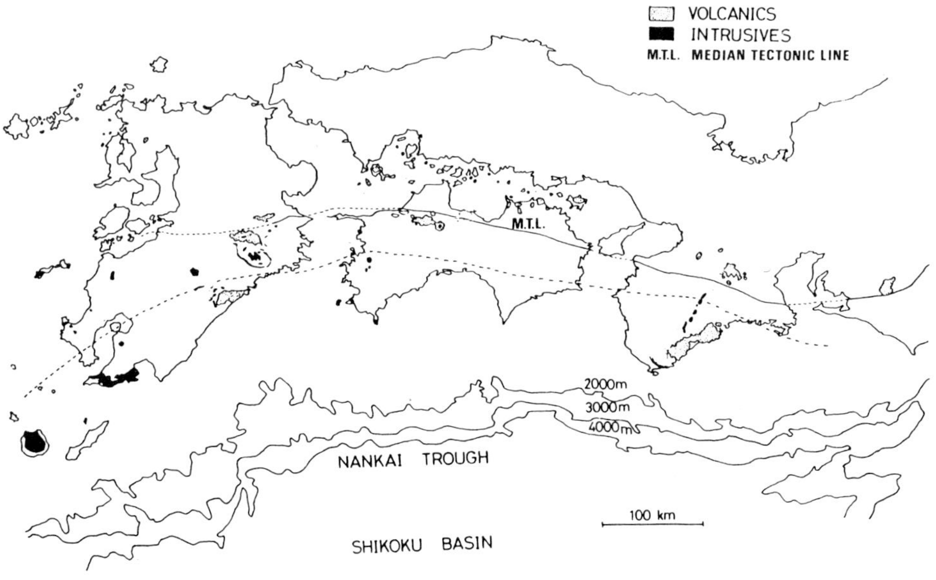


Fig. 9. Distribution of the middle Miocene Outer Zone granitoids and Setouchi volcanics. (After Takahashi et al. 1980)

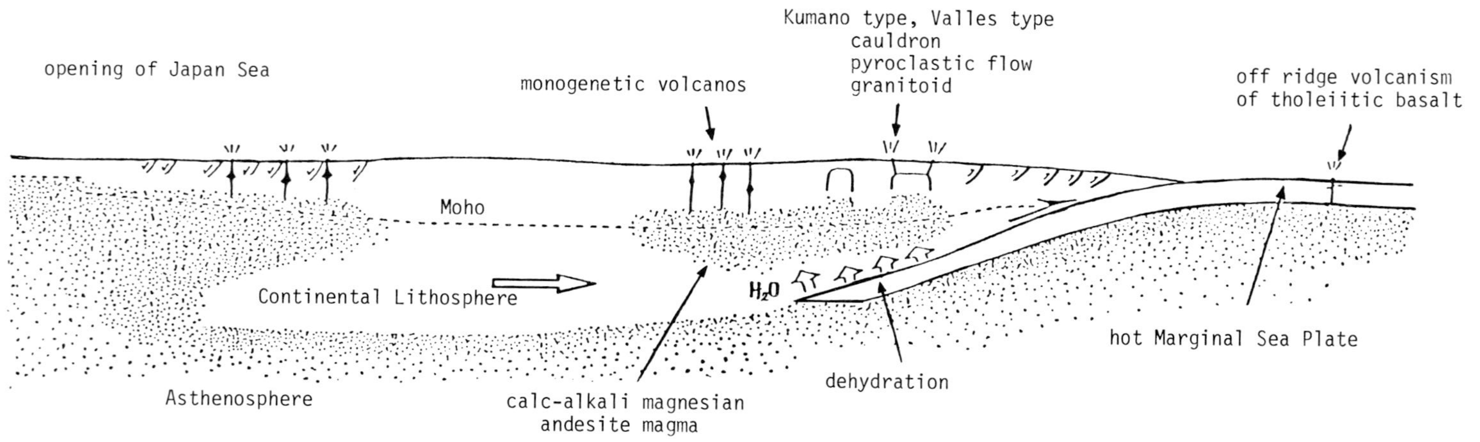


Fig. 10. Schematic north-south cross-section of the Southwest Japan in the middle Miocene and magmatism related with the opening of the Japan Sea. (Takahashi 1980)

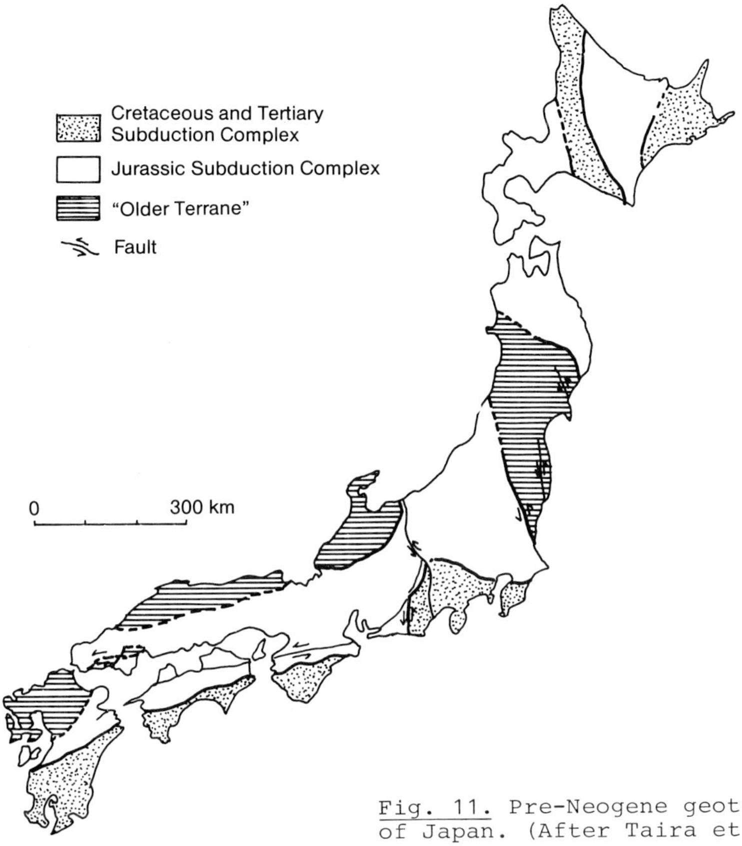


Fig. 11. Pre-Neogene geotectonic framework of Japan. (After Taira et al. 1983)

In Hokkaido Island, Northern Japan, the crust collided at the area of junction between Honshu Arc and Kulile Arc. The collision occurred in the Pliocene, judging from the timing of the folding and thrusting observed in the Pliocene sedimentary layer. In the collision belt, named the Hidaka Mountains, ophiolite are exposing in the West Belt and the island arc or continental crust and mantle are exposing in the Main Belt (Fig. 13). Both belts are separated by the Hidaka Main Thrust. Such a collision at the arcs junction is also found in the Southern Fossa Magna of central Honshu. Here the collision was started in Pliocene, caused by the subduction of the Philippine Sea Plate.

## Proposed Sites for Scientific Continental Drilling

In concluding this paper, I will list the proposed scientific drilling sites based upon the dynamic setting of the Japanese geology and ever evolving knowledge in geophysics (Fig. 14).

1. Continuing the studies of the coral reef drilling, such as Kitadaito-jima drilling, which is aimed at reconstructing the paleoenvironment and evolution of the ocean basin.
2. Opening of the marginal seas, such as the Japan Sea, and rifting process of the crusts of island arc, relating to oil and metal resources.

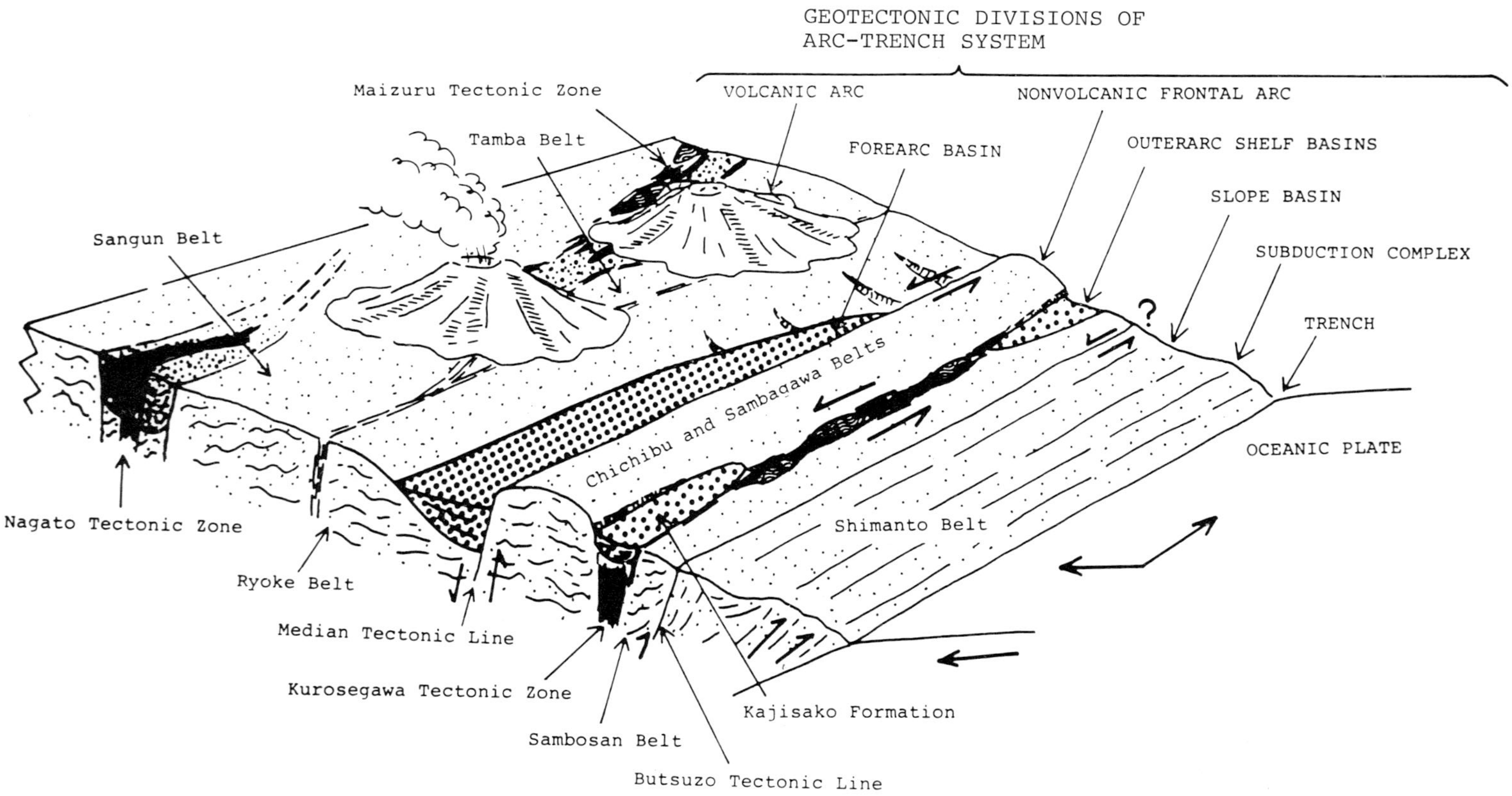


Fig. 12. Reconstruction of Late Cretaceous oblique subduction margin in southwest Japan. (Taira et al. 1983)

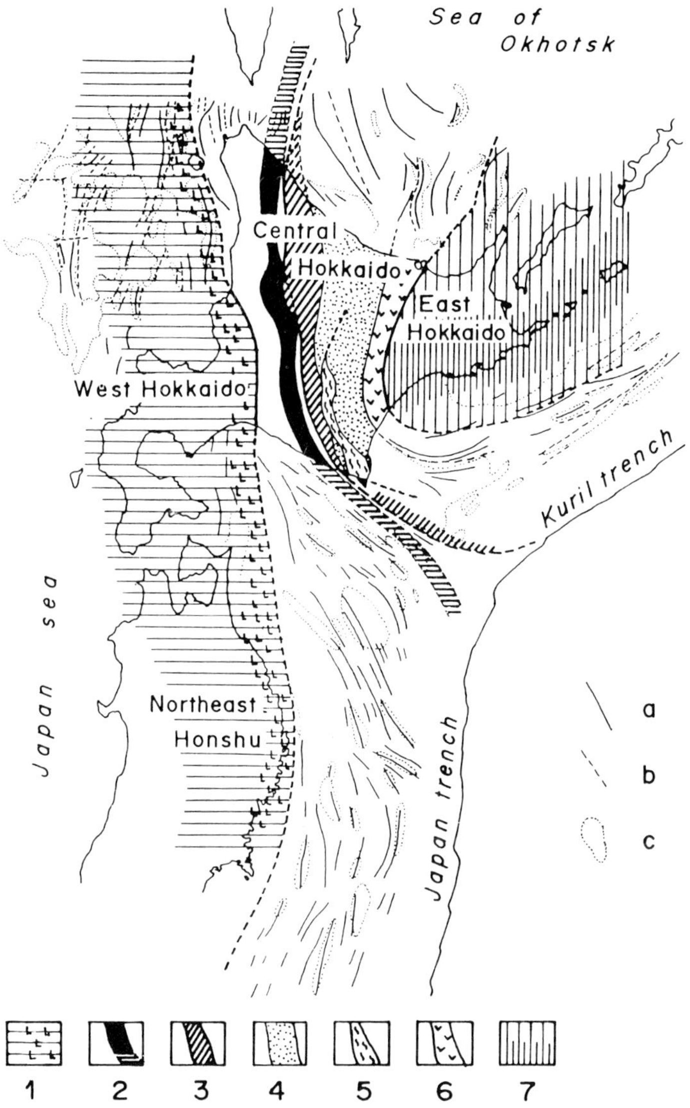


Fig. 13. Tectonic division and structures of Hokkaido and the adjacent areas (Komatsu et al. 1983). *1* Rebun-Kabato range and Iwaizumi belt; *2* Kamuikotan belt and its extension; *3* Hidaka western melange zone; *4* Hidaka sedimentary zone; *5* Hidaka metamorphic belt; *6* Tokoro belt; *7* Kuril arc (outer arc). *a* fold axes; *b* faults; *c* axial part of acoustic basement high

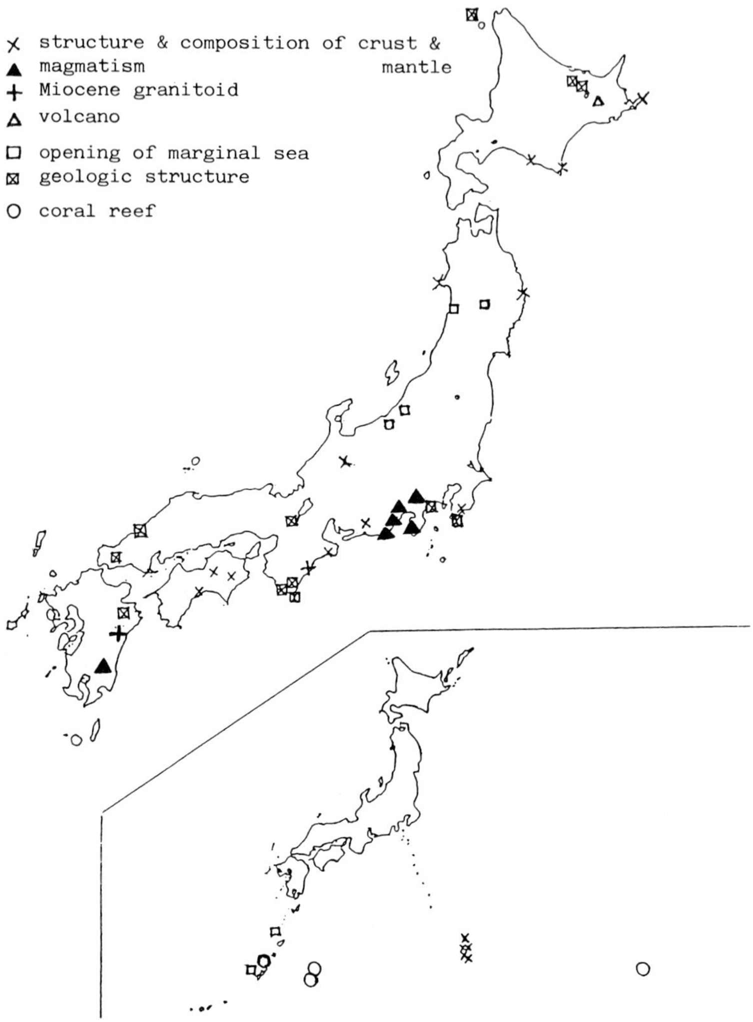


Fig. 14. Proposed scientific drilling sites

3. Formation of the continental crust, which corresponds to granitoid magmatism.
4. Detailed structure of volcanos, magmatism and hydrothermal circulation systems.
5. Structure and composition of crust and mantle, uplifted by the collision processes. Recognition of the gravity high zone along the forearc and outer arc belt.
6. Clarification of the subduction mechanism in ancient subduction zone.
7. Major processes in structural evolution, for example, Decke, Nappe, and Melange complexes.

*Acknowledgments.* I would like to thank B. Raleigh of Lamont-Doherty Geological Observatory for the invitation to the International Symposium on Observation of the Continental Crust Through Drilling and for giving me the international communications. It will be very helpful in realizing a scientific continental drilling project in Japan. I also express my thanks to H. Hase of Geological Survey of Japan and K. Kimura of Japan National Oil Corporation for informations on the recent drilling activities for geothermal energy and oil-gas exploitations, Y. Saito of the National Science Museum, Tokyo, A. Taira of Tokyo University, G. Kimura of Kagawa University, S. Maruyama of Toyama University, M. Hashimoto, K. Amano, and M. Takahashi of Ibaragi University for discussions on the tectonic evolution of Japanese Island Arcs, and S. Honjo of Woods Hole Oceanographic Institution for reviewing the manuscript.

## References

Hanzawa S (1940) Micropaleontological studies of drill cores from a deep well in Kita-daito-zima (North Borodino Island). Jubilee Publication of Prof H Yabe, vol 2, pp 755-802

Ishiwada Y, Ikebe Y, Ogawa K, Onitsuka T (1977) A consideration on the scheme of sedimentary basins of Northeast Japan. Geological Papers dedicated to Prof K Huzioka, pp 1-7

Kitamura N, Onuki Y (1973) Geological and crustal sections of the A-zone, Northeast Japan. The Crust and Upper Mantle of the Japanese Area, part II (Geol Geochem). Jpn Natl Committee Upper Mantle Project. Geol Surv Jpn, pp 38-60

Kitazato H (1983) Submarine topography of the northeast Honshu arc during the early Middle Miocene Nishikurosawa stage, based on the benthic foraminifera. Min Geol Spec Issue 11: 263-270

Komatsu M, Miyashita S, Maeda J, Osanai Y, Toyoshima T (1983) Disclosing of a deepest section of continental-type crust up-thrust as the final event of collision of arcs in Hokkaido, North Japan. Accretion tectonics in the Circum-Pacific regions. In: Hashimoto M, Uyeda S (eds) Terrapub, Tokyo, pp 149-165

Menard HW (1964) Marine geology of the Pacific. McGraw-Hill, San Francisco, p 271

Niitsuma N (1978) Magnetic stratigraphy of the Japanese Neogene and the development of the island arcs of Japan. J Phys Earth Suppl 26: 367-378

Otofuji Y, Matsuda T (1983) Paleomagnetic evidence for the clockwise rotation of Southwest Japan. Earth Planet Sci lett 61: 349-359

Sugi N, Chinzei K, Uyeda S (1983) Vertical crustal movements of northeast Japan since Middle Miocene. AGU-GSA Feodynamics Ser 11, pp 303-316

Taira A, Saito Y, Hashimoto M (1983) The role of oblique subduction and strike-slip tectonics in the evolution of Japan. AGU-GSA Geodynamics Ser 11, pp 317-329

Takahashi M, Aramaki S, Ishihara S (1980) Magnetite-series/Ilmenite-series vs. I-type/S-type granitoids. Min Geol Spec Issue 8: 13-28

Takahashi M (1980) Large scale activity felsic magma and formation of Upper Crust in mobile belt. Chikyu 2: 837-845

Yoshida T, Iizuka S, Igi S (1973) Reconnaissance survey for deep drilling in the Japanese Upper Mantle Project. The Crust and Upper Mantle of the Japanese Area, part II (Geol Geochem). Jpn Natl Committee Upper Mantle Project. Geol Surv Jpn, pp 163-176

# PROGRAMME FOR DEEP DRILLING IN SWEDEN

K. Gösta Eriksson[1]

## Introduction

Increasing interest is being diverted in many countries towards studying the deeper parts of the Earth's crust. Several projects involving deep-earth surveys are either in progress or being planned. The motives for these projects may be briefly summarized as follows:

- the possibility of achieving an overall increase on geoscientific knowledge of the structure, material, and development of the Earth's crust and mantle
- prospecting for valuable substances such as deep-lying oil, gas, minerals, geoenergy, etc.

Deep drilling in Sweden — that means today to a depth of more than 500 m — is or will be of interest mainly for the following fields:

- deep ore exploration
- deep gas prospecting
- radioactive waste disposal
- geothermal energy exploration
- subsurface storage of vis heat, gas, oil, etc.

Deep seismic investigations in Sweden consist today of participating in two international projects:

- European seismic investigation of the Earth's crust
- Project Tornquist (IGCP No. 86)

## Geology

The Precambrian of Sweden is part of the Baltic (Fennoscandian) Shield (Fig. 1). It is characterized by a complex evolution which is still the subject of extensive investigation. The oldest Precambrian rocks in the Swedish part of the shield are of early Precambrian age (>2600 m.y.). Only restricted areas in the northernmost parts of Scandinavia are remnants of this pre-Karelian complex; it contains granitic gneisses dated to be about 2750 m.y. old. In the Russian regions of this very old bedrock a super-deep hole has been drilled (its depth was 38,261 ft or 11,660 m in 1983) in the Kola-Peninsula by geologists from the

[1]Department of Geology, Chalmers University of Technology, Gothenburg, Sweden

Observation of the Continental Crust through Drilling I (ed. by C.B. Raleigh)

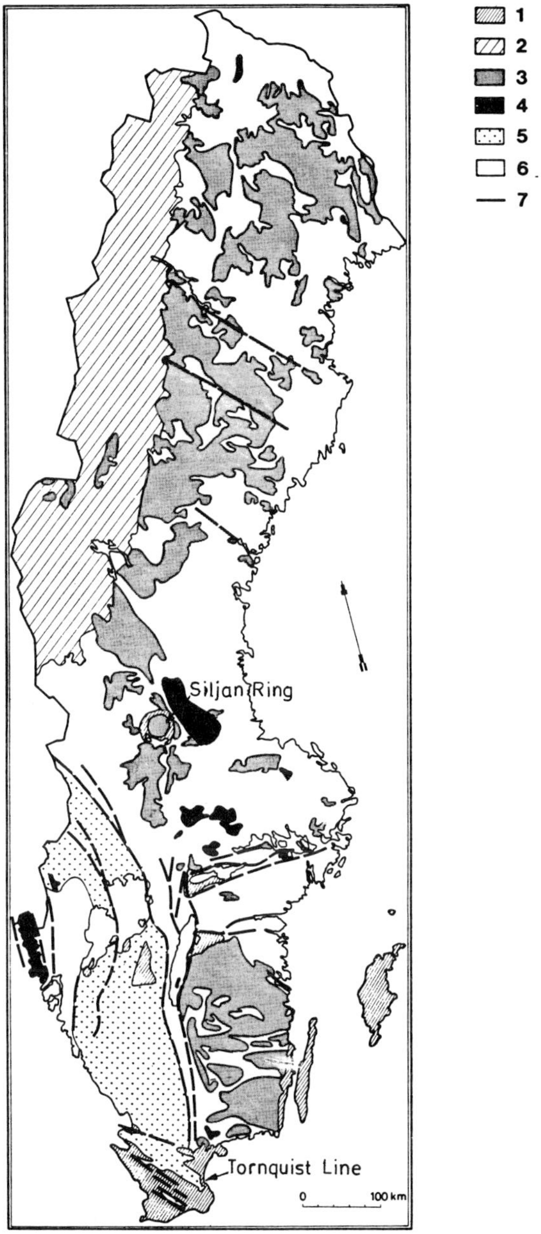


Fig. 1. Sketch map of the bedrock geology of Sweden. *1* Paleozoic-Mesozoic sedimentary rocks; *2* the Caledonides; *3* Precambrian granites which normally have a higher heat flow than the average Precambrian rocks; *4* Precambrian granite areas of a considerably higher heat flow than is normally to be expected; *5* Precambrian gneiss area of southwestern Sweden; *6* other Precambrian rocks; *7* major tectonic lineaments. (Eriksson and Malmqvist 1979)

Soviet Union. The results are very interesting and will probably give base for a reconsideration of previous theories concerning the structure of the Earth's crust (Kozlovsky 1982).

The youngest intrusions of granite in southwest Sweden took place from about 1250 to 900 m.y. ago, the lowest age (910 m.y.) being represented by the Bohus granite. Like other granites with a similar tectonic position, it is characterized by a higher heat flow than the average crystalline rocks in Sweden (Fig. 1).

In northwestern Sweden the eastern parts of the former Caledonian geosyncline are covered by relatively undisturbed old Paleozoic sedimentary rocks. The rocks of the major part of the Caledonian mountain range have been metamorphosed and overthrust in Silurian time. Ancient Paleozoic rocks occur in some other parts of Sweden as platform cover deposits resting on the sub-Cambrian peneplane. Examples are found on the islands of Öland and Gotland in the Baltic, and also in the Bothnian depression.

In the southwestern most part of Sweden - in the province of Scania - the Precambrian crystalline rocks are down-faulted as much as 2000 m along fault lines with a NW-SE direction. This important fault zone (the Tornquist fault) forms the geological boundary between Fennoscandia and the rest of Europe. The southwestern part of Scania is covered by a sequence of Paleozoic, Mesozoic, and Cenozoic sedimentary rocks and sediments about 2000 m thick.

## Deep Ore Exploration

Much of the Swedish mining industry is centred on deposits of complex sulphide ores, particularly in the Skellefte field and the Bergslagen districts. The economically important constituents in these deposits are the base metals copper, zinc, lead, and the precious metals gold and silver.

At the present time complex sulphide ores are being worked at about 15 mines in Sweden. Boliden Mineral AB is the major mining company, but sulphide ores are also mined by LKAB, Stora Kopparbergs Bergslags AB, and Vieille Montagne.

Normal mining depths in Sweden have hitherto been rather moderate, only to depths of 200 m; the Kiruna iron mine was, for example, exploited in an opencast mine until a few years ago. The number of economically interesting ore bodies at shallow depths are, however, radically decreasing.

Worthwhile new discoveries of complex sulphide ores, sufficient to assure the long-term survival of the mining industry in these and other provinces, are therefore likely to be found only through deep ore exploration. In order to stimulate R & D in deep drilling for ore prospecting, the National Swedish Board for Technical Development, STU, has initiated a 5-year programme for development of technical systems and methods to be used in the search for ore deposits, especially sulphide ores, at great depths, i.e., down to 1000 m.

Ore exploration using methods and equipment particularly suitable for locating ore bodies reaching up to the bedrock surface has been carried out for a long time in the main mining districts in Sweden. Simulation studies have, however, shown that the exploration efforts should be concentrated at depths of 400 to 600 m. Potentially economic deposits must be high grade ores of at least some million tons in size.

Exploration at these depths will not be possible without comprehensive drilling. The high cost of diamond drilling is today the limiting factor in deep ore exploration. Therefore, projects aimed at innovations in the field of drilling techniques are given the highest priority within the STU programme. This includes development of new techniques as well as improvements to established techniques and methods.

The development programme is being carried out for the period 1982-1987. For the implementation of the programme STU has appointed a Steering committee with representatives from Swedish mineral exploration companies and authorities. Experiences of ore exploration at depth are limited, in Sweden as well as internationally. So one of the main tasks of the Steering committee is to establish international contacts in order to exchange information about current development with other countries facing similar exploration problems. Of special interest is to learn about corresponding technology development programmes in other countries aimed at making deep ore exploration more efficient.

A pamphlet *Deep Ore Exploration* gives a short programme description which serves as guidelines for the programme. For exchange of information and experience of development in deep ore exploration technology those who are interested should contact the National Swedish Board for Technical Development, STU, Box 43 200, S-100 72 Stockholm, Sweden.

## Deep Gas Prospecting

The established and widely accepted view among petroleum geologists is that oil and natural gas are biological waste products created in sediment under the influence of high pressure and temperature.

During recent years, theories have been advanced propounding a non-biological origin for gas which, if confirmed, offer interesting possibilities from the Swedish point of view.

A limited study, referred to in this paper as the pilot study, has therefore been carried out during the later half of 1983 with the object of elucidating the deep gas theory and its potential for application in the case of Swedish geological structures. The work constitutes a continuation of an initial prestudy, *Deep Earth Gas in Sweden – Background and Possible Prospects*, published at the beginning of 1983.

The principal matters taken up in the study are for example:

- Follow-up deposits of gas and other hydrocarbons in Sweden
- Further evaluation of the areas that had been pointed out as being of interest in the previous study
- Analysis of advanced investigation methods to be used in a possible field investigation phase after the pilot study
- Analysis of drilling techniques in order to provide an indication of the design, costs, etc.
- Description of possible scientific spin-off effects from a deep gas project, i.e., interesting fields apart from deep gas.

It can be judged from the results of the pilot study that there is sufficient justification for continuing the work in a further stage. The work has led to interesting results, even if no long-term conclusions can be reached owing to the limited investments made to date.

Areas given priority for continued investigations are the Siljan Ring, Lake Vättern, and the Bothnian Sea. Among these areas, the Siljan Ring area appears to offer the largest reservoir volume and the greatest scientific interest as a test of the deep gas theory, since the area inside the Ring probably consists entirely of primary rock. On the other hand, it would be wrong to concentrate exclusively on this alternative. The other to areas should also be subjected to continued studies in a further stage.

## The Siljan Area

The Siljan Ring structure (Fig. 2) is the largest impact structure so far known to exist in Europe. The meteorite impact took place in Devonian time (about 360 m.y. ago). It is a complex structure with a central uplift (dome) about 32 km in diameter. The central massif is surrounded by a depression, defined by faults, with an exterior diameter of 44 km.

Within the depression, Ordovician and Silurian sedimentary rocks have been preserved, but are locally uplifted and overturned and can be characterized as a huge megabreccia. As a consequence of the disturbance, the total thickness of the Paleozoic fluctuates widely in different areas, a reasonable maximum value being 300-500 m. The almost circular dome consists mainly of coarse – medium granular Subjotnian Dala granites.

Gravity investigations have been conducted throughout the area, the results of which support the impact theory. The gravity anomalies cover an area twice that of the ring structure itself. A strong peak amplitude is excentrically located close to the periphery of the northern part of the ring. The results indicate a fracturing and brecciation of the uppermost 4 to 5 km of the crust.

Further evidence that the Siljan Ring is a structure created by the impact of a meteorite is provided by finds of shatter cones and the tectonic configuration around the former crater. The origin and appearance of the Siljan Ring can also be explained by a theoretical model for an elastic impact.

Several geophysical investigations have been carried out in the Siljan Ring area using a variety of methods. The investigations have primarily been centred on the north-western part of the area and together they give a good idea of the structure.

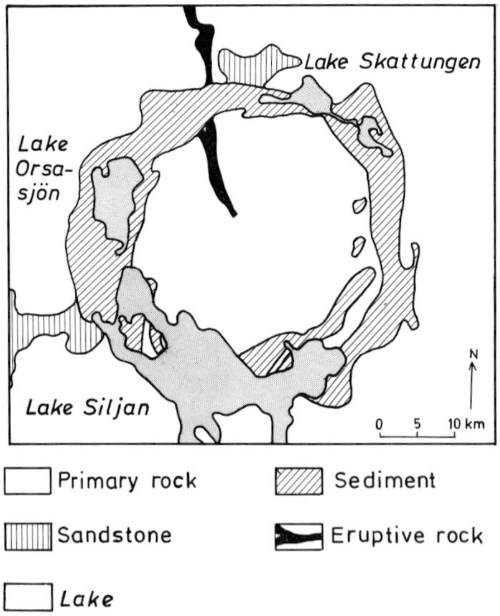


Fig. 2. The Siljan Ring area

The interpretation of satellite photos shows a number of radial shear zones in the rock surface outside the Ring, which supports the theory that the area is the place where a meteorite or other type of object from space struck the Earth.

Research in the impact cratering process has indicated that impact structures could be of interest for hydrocarbon production. The impact process has important implications for petroleum geology in that an impact can instantaneously create porous and permeable rock from non-reservoir material. The fracture pattern emanating from an astrobleme provides an intricate network of migration channels into the breccia lens. The larger the impact event, the deeper the fracture network.

At present, not enough is known about the reservoir parameters of the Siljan structure and the possibility of these being a seal. This means that the possibility of hydrocarbon accumulation in the Siljan structure is a matter for conjecture. Gravimetric and seismic measurements have indicated porosity anomalies, which could be interpreted to mean that additional investigations, including deep pilot-hole drilling, should be carried out before an evaluation and assessment can be made of the potential of the Siljan structure.

## Spin-off Effects

The deep gas study and its possible continuation in the form of advanced field surveys and drilling operations will also result in many other interesting geoscientific aspects alongside the actual prospecting for gas. There are possibilities of acquiring greater knowledge within several fields, for example rock mechanics, tectonics, hydrogeology, geothermy, etc. Historically speaking, it is also the case that significant contributions to the knowledge of the structure and content of the Earth's crust have been obtained in conjunction with the oil and gas prospecting all over the world which, in the course of time, has reached increasing depths. The vast majority of this prospecting has taken place in sedimentary rock. The possibility now exists to obtain increased knowledge within the field of crystalline bedrock. In conjunction with the decision to carry out the detailed preliminary study, it was only natural that it should incorporate elucidation of such spin-off effects.

Deep surveys and drilling are techniques in which Sweden is lagging behind. A Swedish project will provide us with the opportunity to handle, evaluate and develop the extremely advanced technology involved in both logging and drilling to great depths. In this way we will be given the chance to reduce the lead which many industrialized countries have over us in this field, which will in turn provide Swedish manufacturers, contractors, and consultants with export opportunities.

A Swedish deep exploration project would lead to a noticeable upswing within different geoscientific fields. It would provide basic geological research with unique possibilities to study the deep-lying bedrock structure in the country. It would also open the door to deeper scientific and technical cooperation with a number of countries.

The more specific areas of scientific interest in which a deep survey project in crystalline rocks would provide fresh knowledge can be briefly summarized under the following points:

- It would be possible to achieve important contributions to research on the development of the Earth's crust during Precambrian times

- The knowledge and experience gained in connection with deep surveys are of great value in assessing the safety of the design and construction of deep-lying rock caverns, for example, the final storage of radioactive waste
- The existence and behaviour of groundwater at great depth is a subject of considerable interest within geohydrological research
- Deep surveys are of fundamental importance in geothermal research
- Deep surveys in primary rock are of great importance for interpreting the formation and presence of certain ores and minerals
- From the point of view of rock mechanics, research at great depths will carry with it widely diversified spin-off effects within a large number of different fields, including questions concerning the behaviour of minerals and different types of rock under extreme pressure
- Knowledge of the behaviour and origin of earth gases can be of use in connection with ore prospecting
- Light could be shed on questions of the radon content in rock and soil.

## Radioactive Waste Disposal

The development of nuclear power throughout the world has generated several important issues, one of them being the handling and disposal of the radioactive wastes arising from the various operations which involve the nuclear fuel cycle, especially from reprocessing the spent nuclear fuels of the ultimate disposition of the spent nuclear fuels themselves. Problems dealing specifically with the disposal of radioactive wastes into the ground, covering both shallow land burial and disposal in deep continental rocks, have been, and still are, questions of great controversy in Sweden.

Many areas in the country are presently being investigated as possible storage sites. An important aspect of these investigations is drilling and core sampling at considerable depths (> 500 metres). In addition to practical experiences from drilling, valuable geological results can be anticipated, especially in the fields of crustal tectonics, hydrogeology, rock density and rock stress variations. During the next 20 years a series of deep boreholes are planned in the Swedish Precambrian rocks in order to further the research in the previously mentioned fields. It is important here to stress our extreme desire for contact and exchange of information and ideas with other researchers.

## Geothermal Energy Prospecting

Hitherto only a couple of areas in Sweden have been investigated concerning heat production. It is thus too early to make any statement about the geothermal resources of the country. However, the entire knowledge of the geology of Sweden gives some possibilities for certain estimates concerning heat production areas.

Only the granitic rocks may have a sufficient amount of radioactive elements to produce anomalous heat flow in the Swedish Precambrian rocks. Approximately 50 - 70% of Sweden consists of such acid rock material. It is estimated that only younger granites of Postorogenic or Serorogenic age contain anomalous areas.

A briefly outlined mapping of areas with increased gamma-radiation in Sweden has been made by its Geological Survey. In this respect only 25% of the country has hitherto been mapped.

As has been shown during the performed investigations, several granite areas can have gradients of approximately 30° $km^{-1}$. According to the investigations for the Hot Dry Rock method at Los Alamos in the USA, a gradient of this size represents a lower limit for economic extraction of geothermal energy out of crystalline rocks.

If the Hot Dry Rock method is further developed in the fracturing technique, and if the drilling costs are not too high, the method can be used in this country and the geothermal energy can be extracted for low enthalphy production.

Of course, the water temperature can be improved by using a heat pump system, and that might further motivate intensified research in this field.

The objectives and increased interest for deep hole drilling in Sweden is among other things the possibility of using the geological formations for storage purposes and to explore ore deposits and geothermal resources.

In the sedimentary formations in the south of Sweden conventional oil drilling equipment are being used to drill geothermal energy wells to a depth of around 2000 m. Aquifers in Devonian sandstone at a depth of about 800 m are since 1983 producing geothermal heat to the district heating plant of the town of Lund.

Geotechnical exploration drilling down to 1000 m in hard rock is made in order to find safe formations for possible future disposal of waste from nuclear stations.

A large diamond core drilling programme is planned to explore ore bodies at depths of 400 to 600 m in the middle and north of Sweden.

A 1000 m deep well is being drilled in Fjällbacka on the Swedish west coast as a part of a Hot Dry Rock research programme.

## Subsurface Storage of VIS Heat, Gas, Oil, etc.

Deep drilled holes down to more than 2000 m were for the first time in Sweden made in 1939. The pilot project was made through mainly Paleozoic formations, in the southernmost part of Sweden. The sedimentary rocks have here the greatest thickness in the country. A sequence of aquifers exists in these sedimentary rocks and many of them can preliminaryly be used as storage for gas and maybe also heat. If the tests in the project in progress are positive, such a storage of gas will be of an important economic consequence for its supply of energy. There are advanced plans of drilling a sequence of deep holes fur such prospecting in this part of Sweden, if the rocks are enough tight.

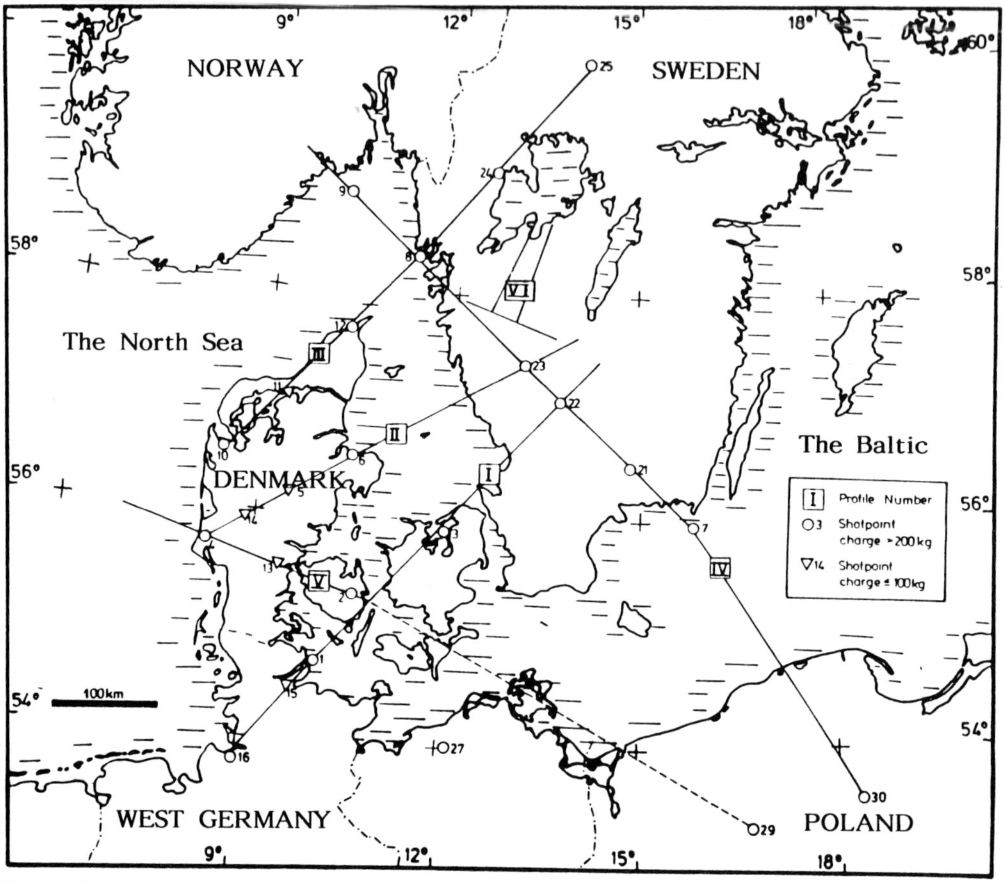


Fig. 3. Deep seismic profile line in western Sweden and Denmark

## Deep Seismic Investigations in Western Sweden

The project called Eugeno-S is sponsored by the European Science Foundation. The aim is to collect seismic data, which will lead to a detailed knowledge of the crustal structure in the area. Recordings will be made along six profile lines (Fig. 3). Dynamite and air gun shots are used as energy sources and all profiles are recorded for refractions and wide angle reflections. The sea profiles are also recorded for near vertical reflections.

The forthcoming interpretation will hopefully show some interesting features of the deep geology in western Sweden, which today is not well understood: the Fennoscandian border zone, the Tesseire-Tornquist line, the Mylonite zone, and the Protogin zone.

## Project Tornquist (IGCP Accession No. 86)

This project has run for a period of 11 years (1975-1985). It principally aims at a general characterization of the Precambrian, Caledonian, Variscan, and post-Variscan stages of geological evolution at the southwestern border of the Eeast European craton margin.

The Swedish project contributions are centred around the Scanian segment of the Tornquist line, a major lineament which can be traced from the Dobrudza at the Rumanian Black Sea via Ukraine, Poland, GDR, South Baltic, Scania, Kattegat, N Jutland into the North Sea.

Concerning geophysical and drilling activities, the project work for 1984-85 will be based on surveys which have been carried out previously (especially in connection with hydrocarbon prospecting), rather than on data from new activities. This work will make use of seismic reflection and refraction surveys carried out in offshore areas surrounding Scania, including Hanö Bay, Bornholm Gat, the south coast of Scania, the Sound (between Denmark and Scania), and Kattegat. Thc major aim is to study the tectonic pattern within the Fennoscandian Border Zone subdividing the crystalline Fennoscandian Shield in the northeast from the sedimentary Danish-Polish Trough in the southwest.

For the interpretation of the seismic records, data from deep wells in and offshore from Scania will be used, i.e., wells drilled for Oljeprospektering AB (OPAB), Stockholm. Of particular interest is to investigate the offshore prolongations of major structural units previously known from the Scanian mainland (horsts, grabens, transverse troughs, etc. and their various stratigraphical representation). In this connection the International Lithosphere Project (ILP) should be mentioned, a project which also has the ambition to investigate the deeper structure of the Tornquist Line among other lineaments.

## Conclusion

There is at the moment no great national deep-drilling programme for the near future in Sweden. Drilling programmes for moderate depths are, however, in progress in many of the fields I have here described.

We are, however, very interested in participating in international cooperation in R & D concerning deep drilling as well as geophysics in large scale to investigate the crust of the Earth.

## References

Eriksson KG, Malmqvist D (1979) In: Čzermák V, Rybach L (eds) A review of the past and present investigations of heat flow in Sweden. Terrestrial Heat Flow in Europe. Springer, Berlin Heidelberg New York

Hjelmqvist S (1966) Description of map of the pre-Quaternary rocks of the Kopparberg County, Central Sweden. Sveriges geologiska undersökning. Ser Ca No 40, Uppsala

National Swedish Board for Technical Development (STU)(1983) Deep ore exploration. A Swedish technical development programme. STU, Box 43 200, S-100 72 Stockholm

Norling E (1983) Geophysical and geological activities 1984/85 for Project Tornquist (IGCP Acc No 86). Sveriges geologiska undersökning, Uppsala

The Swedish State Power Board (1984) Deep Gas – Swedish Premises. Pilot study – report, February 1984. Gas Project G 2, Report No 3, Stockholm

# THERMAL REGIMES

# CHARACTERIZING THERMAL ENERGY AND MASS TRANSPORT IN VOLCANIC CALDERA COMPLEXES; THE ROLE OF SCIENTIFIC DRILLING

John F. Hermance[1]

## Introduction

Of all the geologic features on the earth's surface, the major silicic caldera complexes are one of the most fascinating and fundamental phenomena. Not only do they represent profound thermal anomalies in the lithosphere which potentially may be exploited for energy and economic ore deposits, they also offer grave volcanic and earthquake hazards during major eruptive phases.

The nation's attention, of course, was focused on these hazards by the May 18, 1980 eruption of Mt. St. Helens in the northwest United States. In spite its dramatic nature, however, the total volume of material erupted during this event (2 - 3 km$^3$) was a small fraction of those volumes associated with the caldera forming eruptions of the intraplate silicic volcanic complexes. Three of these centers are thought to be still active in the United States: the Valles Caldera in north-central New Mexico which, during its climax, erupted more than 500 km$^3$ of material; the Long Valley/Mono Craters Volcanic Complex in eastern California which erupted 600 km$^3$; and Yellowstone Caldera in southwestern Wyoming whose most recent climactic eruption released approximately 1000 km$^3$ of material. In short, the geologically recent caldera forming eruptions, associated with what we call the major silicic caldera complexes, are typically several orders of magnitude larger than the Mt. St. Helens class of volcanic eruptions.

Whereas these phenomena have captured the interest of geologists and geophysicists for decades, we are coming to realize that a fundamental understanding of the thermal regime and the associated dynamical processes of the major silicic complexes is of central importance, not only to basic science but to a number of national priorities ranging from resource assessment to the mitigation of volcanic and earthquake hazards. It is clear that of all classes of volcanic phenomena within the conterminous Unites States, the active caldera complexes have the highest accessible geothermal resource base, as well as the greatest destructive power during major eruptive phases. In addition, the exhumed fossil analogs of these systems are associated with extensive mineralization and economic ore deposits. What is lacking however is a predictive scientific theory describing the fundamental physiochemical processes responsible for the development and long-term sustenance of these major volcanic centers in space and time; a theory which is testable.

Therefore, in response to a growing interest among earth scientists, geotechnologists and government policy-makers, a coordinated research

[1]Department of Geological Sciences, Brown University, Providence, RI 02912, USA

Observation of the Continental Crust through Drilling I (ed. by C.B. Raleigh)

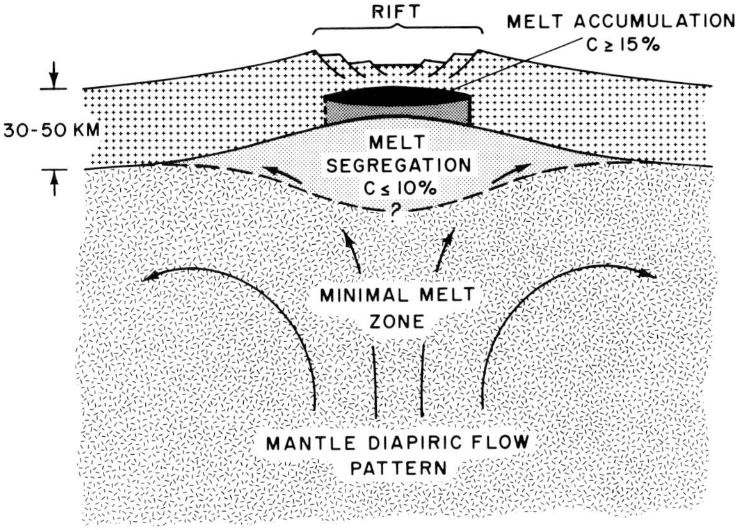


Fig. 1

effort is being mobilized by the DOE National Laboratories, the USGS, NSF, industry and universities to develop a comprehensive understanding of the morphology and dynamical evolution of these first-order tectono-magmatic features (EOS; June 26, 1984). Of particular interest are questions regarding the transfer of energy and mass between magma reservoirs deep-seated in the crust, and the shallower hydrothermal systems which they drive. A key tool in these studies has been and will continue to be scientific drilling.

## Conceptual Models for the Thermal Regime of Major Silicic Centers

### Regional Processes

In order to focus the following discussion, we summarize some of the current geophysical and geological constraints on these systems as reported by Hermance (1982). It is well known that centers of silicic volcanism are regionally associated with zones of extensional tectonics. It has also been suggested by a number of workers that these silicic centers have developed through partial melting (anatexis) of crustal material through interaction with more primitive basaltic magma at depth. Recent geophysical investigations support the idea that beneath zones of extensional tectonics (e.g., mid-oceanic and intracontinental rift systems), ascending masses of material from the mantle are intimately coupled with regional doming and the morphotectonic development of rift features (Fig. 1). Since crustal temperature gradients are on the order of 40 to 100°C $km^{-1}$ in these regions, and gradients in the upper mantle appear to be less than a few °C $km^{-1}$, heat transfer mechanisms in the mantle must be extremely efficient. This implies mass transport. In addition, electrical resistivities are such that minimal melt is thought to be present at depths greater than 50 km or so. In Figure 1 this is indicated as the "minimal melt zone".

At the top of the mantle diapir, lowered values of the bulk resistivity indicate, in some areas, a slight enhancement of a partial melt fraction

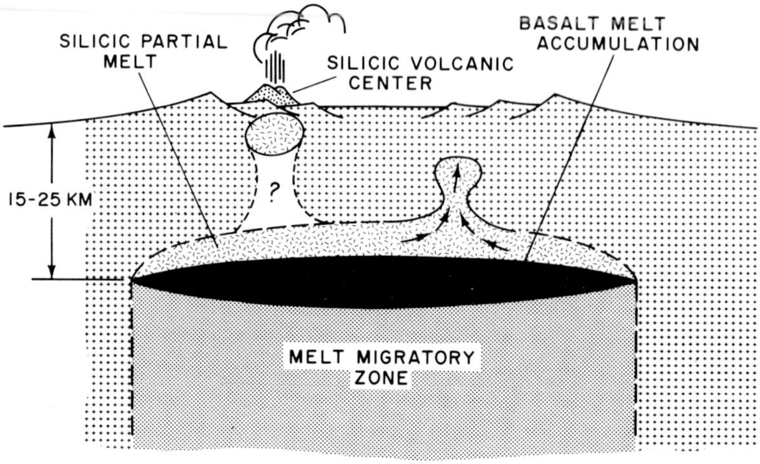


Fig. 2

which may be associated with a seismic low-velocity pillow seen beneath many rifts. In Figure 1 this is indicated as the zone of "melt segregation".

It does not seem plausible that melt, having segregated and formed a concentration of greater than a few percent of the total volume, can be dynamically stable (Walker et al. 1978). Therefore, melt in the upper mantle probably migrates either laterally as part of the diapiric flow pattern, or vertically in response to a combination of buoyancy and lithostatic pressures, and accumulates at a preferred level which is presumably hydrostatically controlled at the base of, or within, the crust. This is shown as the zone of "melt accumulation" in Figures 1 and 2. Geophysical interpretations place this zone of magma accumulation at the base of the crust beneath some areas whereas it seems to be at intracrustal levels beneath others.

Implicit in the above discussion is the concept that basalt melt *segregates* from a source region in the mantle, which at any given time has a low concentration of liquid, and *accumulates* at some level in the lithosphere.

Figure 2 represents an enlarged version of the uppermost section of Figure 1 and shows the presence of a melt migratory zone. The nature of such a zone separating the region of melt segregation in the mantle from the zone of melt accumulation at the base of, or within, the crust is obscure at the present time, and its character undoubtedly varies from region to region. In some areas, such as the central and north-central Rio Grande rift, the zone of melt accumulation seems to be a discrete layer at intermediate levels in the crust which can be clearly distinguished from its anomalous roots in the mantle. In other regions, such as Iceland, the zone of melt accumulation may be more intimately coupled to its mantle origins.

The level at which basaltic magma from the upper mantle accumulates must depend to a marked degree on its depth of origin as well as on the overall character of temperature and density gradients in the lithosphere and asthenosphere. Moreover, the frequency and mass flux with which the magma zone is replenished and sustained determines the extent to which the thermal regime of the lower crust is perturbed from a conductive environment to a convective environment.

## Relationship to Extrusive Silicic Volcanism

We suggest (Fig. 2) that the emplacement of a molten (1200 °C) basaltic lens at depth will thermally perturb the surrounding country rock, which may undergo local secondary remelting (to temperatures apparently as low as 750 - 800°C; Bailey et al. 1976). In this case, up to 400 cal $g^{-1}$ of excess heat is available in the basaltic intrusion to locally remelt the crust. Assuming for the sake of illustration that the latent heat of fusion for basalt and granite are approximately equal (i.e., 100 cal $g^{-1}$), then one unit of basalt can mobilize up to four units of silicic material.

The partial melt will segregate as a rhyolite magma (Fig. 2) and, because of its lower density, will tend to rise as a silicic diapir, which, if it reaches the surface, will lead to extrusive volcanism. We submit that episodic silicic volcanism at the surface is associated with the episodic replenishment of the basaltic magma layer at depth. Moreover we emphasize, as a corollary to this observation, that long periods of dormancy in which the magma reservoir is unreplenished – and may even approach a solidified state – is a fundamental feature of the known thermal regime of these systems.

Consider, for example, the expected scale-size of the ultimate magma reservoir. Smith and Shaw argue that even during a *major* eruption, only 10% of the magma present is erupted. Therefore, for a typical erupted volume of 500 $km^3$, this implies a reservoir capacity of 5000 $km^3$, or a totally molten body having a typical diameter of 20 km (for a sphere-like body), emplaced at midlevel in the crust (as implied by certain fluid inclusion and other data). Such thermal perturbations do not seem to be a steady-state feature of the major caldera systems.

Thus the model in Figure 2, driven by episodic mobilization and upward migration of basalt from the upper mantle, provides a basis for relating geophysically delineated structures in the deep earth to the genesis of major centers of silicic volcanic activity.

## Coupling of Hydrothermal Systems to Deeper Magma Bodies

Insight into the dynamics of how hydrothermal systems in the upper crust are driven by deeper magma bodies has relied on both direct and indirect measurements to refine conceptual and physio-mathematical models. Approaches by previous workers have included extrapolation of surface geology and geophysics, direct measurements in a number of shallow and a few intermediate-depth drillholes, inferences from fluid geochemistry, and comparison with fossil magma-hydrothermal systems which have been exhumed and dissected by erosion.

Scientific drilling by the USGS in the 1950's and 60's laid much of the foundation for understanding ground water circulation in hydrothermal systems. White (1973) proposed a model in which low temperature meteoric water percolates downward along fractures and is driven laterally by a hydraulic pressure imbalance while being heated from below. An uprising plume develops and the hot water rises essentially adiabatically until it encounters its boiling point and flashes to steam. A two-phase mixture of fluid and gas then persists to the surface.

This model, while conceptually satisfying, remains essentially untested in the major caldera systems. In particular the region of heating at depth, and the degree to which the fluid interacts with the country rock and the molten magma reservoir itself (in dissolving and redistributing mineral components, some of which have economic value), is completely unstudied.

## Rationale for Scientific Experiments in Intermediate and Deep Drillholes

Our understanding of the overall system is limited primarily by our being able to sample only the upper, lower temperature portions of the active hydrothermal system itself. Although geothermal wells have been drilled to depths greater than 4 km and temperatures greater than 400°C, meaningful measurements are presently restricted to temperatures less than 250°C. Ideally, one would like to drill and carry out observations in the entire magmahydrothermal system, to magmatic temperatures (>900°C) and to depths well within the crust (>10 km). Although this may be technically feasible in the near future, at present it seems realistic to restrict our objectives to temperatures of less than 400°C and to depths of less than 4 km – in many cases, this would allow one to study the *roots* of hydrothermal systems. Direct sampling of this environment through drilling, while representing a distinct challenge to present technology, would represent a dramatic improvement in our understanding of active physio-chemical processes in this regime not obtainable in any other way.

Information from scientific drillholes to temperatures of 400°C in magmahydrothermal systems would serve a number of purposes:

1. An overall characterization – from top to bottom – of the natural hydrothermal system.
2. Evaluation of conceptual models for the evolution of the overall magmahydrothermal system in space and time.
3. Quantitative parameterization of energy and mass transfer mechanisms throughout the total system.
4. Evaluation of interpretations from surface geophysical and geological observations. The drill-hole offers an opportunity to validate and to refine surface techniques in what is essentially a "calibrated" environment. This would go a long way toward optimizing pre-drilling exploration activities in other less-studied systems elsewhere.

## Background on Potential Drilling Sites

Each of the three young, large silicic volcanic complexes in the western U.S. – the Valles caldera, New Mexico; the Yellowstone caldera; Wyoming; and the Long Valley caldera, California – has an associated hydrothermal system and has been subjected, in some degree, to a wide variety of earth science investigations, including in some cases commercial drilling to depths on the order of 2 km. However, in none of the three areas do we have direct drill-hole knowledge of the *roots* of the hydrothermal systems (2 - 5 km), and how these hydrothermal systems derive energy from molten rock sources within the earth's crust (from depths greater than 5 km). A number of workers concur that, in choosing one or more of these caldera complexes for scientific drilling, the following criteria should be considered:

1. The system should represent an active counterpart of fossil caldera systems.
2. A well-defined magma body should be present.

3. The target should represent a clearly defined stage in the evolution of silicic centers.
4. A complete, compatible set of geological, geophysical, and intermediate-depth drilling data should exist.
5. A significant area of the caldera should be available to drilling in terms of both geographic accessibility and environmental sensitivity.
6. Actual siting of scientific drill-hole(s) should be based on a reasonable certainty of encountering temperatures of 400°C or greater at depths of 5 km.
7. Drilling and maintaining the drill-hole(s) should be technically feasible.
8. Consideration should be given to the benefits from add-on commercial drilling.

## The Valles Caldera

The Valles Caldera resulted from a climactic eruption of the Jemez Mts. Volcanic Complex in north-central New Mexico 1.1 m.y. ago. This eruption was preceded at 1.4 Ma by the formation of the Toledo Caldera which is now largely obscured by the most recent event. The eruptives from both calderas are generally known as the Bandelier Tuff, which has an estimated volume of approximately 600 $km^3$ (Smith 1979).

Volcanism in the Jemez Mts. spans the time interval from 13 m.y. to less than 0.1 Ma ago, and by most standards (including recent seismicity), the complex is likely to be still active.

The Valles Caldera was the classic type-example used by Smith et al. (1961) to develop their concept of resurgent calderas (Fig. 3). The fundamental features of the Valles system are extraordinarily well developed, from the physiographic boundary fault to the spectacular resurgent dome and a number of discrete post-climactic rhyolite dome extrusions along the intracaldera ring fracture.

Smith and Bailey (1968) visualized the following series of stages in the development of resurgent calderas (Fig. 3):

1. Regional tumescence and the development of ring fractures culminating in caldera forming eruptions (Fig. 3A).
2. Caldera collapse and post-caldera volcanism (Fig. 3B).
3. Resurgent doming and major ring fracture volcanism (Fig. 3C).
4. Continuing ring fractures volcanism, primarily in the form of extrusive rhyolite domes. The development of keystone graben-like features on the resurgent dome (Fig. 3D).

It is important to note that each of the three major silicic centers with which we are concerned have resurgent calderas, and generally have progressed through the above sequence – some, perhaps, several times.

On the basis of available data, the Valles caldera appears to be a highly favorable site for scientific drilling, primarily because of the already demonstrated high-temperature geothermal system at the Union Baca hydrothermal site, the large amount of intermediate-depth drilling by industry, drilling at Fenton Hill by Los Alamos National Laboratory, and the possibility of good access logistically because of a single land-owner for most of the caldera and a single commercial leasee. At present, however, access to this site is restricted because of pending legal and environment issues.

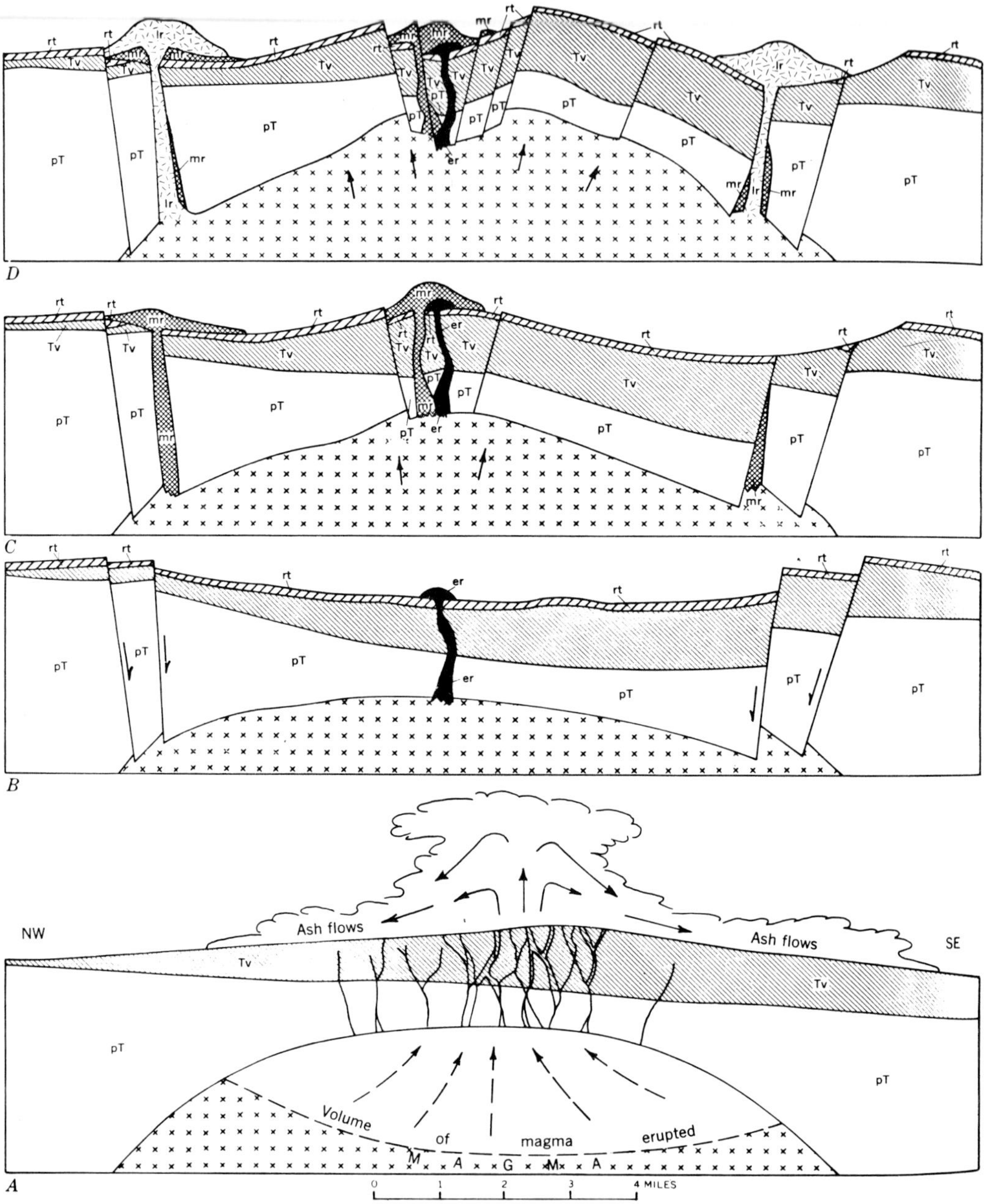


Fig. 3

A recent workshop hosted by Los Alamos National Laboratories and the USGS (EOS June 28, 1983) underscored the attraction of the Valles caldera as a site for continental scientific drilling. A number of participants argued that there already exists a considerable background in regional and local geology, geophysics, and geochemistry. In addi-

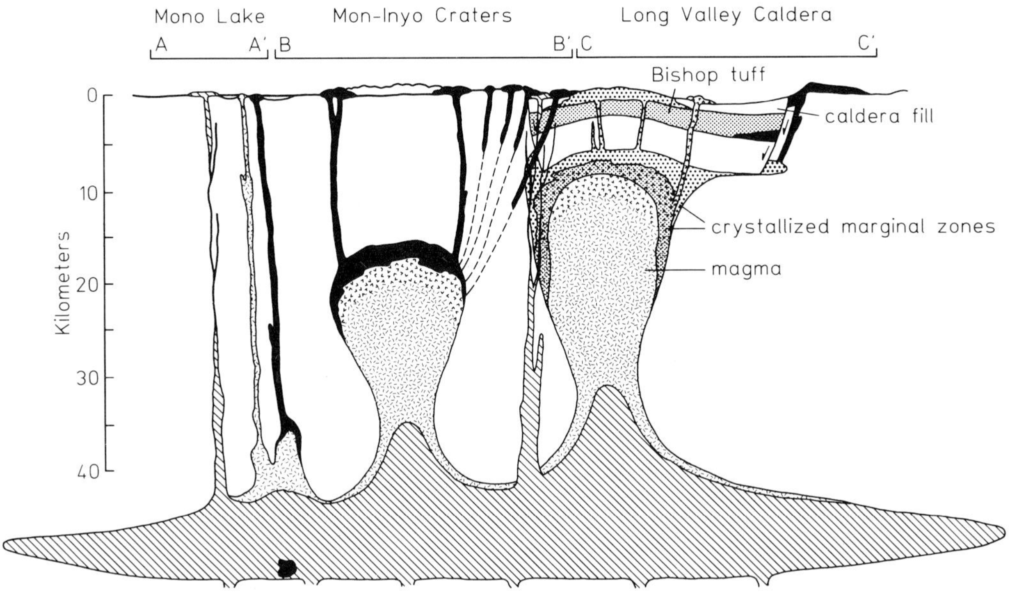


Fig. 4

tion, lithologic, geochemical and thermal data have been obtained from a number of intermediate depth holes within and around the Baca geothermal field, as well as from the Hot Dry rock project on Fenton Hill (at a location immediately outside the caldera), where a hole has already been drilled to 4.5 km in basement, encountering temperatures of 325°C. To supplement these data, it has been recommended that a number of intermediate depth holes (on the order of 1 km, with one perhaps going as deep as 3 km) be drilled to better qualify (a) the magma-hydrothermal model, (b) features within the intrusive aureole of the principal magma chamber, (c) the stratigraphic record within the caldera structure, (d) the possibility of interstitial melt being still present at upper levels in the crust (i.e., above 10 km).

Although it is not likely that a major magma chamber presently exists in a molten state at shallow levels in the crust, there is very good reason to expect that one did exist in the recent past, but is now largely solidified. Scientific drilling in the Valles Caldera therefore offers an opportunity, in terms of presently available technology of drilling into and even *through* a high level magma chamber which has been active so recently that significant retrograde metamorphism has not obscured the basic petrologic signatures of processes which were in effect when the system was still active.

## The Long Valley/Mono Craters Volcanic Complex

Long Valley Caldera is associated with a structural embayment in the northerly trending frontal fault system of the Sierra Nevadas, and is one of several young volcanic centers in the area. Volcanism in the vicinity of Long Valley began about $3 \times 10^6$ yr ago with scattered eruption of basalt and andesite, becoming localized about $2 \times 10^6$ yr ago with the eruption of the Glass Mountain rhyolite, and culminating with the tremendous eruption of 600 $km^3$ of rhyolitic Bishop Tuff, $0.7 \times 10^6$ yr ago. The subsequent collapse of a shallow, zoned magma chamber produced the Long Valley caldera, 17 × 32 km in size (Bailey et al. 1976). This was followed by resurgent doming of the caldera floor and episodic intracaldera volcanism as recently as 0.05 m.y. ago. Bailey (1982) has suggested that the volcanism responsible for forming Long Valley caldera itself has probably completed the cycle usually associated with resurgent activity.

Long Valley Caldera is the southern-most and oldest member of a succession of three progressively younger volcanic complexes each in different stages of evolution (Bailey 1980). These are in decreasing age, Long Valley (3.0 - 0.05 m.y.), Mono Craters (40 k.y.-1 k.y.), and the centers in Mono Lake (<2000 yr).

According to a conceptual model of Bailey (1982), Mono Craters (1300 yr) may have evolved through stages 1 and 2 described for Long Valley above, and the centers in Mono Lake form the smallest and youngest complex, and seem to be in stage 1 (Fig. 4). Mono Craters are chemically homogenous which, along with their recent age and frequency of eruption, suggests that they were extruded from a single magma chamber, largely molten and perhaps still rising to the surface. The crescent-shaped trend of Mono Craters has been used to infer the outer margin of this inferred magma chamber, and its center may be beneath Pumice Valley.

In contrast to the chemical *homogeneity* of Mono Craters, the chemical *heterogeneity* of the underdeveloped volcanic centers in Mono Lake seems to suggest that they may have issued from small, multiple, and relatively deep magma bodies that have not yet coalesced (Fig. 4). We therefore have available in these three volcanic complexes a set of field examples, each of which represents a different stage in their tectonomagmatic evolution.

## The Inyo Domes and Craters

The youngest volcanic features in Long Valley caldera are the Inyo craters and domes which lie on a linear trend extending north from the northwest sector of the caldera, across the caldera rim, to Mono Craters (Bailey et al. 1976). The five Inyo domes are roughly rhyolitic in composition, and the youngest are less than 550 yr old (Miller 1984). Early studies indicate that Inyo domes are chemically heterogeneous, which suggests that they may be a product of mixing magmas from the Long Valley chamber and the possible chamber beneath Mono Craters to the north (Bailey et al. 1976). Recently, however, Bailey (pers. commun. 1982) found evidence that the Inyo volcanoes may have derived from a single source associated with the Mono Craters. Thus, geological evidence suggests the possibility that a rather extensive magma reservoir (or system of reservoirs) may exist beneath the northwestern section of Long Valley caldera and extend 30 - 35 km north to Mono Lake (Hermance 1983).

Following a major earthquake sequence in 1980, a significant increase of tectonic activity in Long Valley caldera has lead some workers to infer that magma has intruded the upper crust, and there has been concern about possible renewed volcanic activity at the surface. The three most compelling pieces of evidence for this are the recency of volcanism along Inyo Domes (550 yr), the localization of spasmodic tremor in the southwest moat, and the recent uplift of the resurgent dome (Fig. 5).

All these phenomena are particularly striking when it is realized that volcanism in the Long Valley/Mono Craters complex is the most recent of any of the major silicic centers we are discussing. In fact there

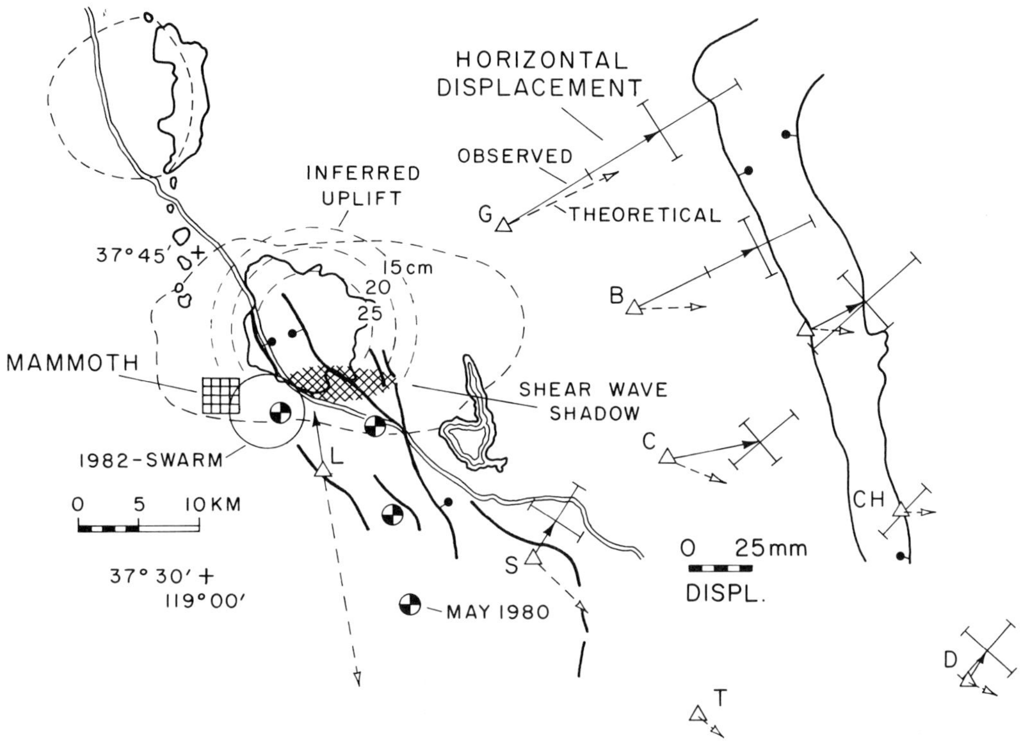


Fig. 5

may have been a historic eruption beneath Mono Lakes in the late 1800's. Volcanism in both the Yellowstone and the Valles Calderas is most likely older than 50,000 years.

## Yellowstone Caldera

Yellowstone Caldera is the youngest major element of a 7500 $km^2$ caldera complex at the terminus of what many workers feel is a mantle hot-spot track (Christiansen 1984). The volcanic center occurs at the confluence of three major mega-tectonic stress trends: the downwarped northeast trending Snake River Plain (the hot-spot trace), and Basin-Range type faulting which trends both southwards to the south of the Caldera and northwestward to its north. Seismicity occurs throughout the region (smaller events are shown as open circles in Figure 6 with some of the largest earthquakes in this region of the Rockies occurring close to the northern rim and shown by stars; Smith and Christiansen 1980).

The volcanic complex, itself, formed over three major eruptive cycles at 2.0, 1.3, and 0.63 m.y. ago. The latter event is the best understood because it is the least obscurred by subsequent eruptions. It produced 1000 $km^3$ of the Lava Creek Tuff and formed the 45 × 70 km Yellowstone Caldera proper. Resurgence of the caldera floor followed the main eruptive phase, with intracaldera eruptions occurring up to 0.075 m.y. ago. Many workers feel that a massive magma reservoir still persists beneath a portion of the caldera, but this is not conclusive. Clearly, however, some high temperature source is needed to sustain the vigorous hydro-

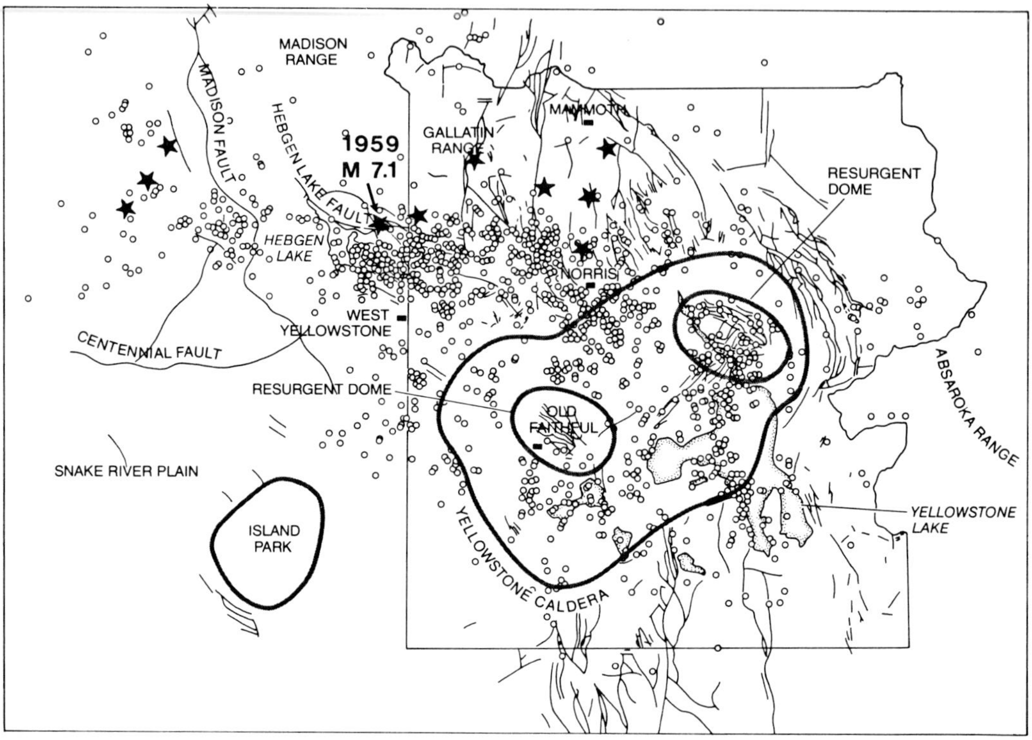


Fig. 6

thermal system which accounts for the present average advected heat flow of 1700 mW $m^{-2}$ (60 mW $m^{-2}$ is considered normal for a stable geologic province). Christiansen (1984) argues that only magma can account for such a high sustained thermal discharge.

There is little argument, therefore, that Yellowstone Caldera represents the most intense magmatic and geothermal anomaly in the conterminous U.S., but at the same time, being in a National Park, this area is environmentally sensitive to studies using scientific drilling. Even for drill-holes dedicated to purely scientific objectives, scientists and environmentalists are concerned regarding the potential hazard to geyser activity from any hydrologic disturbance. The Continental Scientific Drilling Committee has recently established a task group under the direction of Bob Fournier of the USGS to study these issues further and to identify unique scientific questions which can only be addressed through drilling in Yellowstone. If drilling is recommended for this area, it will, of course, be strictly limited geographically and for purely scientific reasons with full regard for mitigating any negative impact whatsoever on one of our finest national parks.

In summarizing the characteristics of these three silicic calderas we note that:

1. Each is associated with a locally complex stress field.
2. Each is characterized by episodic volcanism preceding and following major caldera forming eruptions.

3. Each is associated with a hydrothermal system which is largely fault-controlled at the surface.
4. But, we emphasize, each is in a different stage of their tectono-magmatic evolution which makes them each uniquely and equally attractive for scientific drilling in the type of generic research program we are describing.

## Priorities and Tradeoffs in Terms of Cost Versus Recovered Information

An integral step in the foreplay that goes into planning a program of scientific drilling needs to involve a careful assessment of the progressive tradeoffs in cost and technological development on one hand and the quantity and quality of information recovered on the other. Planning a scientific drillhole will involve considering the following stages:

1. Drilling
2. Temperature logging
3. Recovery of cuttings
4. Pressure measurements
5. Reservoir testing
6. Core recovery
7. Conventional logging
8. Short-term downhole instrumentation
9. Long-term (observatory?) downhole instrumentation

Are each of these activities as critical to the success of the experiment as the others? As we go down this list we progressively increase the total cost of the experiment, while at the same time, of course, we hopefully recover more information. The question we have to ask is, where in the cost structure does the rate of information returned become offset by the enormous cost of retrieving that last data point?

## Can One Constrain in-Situ Physical Processes with Minimal Information?

Perhaps a more positive way to view the above exercise is to ask what minimal experiment justifies the cost of drilling a given hole? One might argue that in some cases simply temperature and perhaps cuttings might be sufficient. Anything else that one can cram down or pull out of the hole would be frosting on the cake.

Temperature, after all, can be directly and readily measured – in fact to temperatures in excess of 1000°C. Knowing the in-situ temperature provides a great deal of insight into physical processes. This can be illustrated with a set of simple models. It is important to recognize that constraints on dynamical processes do not always require exotic instrumentation and/or sample recovery methods downhole. Fundamental constraints often come from the simplest measurements.

Figure 7 illustrates a set of possible temperature profiles following the intrusion of magma at a temperature of $T_m$ at a depth D. The left hand panel shows the initial transient condition at an early time after the intrusion so that temperatures at shallow depth are unperturbed from their "normal" value. (To simplify our present discussion, we neglect the contribution of the normal background temperature gradient and the fact that the boiling point of water increases with pressure

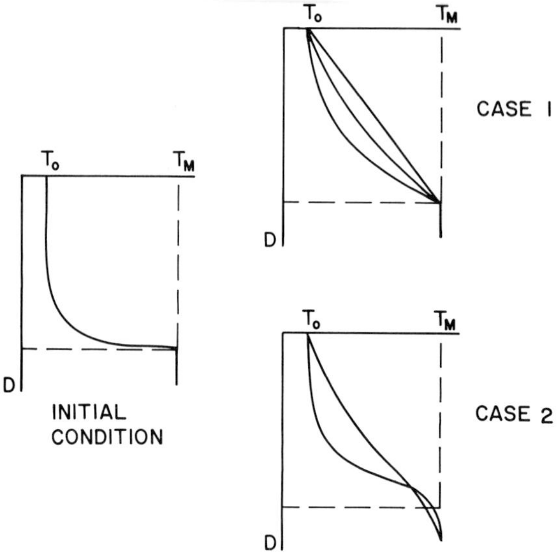


Fig. 7

INTRUSION - SOLID CONDUCTION

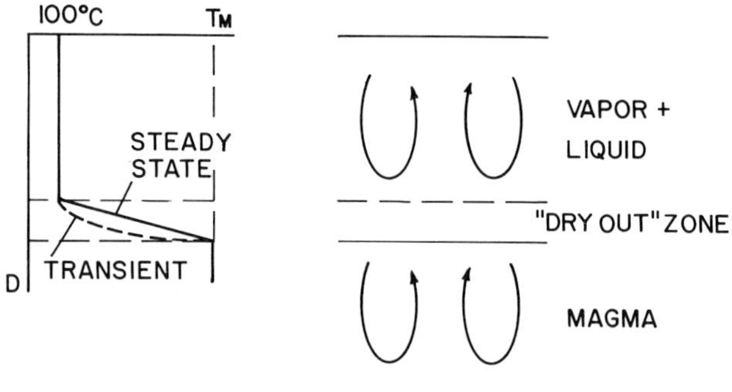


INTRUSION - CONVECTION DOMINATED

Fig. 8

and depth.) Case 1 in the right hand set of panels shows the temperature stabilizing toward some steady-state temperature gradient – this of course implies that $T(D) = T_m$ for all time, which in turn implies that the magma body is large and is actively convecting.

Case 2 illustrates the case where temperatures in the magma systematically *decrease* with time while they *increase* at shallow levels in the crust – at least for a while. Such a situation implies that the magma body is quite small, is quite viscous, or is not convecting at all.

The point we would make is that the physical process is reflected in the nature of the temperature distribution.

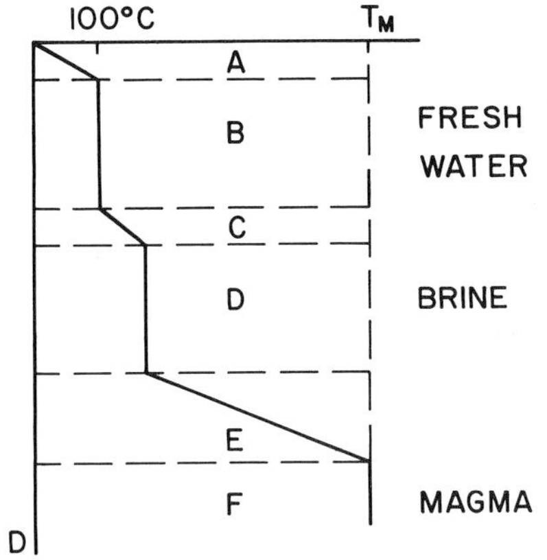


Fig. 9

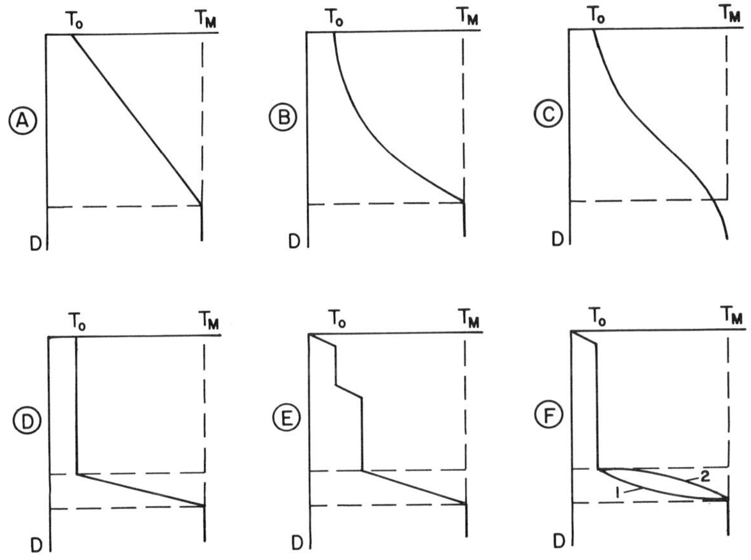


Fig. 10

Next we consider in Figure 8 the case of an intrusive in which heat transfer is convection-dominated both in the magma and in the overlying crust. Following an initial transient stage, a quasi steady state is reached where temperatures are essentially constant with depth in the overlying crust, rise along a constant gradient in a thin boundary layer, and are again constant in the magma at the magmatic temperature.

An example of "stacked" convection cells is illustrated in Figure 9 (following Fournier 1983). Here a high density brine will convect at a higher temperature than a lighter freshwater layer above. The interface at C may be simply a thermal boundary layer or may in fact be a physically impermeable zone generated by deposition of minerals exsolved from the brine below.

These situations are resummarized in Figure 10. A, B and C represent conduction-dominated situations. A is a steady state constant gradient requiring a considerable reservoir of convecting magma at depth. B reflects a transient temperature build-up due to a recent intrusion. C represents a cooling magma reservoir and a heating crustal layer above. This will continue to be a transient situation until all the heat from the magma reservoir is depleted. D, E, and F represent the effects of convection-dominated heat transfer in an aquifer overlying the magma reservoir.

Case F is particularly interesting because the two profiles in this panel (1 and 2), while their difference is fairly subtle in detail, can in fact be resolved in actual field studies and represent dramatically different histories for the intrusion. Case F1 illustrates a transient "start-up" condition in which the intrusion is very recent and the heat flux above the magma source is still increasing.

On the other hand, Case F2 represents a transient *cooling* condition which has been well-documented in studies of Kilauea Iki Lava Lake on Hawaii (Hardee 1980; Helz 1980; Hermance and Colp 1982). The lake was caused by a lava flow from the November-Dezember 1959 eruption of Kilauea volcano which ponded in Kilauea Iki pit crater. The lava flow was approximately 120 m deep and had a diameter of 750 m. Over the years a number of scientific drillholes have recorded its cooling history and a variety of geophysical measurements have been employed to remotely sense the status of its molten core (see, for example, the review by Hermance and Colp 1982).

Hardee (1980) noted that over a significant portion of the colling history of the lava lake, temperatures in its lower conduction-dominated zone showed a systematic concave-downward curvature with depth (Case F2 in Figure 10). In a simplistic sense, curvature of the geotherm represents either a change of thermal conductivity or a non-zero divergence of the heat flux. Hardee argues, however, that in this case the change of conductivity is too small, and may even have the wrong sense, to account for the curvature. Hence, it seemed reasonable to explain this feature as divergence of the heat flux.

In the conventional one-dimensional heat flow equation

$$\partial^2 T/\partial z^2 = -H_s/k + (\rho C_p/k)\,\partial T/\partial t + (\rho C_p/k)\,v \cdot \nabla$$

the curvature of the temperature field is embodied in the second derivative on the left-hand side. The heat productivity $H_s$ due to radioactivity is small for basalt, therefore only the transient cooling effect $\partial T/\partial t$ and the advection term $v \cdot \nabla T$ were possible contributors to the negtive curvature in the geotherm of Kilauea Iki Lava Lake. Originally it was suggested that advection of steam through a disconnected zone of molten inclusions (Fig. 11, top panel) could explain the temperature field above the melt zone. But the results of electromagnetic sounding experiments were incompatible with a discontinuous melt phase for the molten zone; its conductivity was too high. Geophysical considerations, therefore, required a macroscopically continuous zone of thoroughly interconnected melt for the molten core.

Hardee (1980) presented a model in which the transient cooling term dominates the curvature of the temperature field. He argued that although the solidification of the crust followed the square-root-of-time law early in the history of the lake (e.g., proposed by Peck et al. 1964 for Alae), the crust then began solidifying at a nearly constant rate (Fig. 11, bottom panel). The observed constant solidifications rate appeared to result from a natural heat flux balance between the two zones of the crust, the upper two-phase convection-dominated zone and the lower zone dominated by transient conduction effects due to the moving solidification front. Hardee argued that the curvature

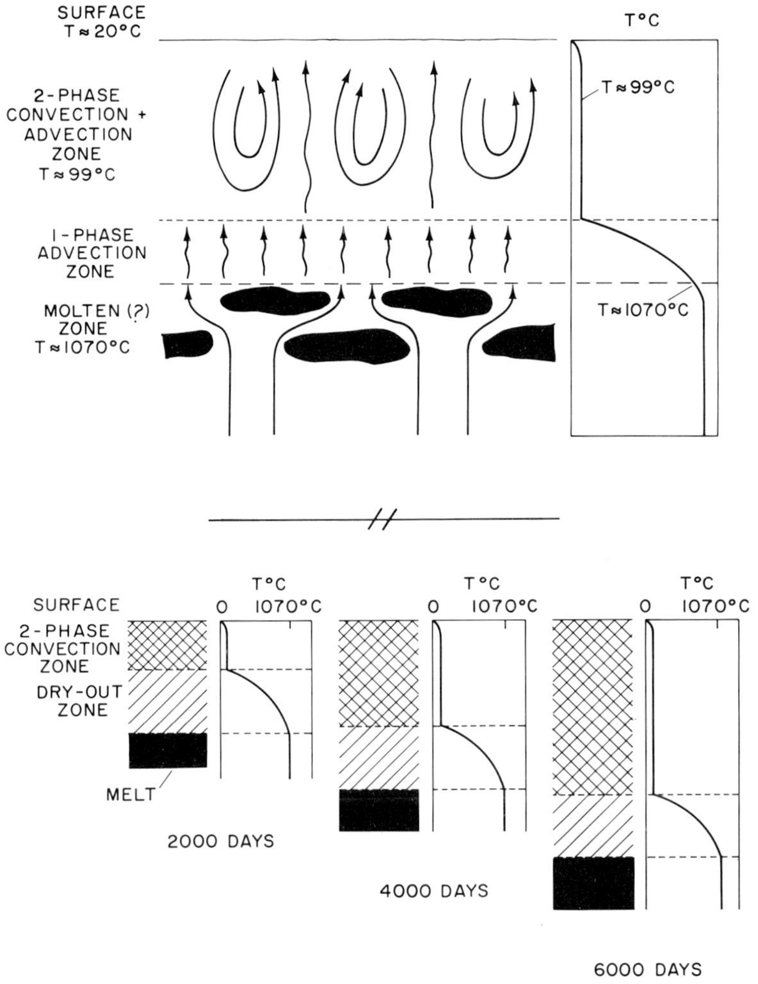


Fig. 11

of the temperature profile is fundamentally related to the rate of solidification, and that the rate of solidification can be determined directly from the temperature profile.

The lessons to be learned here are that physical processes do, in fact, have distinctive signatures in the thermal field. Much can be learned about a system from simple measurements of temperature. Nevertheless the ultimate interpretation of the temperature profile was ambiguous — did the cooling term ($\partial T/\partial t$) or did the advective term ($v \cdot \nabla T$) dominate the divergence of the heat flux? In this case a single drillhole was insufficient to discriminate between these two possibilities and recourse was made to surface geophysical-electromagnetic sounding measurements.

The underlying theme of the Kilauea Iki experiment was based on recognizing the basic ambiguities in proposed models, then sharpening the questions to be asked to the point where drilling becomes a cost-effective, in fact the *only*, solution.

Fig. 12

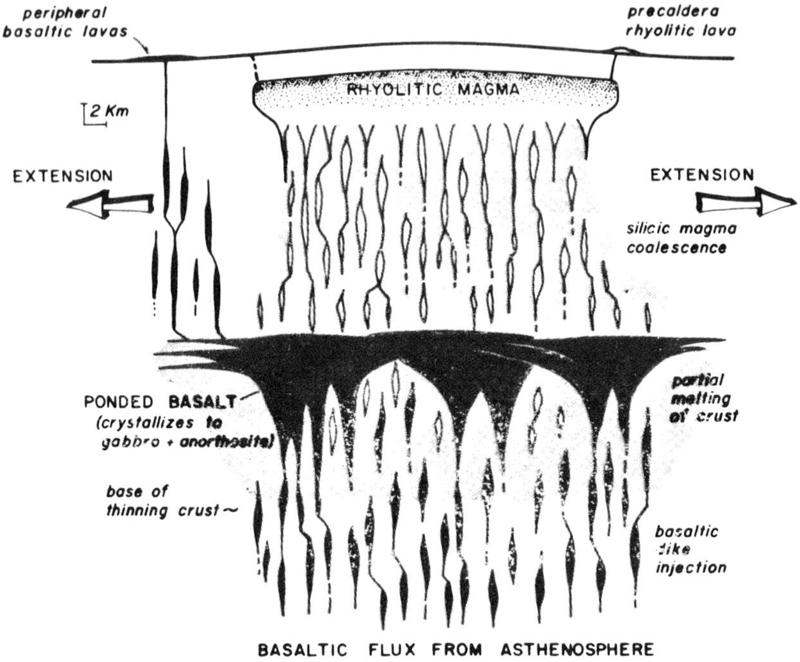


## Some Testable Concepts for Scientific Drilling

The above discussion argued that a great deal of insight into physical processes in magma-hydrothermal systems can be obtained through relatively simple measurements in drillholes using present-day technology. It was also argued that the interpretation of such measurements can be ambiguous – supporting the position that while a single suite of measurements from a drillhole may be acceptable, this is generally not the most desireable nor cost-effective use of limited funds. The prospect of a single, super-deep hole into such systems is particularly inadvisable.

We now consider a final example which is not only concerned with *physical* processes associated with magma-hydrothermal systems but *chemical* processes as well.

Hildreth (1981) has proposed a model for the evolution of zoned silicic magma chambers (Fig. 12) based on an impressive synthesis of data from the class of caldera complexes described above. Through a process described earlier, basalt flux (T approx. 1200°C) from the asthenosphere ponds at the base of the crust causing partial melting of silicic material having a low melting point (700° - 800°C). The silicic magma thus formed rises and collects in a magma chamber at high levels in the crust. Such a shallow depth appears to be compatible with gas inclusion studies of the erupted products. Needless to say, if such bodies exist at such shallow levels they are indeed exciting geophysical and scientific drilling targets.

A conceptual model is shown in Figure 13 (following Hardee 1982) for the hydrothermal system associated with such a magma chamber at a depth of 5 km or so. A zone of conduction dominated heat flow (Zone B in

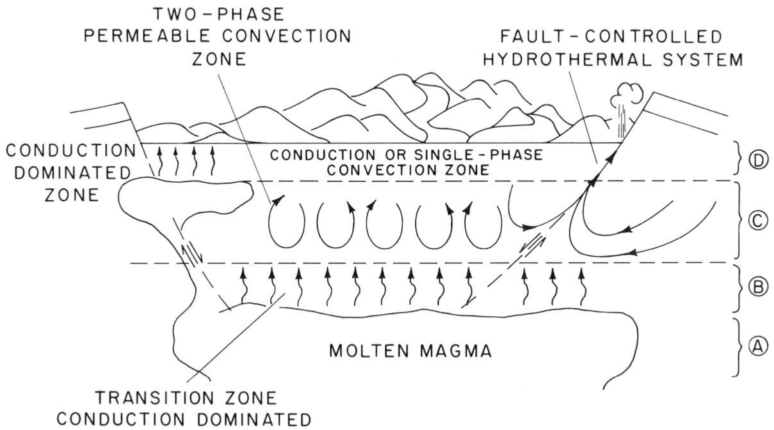


Fig. 13

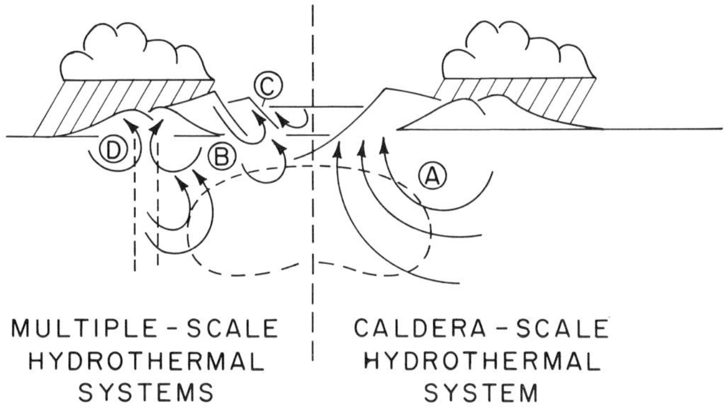


Fig. 14

Figure 13), probably highly resistive and somewhat ductile, is thought to be present directly above the molten magma chamber (A in Figure 13). The high heat flux from the parent magma body forces the hydrothermal system to convect as a two-phase flow (liquid water plus steam) in a permeable zone (C). Some of the fluids may leak to the surface along faults and fractures. Above the convection zone (C), heat may be transferred by solid conduction or by single-phase advection or convection along systems of fractures (Zone D, Fig. 13).

The essential features of this model is that during a significant portion of its lifetime, a vigorous hydrothermal system is maintained through a mechanism in which the magmatic source is cooled through a downward moving solidification front, in analogy with the process described for Kilauea Iki Lava Lake.

It seems that hydrothermal systems which totally penetrate an intrusive body, as visualized a number of years ago by Taylor and Forester (1971), occur in a late stage of the evolution of these systems, when the intrusive has essentially solidified. It is quite likely, in fact, that the pattern of hydrothermal circulation is somewhat more complicated than visualized by Taylor and Forester. In Figure 14 we show two possible modes for penetrative convection of hydrothermal fluids. On the right hand side, A indicates a single caldera-scale hydrothermal cell which geochemists seem to readily visualize, but of which we do not know of a single documented example. On the left hand side of Figure 14, we see hydrothermal systems having multiple-scales: B represents a small convective system penetrating the periphery of a much larger scale solid pluton; C is a fault-controlled high level hydrothermal system, driven essentially by high heat flow from below; D represents convection totally penetrating a local stock. Most electromagnetic workers who have surveyed specific geothermal targets with a dense coverage of sites will be all too familar with the small scale electrical anomalies which are implied by the phenomena illustrated on the left hand side of Figure 14.

The fundamental questions we are posing here – which are only several of many which workers might pose – are at what stage does penetrative hydrothermal convection occur in the evolution of massive magma chambers? – and what is the scale-size of the deep-lying convection cells? Is it reasonable to consider a model in which the hydrothermal system progressively evolves from being dominated by a moving solidification front (Figure 13) during its early history, to a caldera-scale convection cell which totally penetrates the parent magma body during its later history (Fig. 14, left side)? Or does a caldera-scale hydrothermal system *ever* exist in the evolution of these complexes? – does the hydrothermal system always consist of multiple, randomly placed small scale units?

We would argue that our models are becoming sufficiently quantitative so that proposed hypotheses can be tested. The generic program we are describing here will attempt to characterize the roots of these hydrothermal systems, particularly emphasizing the transfer of mass and energy throughout the magma-hydrothermal system. What is the effect of the magnitude and recency of volcanism? Is a brine layer present as a ubiquitous feature in all of these systems? What is the nature of processes at the magma-hydrothermal interface as it evolves in space and time?

These are first-order questions which can only be addressed through scientific drilling..., and only through drilling in a number of target areas at different stages in their evolution.

*Acknowledgements*. Many of the ideas presented here have evolved through discussions with my colleagues on the Thermal Regimes Panel of the CSDC and elsewhere. In particular I am grateful for the interaction I have had with Patrick Muffler, Bob Decker, Dick Dondanville, Hugh Taylor, John Sass, Sam Varnado, Terry Gerlach, Harry Hardee, John Colp, Bill Luth, Fraser Goff, and Bob Fournier. Although I have gleaned ideas from these and other sources, the comments presented here are not to be taken as an official position of the CSDC nor of the National Academy of Sciences. The preparation of this manuscript was supported by the Department of Energy Office of Basic Energy Sciences through Contract No. DE-AC02-79ER10401 to Brown University.

## References

Bailey RA (1982) Other potential eruption centers in California: Long Valley-Mono Lake, Coso, and Clear Lake Volcanic Fields, Status of Volcanic Prediction and Emergency Response Capabilities in Volcanic Hazard Zones of California, Special Publication No 63, California Dept. of Conservation, Division of Mines and Geology, pp 17-28

Bailey RA (1980) Structural and petrologic evolution of the Long Valley, Mono Craters, and Mono Lake volcanic complexes, eastern California, EOS 61:1149

Bailey RA, Dalrymple GB, Lanphere MA (1976) Volcanism, Structure and Geochronology of Long Valley Caldera, Mono County, California. J Geophys Res 81:725-744
Christiansen RL (1984) Yellowstone magmatic evolution: Its bearing on understanding large-volume explosive volcanism, In: Boyd FR (ed) Explosive Volcanism. National Academy of Sciences, Washington
Decker R, Decker B (1981) The eruptions of Mount St. Helens, Sci Amer 244:68-80
Fournier RO (1983) Active hydrothermal systems as analogues of fossil systems, Geothermal Resources Council, Special Report No 13:263-284
Hardee HC (1982) Permeable convection above magma bodies. Tectonophysics 84:179-195
Hardee HC (1980) Solidification in Kilauea Iki lava lake. Jour Volcanol and Geotherm Res 1:211-223
Helz RT (1980) Crystallization history of Kilauea Iki lava lake as seen in drill cores recovered in 1967-1979. Bull Volcanol 43:675-701
Hermance JF (1983) The Long Valley/Mono Basin volcanic complex in Eastern California; Status of present knowledge and future research needs. Rev Geophys Space Phys 21:1545-1565
Hermance JF (1982) Magnetotelluric and geomagnetic deep-sounding studies in rifts and adjacent areas: Constraints on physical processes in the crust and upper mantle. Int Geodynam Project Sci Vol on Continental and Oceanic Rifts, Geodynam Ser, A.G.U.
Hermance JF, Colp JL (1982) Kilauea Iki lava lake: Geophysical constraints on its present (1980) state. J Volcanol Geotherm Res 13: 31-61
Hildreth EW (1981) Gradients in silicic magma chambers: Implications for lithospheric magmatism. Jour Geophys Res 86:10153-10192
Miller CD (1984) Holocene Eruptions at the Inyo Volcanic Chain, California - Implications for Possible Eruptions in Long Valley Caldera, Proceedings of Workshop XIX, Vol 1, Open-File Report 84-939, p 76-95
Peck DL, Moore JG, Kojima G (1964) Temperatures in the crust and melt of Alae lava lake, Hawaii, after the August 1963 eruption of Kilauea volcano: a preliminary report. US Geol Surv Prof Pap 501-D:1-7
Smith RB, Christiansen RL (1980) Yellowstone Park as a window on the Earth's interior. Sci Am 242:104-117
Smith RL (1979) Ash-flow magmatism. Geol Soc Amer Spec Paper 180:5-27
Smith RL, Bailey RA (1968) Resurgent cauldrons. In: Coats RE, Hay RL, Anderson CA (eds) Studies in Volcanology, Geol Soc Amer Mem, 116, pp 613-632. Geol Soc Amer, Boulder, Colorado
Smith RL, Bailey RA, Ross CS (1961) Structural Evolution of the Valles Caldera, New Mexico, and its Bearing on the Emplacement of Ring Dikes: US Geol Surv Prof Pap 424-D:D145-D149
Taylor HP Jr, Forester RW (1971) Low-$O^{18}$ igneous rocks from the intrusive complexes of Skye, Mull, and Ardnamur Chan, Western Scotland. J Petrol 12:465-497
Walker D, Stopler EM, Hays JF (1978) A numerical treatment of melt/solid segregation: Size of the eucrite parent body and stability of the terrestrial low-velocity zone. J Geophys Res 83:6005-6013
White DE (1973) Characteristics of geothermal resources. In: Kruger P, Otte C (eds) Geothermal Energy: Resources, Production, Stimulation. Stanford Univ Press, Stanford, California, pp 69-94

# MODELS FOR VOLCANIC PROCESSES IN LONG VALLEY CALIFORNIA: TESTING BY CONTINENTAL DRILLING

John B. Rundle[1]

## Seismicity

The occurrence of a local magnitude $M_L$ = 5.8 earthquake on the Wheeler Crest fault on 4 October 1978 (Fig. 1) signaled the onset of significant seismic activity in the Long Valley, California, volcanic region.

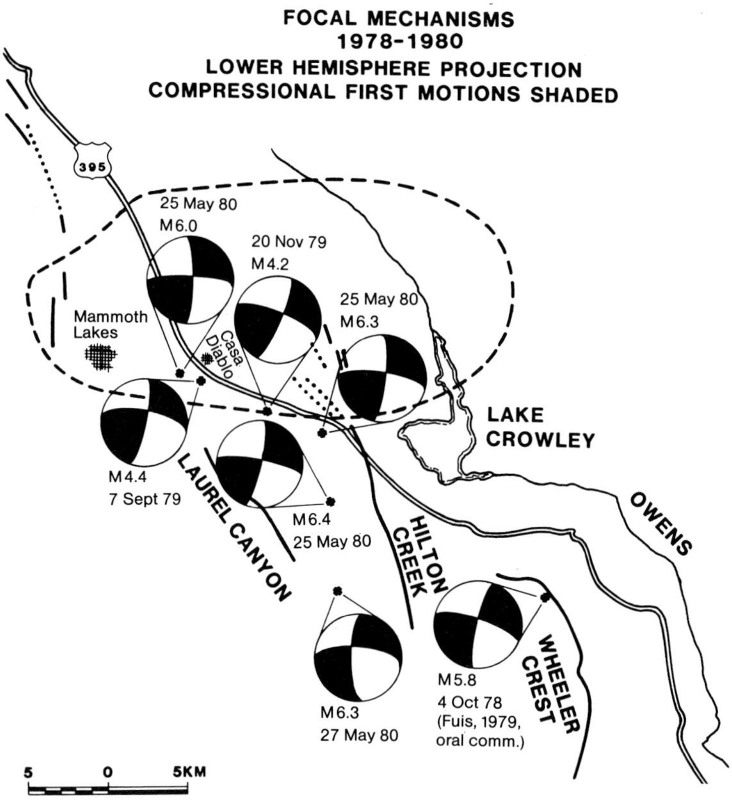


Fig. 1. Map showing Long Valley, together with major faults and focal mechanisms for largest events from 4 October 1978 to 25 May 1980. (Cramer and Toppozada 1980)

---

[1]Division 1541, Sandia National Laboratories, Albuquerque, NM 87185, USA

Observation of the Continental Crust
through Drilling I (ed. by C.B. Raleigh)

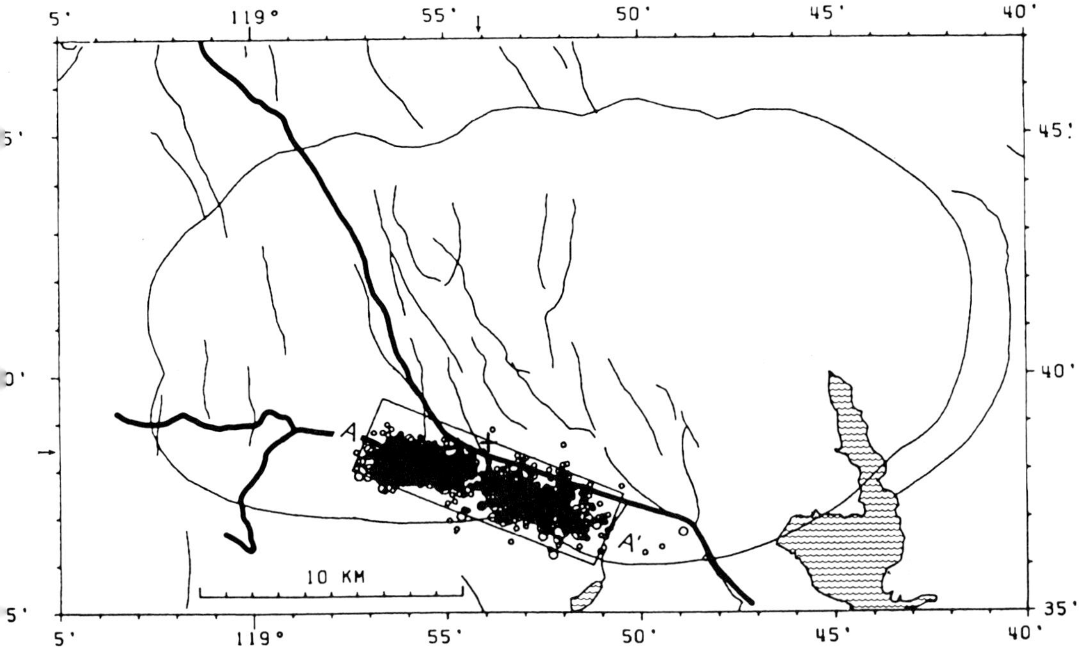


Fig. 2. Map of epicenters for the period January 7 - January 31, 1983. (Savage and Cockerham 1984)

This event followed a period of relative quiescence lasting about 37 years, which began upon culmination of the 1941 sequence (Cramer and Toppozada 1980). The 1941 earthquakes may have been associated with the filling of Crowley lake (Toppozada 1979), and included an event with a magnitude $M_L = 6.0$. Evidently, the 1941 sequence involved little ground uplift (Bennett 1980; Castle et al. 1984), at least along highway 395 (Fig. 1). By contrast, the 1978 event was followed by an extremely unusual swarm of four $M_L \geq 6.0$ events on May 25 - 27, 1980 (Fig. 1). Interpretation of these events in terms of tectonic processes has caused a great deal of controversy. Whereas the short period data imply almost pure strike slip, focal mechanisms derived from longer period data show an additional component of normal slip. Hence, while Julian (1983) and Julian and Sipkin (1984) argue that at least two of the events may have been caused by injection of magma into a dike complex, Given et al. (1982), Wallace et al. (1982), and Barker and Langston (1983) believe that the earthquakes were complex faulting events. All of the events generally show tension axes oriented ENE-WSW.

Using aftershocks of these events in the southern part of the caldera, and south into the Sierra Nevadas, Ryall and Ryall (1981) examined waveforms recorded at a variety of seismic recording stations located to the northwest, north and east. They found that for raypaths which crossed a region in the south-central part of the caldera, S wave energy was almost totally absent on the records, implying passage across a zone of partial melt. Depth to this zone was about 7 - 8 km. Following up on this work, Sanders (1984) did a more detailed S-wave shadowing study and contoured the depth below the surface to several magma bodies within the caldera. Minimum depth to the body was 5 km.

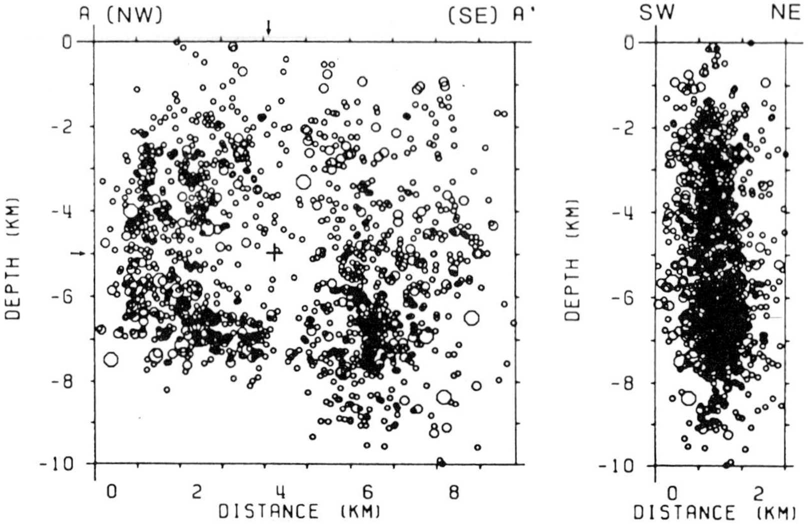


Fig. 3. Hypocenters of events shown in Figure 2. Note distinct gap in seismicity centered on 5 km depth, about 4 km from the western-most end of the box in Figure 2. (Savage and Cockerham 1984)

Following the events of May 1980, activity has continued at a relatively high level both within and south of the caldera. The seismicity is generally characterized by periods of intense swarms, separated by periods in which the background activity averages several events per day with $M_L \geq 2$. The most intense of these swarms began on January 6, 1983, and included two events of magnitude $M_L > 5$. These events were centered in the southwest moat (Bailey et al. 1976) of the caldera, between the southern end of the resurgent dome and the caldera bounding fault (the resurgent dome is the area bounded roughly by benchmarks CONVICT-D916-23DOR-43DOR in Figure 5). Focal mechanisms showed strike slip with T axes again directed ENE-WSW. The seismicity associated with this swarm (Figs. 2 and 3) during the month of January and subsequently (Savage and Cockerham 1984) shows a two-lobed pattern in plan view (Fig. 2). Hypocenters (Fig. 3) show a "gap" or "hole" in seismicity centered on a depth of about 5 km, and at a lateral position of about 4 km from the left hand edge of the box defined in Figure 2. Moreover, the seismicity in the western half of the box in Figure 2 is characterized by focal mechanisms consisting of both strike slip and normal slip events (Savage and Cockerham 1984). By contrast, the focal mechanisms for events in the eastern half of the box are almost purely strike slip in character. These features of the seismicity distribution evidently continue to dominate.

## Deformation

Beginning in 1905 with leveling (Castle et al. 1984), 1979 with horizontal strain, and 1982 for gravity change observations, crustal deformation data have been collected which unambiguously show that large uplift and strain has accompanied the seismicity discussed above. From 1975 to 1983, uplift has been in excess of 400 mm along Highway 395, with a maximum centered near the Casa Diablo area (Figs. 4-5, Castle

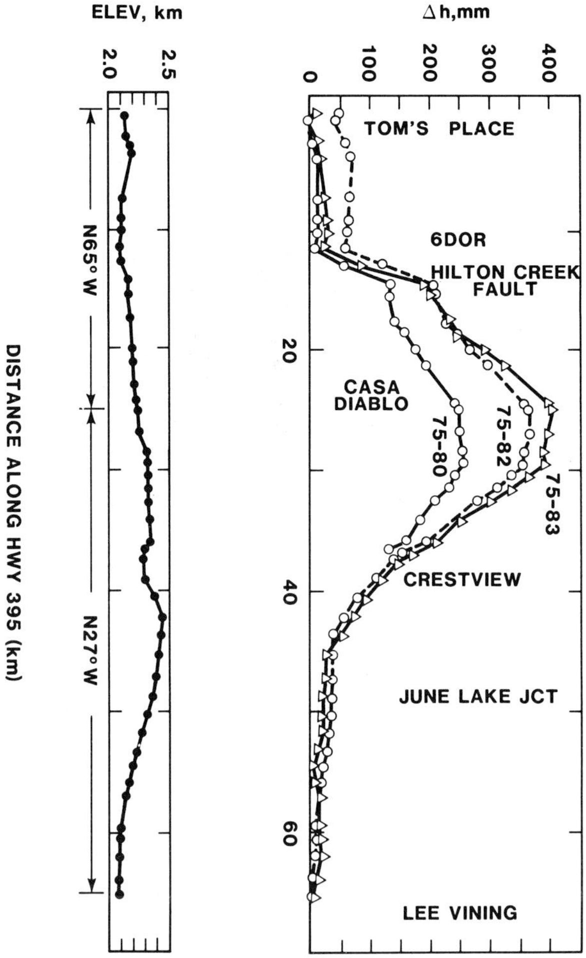


Fig. 4. Change in elevation along Highway 395, from Tom's Place south of Lake Crowley, to Lee Vining, west of Mono Lake, during the years 1975 - 1983 (Castle et al. 1984)

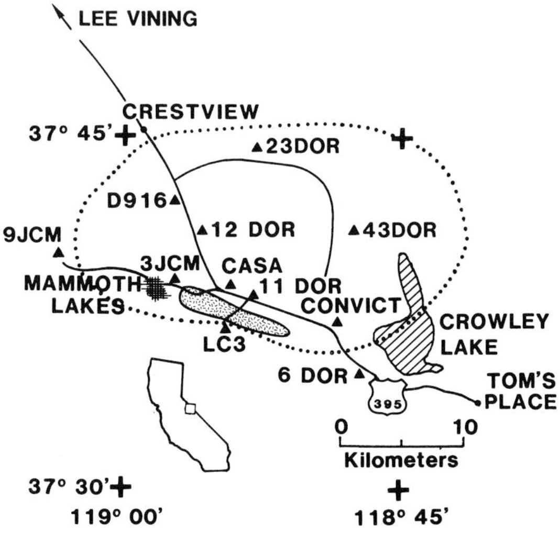


Fig. 5. Locations of major benchmarks in the Long Valley area. *Stippled area* is the location of the January, 1983 swarm

et al. 1984). Other leveling observed along routes off Highway 395 supports the picture of substantial uplift centered on the old resurgent dome of Long Valley caldera, suggesting the recent re-injection of magma into the system. Additionally, horizontal strain data collected since 1979 (Savage and Clark 1982; Savage and Cockerham 1984; Savage and Lisowski 1984) show clearly that significant deformation occurred in association with the 1980 events, and has continued at least through the summer of 1983. In particular, significant strain changes were observed at the time of the January, 1983 earthquakes, and for the succeeding 6 months into the summer of 1983. Finally, gravity changes have been observed by at least two groups of investigators (R. Jachens, pers. commun. 1984; Whitcomb and Rundle 1984) since June, 1980. These data have shown that substantial gravity changes occurred in conjunction with the other deformation changes, confirming the fundamental picture of the caldera actively undergoing uplift and horizontal extension. Interpretations of these data have been published by Savage, Castle and coworkers (Savage and Clark 1982; Savage and Cockerham 1984; Castle et al. 1984; Savage and Lisowski 1984), and by Rundle and Whitcomb (1984) and Whitcomb and Rundle (1984). All interpret the 1980 deformation data as a result of magma injection into a chamber beneath the central part of the resurgent dome, at a depth of 8 - 11 km. Right lateral strike slip faulting in the southwest moat contributed to the observed deformation as well. However, for the January 1983 events, significant differences of opinion exist between the two groups as to the causal process. While Savage and Cockerham (1984) believe that the deformation was caused by the formation of a WNW-ESE dike in the south moat, with associated strike slip, Rundle and Whitcomb (1984) proposed that magma from the deep chamber under the central resurgent dome made its way to shallower depth and produced inflation of a shallow chamber at 5 km under the Casa Diablo hot spring area (Fig. 1). Savage and Cockerham's (1984) model is based on the concept that the activity in the south moat is due to stress relief on the fault zone in the south moat. Thus, in their model, the "gap" or "hole" in seismicity seen in Figure 3 is due to a low shear stress environment. By contrast, Rundle and Whitcomb (1984) show that the seismicity distribution agrees with predictions of their model, and that the hole in the seismicity is due to a very high normal stress generated by the injection of magma into the shallow chamber underneath Casa Diablo hot springs. The debate between these two groups therefore revolves around the question of whether the stress in the south moat of Long Valley is high or low. Measurements of in situ stress can thus in principle resolve the question, and help to determine the correct causal process for the deformation and seismicity.

## Site Assessment for Thermal Regimes Drilling

During June, 1982, the Long Valley area was placed under a U.S. Geological Survey Notice of Potential Volcanic Hazard as a result of the intense level of tectonic activity there. This in turn has stimulated many geological and geophysical investigations oriented at determining the level of potential hazard that may exist. In addition, an intensive investigation has been underway since the summer of 1982 to determine the suitability of the Long Valley-Mono Craters system as a site for scientific drilling (Rundle and Eichelberger 1983). Teleseismic P-delays (Steeples and Iyer 1976) had first suggested that there might yet be a substantive body of magma underlying the caldera proper, at depths greater than 10 km. Smaller bodies of magma were also found within the caldera at depths as shallow as 7 km (Hill 1976; Ryall and Ryall 1981). Moreover, changes in the caldera hydrothermal system observed at the time of the 1980 events and subsequently (U.S. Forest Service, pers. commun. 1984), suggested that the thermal regime within the system was participating in the tectonic activity. It was therefore proposed, by a group of investigators from the National Laboratories, the U.S. Geological Survey, private industry, and some universities, to conduct a systematic investigation of the Long Valley-Mono Craters volcanic system. The purpose of the study was to characterize, in as much detail as possible, the geologic evolution of the system, and to investigate the present seismic, tectonic, and thermal activity. Funding was subsequently provided by the U.S. Department of Energy, Office of Basic Energy Sciences. Included within the project was a large scale seismic refraction experiment to be conducted by the U.S. Geological Survey over the years 1982 - 1984. Also included was a program to establish and regularly observe a dense network of high precision gravity change stations, heat flow measurements, geologic and petrologic investigations, and a geophysical synthesis of all available data. Much of this work is now complete, and first results are now becoming available (e.g., Meador and Hill 1983; Rundle and Whitcomb 1984; Whitcomb and Rundle 1984, Kissling et al. 1983; Eichelberger and Reece 1983).

One of the most interesting results to come out of the U.S. Geological Survey data collection has been obtained using a new data analysis technique, tomographic inversion for seismic structure. Using this technique, a narrow, 3 - 4 km wide linear NNW-trending low velocity zone has been identified at moderate crustal depths, 7 - 14 km. This feature extends from Mono Craters in the north, through the caldera under the west-central resurgent dome, and exists under the Sierran block bordering Long Valley to the south (E. Kissling, pers. commun. 1984). At shallower depths, perhaps 5 km or less, smaller, more local low velocity regions may exist under the northwest part of the resurgent dome, under Casa Diablo hot springs, and under Hot Creek, near the eastern extent of the resurgent dome. These shallow features offer the prospect of readily drillable targets at reasonable depths. Combined with the other site assessment work to date, it seems clear that the Long Valley caldera offers several drilling targets, whose investigation will resolve a number of important scientific thermal regimes problems. These problems include recent caldera evolution, origin of volcanic earthquakes, nature of heat flow and cooling of high level intrusions within the shallow crust, structure and evolution of the hydrothermal system, physical and chemical alteration processes active within the caldera, and stress and stress changes within an actively evolving volcanic system.

## Crustal Imaging Experiment

One of the goals of the newly-created Incorporated Research Institutions for Seismology (IRIS) is to establish a vigorous research program aimed at a systematic investigation of the lithosphere. It is anticipated that eventually, as many as 1000 portable seismographs will be available, together with vibrator sources, for use in three dimensional imaging of the crust and lithosphere. Prior to that time, however, it will be necessary to take advantage of opportunities that may arise, and to combine existing instrumentation and capabilities into unified experiments. Sandia National Laboratories has recently been granted access to a 3000' well located about 8 km NNW of Casa Diablo hot springs (Figs. 1 and 5). This well, which has a nominal diameter of 6", was originally drilled by Occidental Geothermal, Inc., for use as a temperature gradient hole. Tubing was subsequently pulled, and the hole is apparently in very good condition. Current plans call for

Sandia to contract with a commercial driller to clean out the hole, as its location is particularly well suited for a variety of passive and active seismic experiments. One such experiment, currently being funded by the Magma Energy Technology Program of the Department of Energy's Geothermal and Hydropower Technology Division, involves passive recording of waveforms from earthquakes occurring in the south moat of the caldera.

By emplacing the geophone at the bottom of the well, we anticipate several unique benefits, not obtainable with surface instruments. First, the recording site will be extremely quit, allowing detection of very small events, and recording of high frequencies in the tens of Hertz range from relatively distant events. Second and most important, we expect to see phases on the record not normally recorded, which involve upgoing rays bouncing off the ground surface, and travelling down to be recorded at the geophone site. These pulse shapes can then serve as a model for waveform shapes of energy reflected off an upper or mid crustal low velocity magma body. The ground surface reflection can thus be used as a matched filter for the magma body reflection. Alternatively, an autocorrelation on the waveforms should give two negative peaks at nonzero time lag, in addition to the positive peak at zero lag, if indeed a magma body reflection is present in the waveform. The first negative peak should correspond to the ground surface reflection, the second peak to reflection off the magma. And finally, complexities in the waveform due to near-surface inhomogeneities ought to be absent. This, for example, ought to make shear wave shadowing effects easier to see and interpret.

The passive experiment planned for the summer of 1984 makes use of the above advantages, in that $S_H$ waves are expected to show clear phase reversals upon reflection off the top of the magma body postulated to exist underneath Casa Diablo hot springs. The source-receiver geometry is also very favorable for this experiment. However, the hole location itself makes it an attractive candidate for pursuing an active seismic experiment as well. By coincidence, the hole lies almost directly over the bottoming point of rays showing an anomalous waveform in the U.S. Geological Survey refraction profiles of 1982 (Hill 1976) and 1983 (D. Hill, pers. commun. 1984). These waveforms appear to be similar to what might be expected from waves reflecting off a low velocity magma body at 7 - 8 km depth. Thus, two postulated magma bodies will be investigated: one at 5 km depth under Casa Diablo Hot Springs via the passive experiment, and one in the NW part of the caldera at 7 - 8 km depth via the active experiment.

The active-source, crustal imaging experiment planned for summer 1984 includes a near-vertical incidence Vertical Seismic Profile (VSP), a wide angle VSP, and a reflection survey around the borehole. Since the hole is located just at the intersection of the two refraction profiles, the VSP will aid an interpretation of both. Participants in the active experiment include personnel from Sandia National Laboratories; the U.S. Geological Survey; Lawrence Berkeley Laboratory; Lawrence Livermore National Laboratory; the University of Southern California; the University of California at Santa Barbara; the University of Wyoming; the California Institute of Technology; the University of Nevada, Reno; and the Massachusetts Institute of Technology. The goal of the experiment is to obtain as detailed a picture as possible of the crustal structure in the upper 10 km of the earth around the borehole site. Funding for these experiments is coming from a variety of agencies and offices within each agency. The organization and structure of this experiment will be used as a prototype for future experiments conducted under the auspices of IRIS. First results are expected to be available in time for a special symposium to be held at the Spring 1985 American Geophysical Union meeting.

## CSDP: Some Plans for the Future

One of the necessary pieces of physical data needed in order to understand the tectonic and volcanic evolution of the Long Valley caldera is the state of in situ stress and the magnitude and orientation of the time dependent stress changes. Both the stress and the stress changes are responsible for a diversity of phenomena observed in the caldera, including the occurrence of the major earthquakes and swarms; spatial seismicity patterns; changes in the hydrothermal system; ground deformation; mechanism of magmatic injection; and mode of ultimate rock failure and time and location of ultimate eruption. Hence it is vital to obtain as much information as soon as possible, and in as much detail as possible, about the stress patterns within the caldera. Frequent measurements are also of great interest in tracking the temporal details of the ongoing physical processes. Clearly, these data may only be obtained by drilling. Accordingly, a group of investigators from a variety of institutions has come together with the common interest of understanding the physical processes operating in the south moat of the caldera. Several proposals were written and have now been submitted to the Department of Energy, Office of Basic Energy Sciences. With a diversity of objectives, the proposals all have the common need for several 1 to 1.5 km deep wells in the south moat, and on the resurgent dome, of the caldera. In particular a group from Sandia, the University of Colorado, and the California Institute of Technology has written a proposal directed at measuring stress along the south end of the resurgent dome. This should allow resolution of the question, discussed in section 2 above, of whether the January 1983 activity was caused by dike injection in the south moat, or magma injection into a shallow chamber underneath Casa Diablo hot springs. While these rather expensive proposals are under consideration, other opportunities are becoming available to pursue in a limited way some of the objectives desired. A prime example of this approach is the crustal imaging experiment to be conducted in the Occidental Geothermal hole. As other holes become available in the caldera for similar types of work, it may be possible to accomplish a great deal on a relatively limited budget. In any case, all of the opportunities which present themselves can and will be pursued.

*Acknowledgements.* I would like to thank Yolanda Miller for typing the text of this manuscript. This work was performed under contract DE-AC04-76DP00789 from the Department of Energy to Sandia National Laboratories.

## References

Bailey RA, Dalrymple GB, Lanphere MA (1976) Volcanism, structure, and geochronology of Long Valley caldera, Mono County, California. J Geophys Res 81:725-744

Barker JS, Langston CA (1983) A teleseismic body-wave analysis of the May, 1980 Mammoth Lakes, California, earthquakes. Bull Seism Soc Am 73:419-434

Bennett JH (1980) Vertical crustal movement from 1932 and 1957 geodetic surveys in the Mammoth Lakes-Long Valley area, Mono County, California. Spec Report 150, California Div Mines Geol

Castle RO, Estrem JE, Savage JC (1984) Uplift across Long Valley Caldera, California. J Geophys Res (submitted)
Cramer CH, Toppozada TR (1980) A seismological study of the May 1980 and earlier earthquake activity near Mammoth Lakes, California. Spec Report 150, California Div Mines Geol
Eichelberger JC, Reece M (1983) Degassing of magma within a conduit. Trans Am Geophys Un EOS 64:895
Given JW, Wallace TC, Kanamori H (1982) Teleseismic analysis of the 1980 Mammoth Lakes earthquake sequence. Bull Seism Soc Am 72:1093-1109
Hill DP (1976) Structure of Long Valley Caldera, California, from a seismic refraction experiment. J Geophys Res 81:745-753
Julian BR (1983) Evidence for dike intrusion earthquake mechanisms near Long Valley caldera, California. Nature (Lond) 303:323-325
Julian BR, Sipkin SA (1984) Earthquake processes in Long Valley caldera area, California. J Geophys Res (submitted)
Kissling E, Cockerham RS, Ellsworth WL (1983) Structure of the Long Valley caldera region as interpreted from seismic data. Trans Am Geophys Un EOS 64:890
Meador PJ, Hill DP (1983) Data report for the August 1982 seismic-refraction experiment in the Mono Craters-Long Valley region, California. US Geol Surv Open File Report, pp 83-708
Rundle JB, Eichelberger C (1983) Continental scientific drilling at Long Valley-Mono Craters. Trans Am Geophys Un EOS 64:12-15
Rundle JB, Whitcomb JH (1984) A model for deformation in Long Valley, California, 1980-1983. J Geophys Res (in press)
Ryall F, Ryall A (1981) Attenuation of P and S waves in a magma chamber in Long Valley caldera, California. Geophys Res Lett 8:557-560
Sanders CO (1984) Location and configuration of magma bodies beneath Long Valley, California, determined from anomalous earthquakes signals. J Geophys Res (in press)
Savage JC, Clark MM (1982) Magmatic resurgence in Long Valley caldera, California: Possible cause of the 1980 Mammoth Lakes earthquakes. Science (Wash DC) 217:531-533
Savage JC, Cockerham RS (1984) Eartquake swarm in Long Valley caldera, California, January 1983: Evidence for dike inflation. J Geophys Res (in press)
Savage JC, Lisowski M (1984) Deformation in the White Mountains seismic gap, California-Nevada, 1972-1982. J Geophys Res (in press)
Steeples DW, Iyer HM (1976) Low-velocity zone under Long Valley as determined from teleseismic events. J Geophys Res 81:849-860
Toppozada TR (1979) Seismicity, possibly induced by Lake Crowley, on the eastern front of the Sierra Nevada, California. Trans Am Geophys Un EOS 60:882
Wallace T, Given J, Kanamori H (1982) A discrepancy between long- and short-period mechanisms of eartquakes near the Long Valley caldera. Geophys Res Lett 9:1131-1134
Whitcomb JH, Rundle JB (1984) Vertical distortion of the Long Valley, California, caldera associated with the January 1983 earthquake swarm. J Geophys Res (submitted)

# GEOCHEMISTRY AND MINERAL RESOURCES

# CONTINENTAL SCIENTIFIC DRILLING TO INVESTIGATE BRINE EVOLUTION AND FLUID CIRCULATION IN ACTIVE HYDROTHERMAL SYSTEMS

Robert O. Fournier[1]

## Introduction

A major objective of studying active hydrothermal systems is to gain information that can be used to help formulate and test models of coupled hydrothermal-magma systems, and mineral genesis in shallow magmatic environments. Many researchers have discussed the evolution of aqueous fluids from crystallizing magma in relation to ore deposition, including Bowen (1933), Emmons (1933), Fenner (1933), Lindgren (1937), Neuman (1948), Burnham (1967, 1979), Fournier (1968, 1972), Holland (1972), Phillips (1973), Norton and Cathles (1973, 1979), Whitney (1975), Cunningham (1978), Henley and McNabb (1978), and Henley and Ellis (1983). Models for fluid flow in hydrothermal systems have been developed by Elder (1965), Lister (1974, 1980, 1983), Ribando and others (1976), Cathles (1977, 1980, 1983), Norton and Knapp (1977), Norton and Knigth (1977), Norton (1978), Torrance and Sheu (1978), Fehn and Cathles (1979), Norton and Taylor (1979), Hartline and Lister (1981), Knapp and Norton (1981), Hardee (1982), and Sleep (1983).

Important parameters of models of hydrothermal-magma systems should include (a) the factors that control the maximum depth of circulation and maximum temperature likely to be attained by hydrostatically pressured liquids, including the factors that create and decrease permeability in rocks, (b) the evolution of brines in and around shallow magmatic intrusions, (c) the impact of those brines on fluid circulation and transport of metals, (d) the nature and distribution of sulfur species in the system, and (e) the mechanisms of water-magma interactions, in general. This paper focuses on the neeed for scientific drilling in active hydrothermal systems to gain information about coexisting hydrologic factors that cannot be otained from studies on fossil hydrothermal systems: temperatures, pressures, permeabilities, and salinities from the tops to the bottoms of systems at given times.

## Heat Transfer Regimes and Circulation of Hydrothermal Fluids

To explain hydrothermal activity and heat flow relationships at oceanic spreading centers, a model has been proposed in which heat is transferred by conduction through hot rock from magma to the base of a convecting hydrothermal system that is hydrostatically pressured (Lister 1974). In that model, as downward convecting water extracts thermal energy and cools the underlying rock, thermal cracking allows the water

[1]U.S. Geological Survey, Menlo Park, CA 94025, USA

Observation of the Continental Crust
through Drilling I (ed. by C.B. Raleigh)

to circulate ever deeper while maintaining a relatively constant flux of thermal energy from the system.

The thermal interaction of water with hot rock might be limited by decreasing permeability caused by deposition of minerals, such as quarz, if solutions are heated to temperatures above about 350° to 400°C (Fournier 1977, 1983a; Sleep 1983). It has been suggested that deep circulation of dilute water at Yellowstone National Park (Truesdell and Fournier 1976) and sea water at oceanic spreading centers (Bischoff 1980) might be limited by the rapidly increasing buoyancy of liquids as critical conditions are approached. Dunn and Hardee (1981) cite experimental data that show that water becomes "super convecting" when the temperature and pressure are in the vicinity of the critical point. Such super convection would allow fluid flow and relatively rapid transfer of heat through rocks with relatively low permeabilities. But super convection has not been demonstrated in systems in which the fluid is a brine.

Fluid circulation may also be affected by the mechanical properties of rock. The strength of rock should decrease with depth and, if temperature and rock type permit, the style of deformation may change from mainly brittle fracture to mainly ductile flow. Ductile flow of rock will decrease fracture permeability and limit the convective flow of fluids at hydrostatic pressures. However, where slow ductile flow of rock has caused fractures to close, newly imposed and relatively rapidly accumulation local stresses may cause other fractures to open. Thus, the maximum depth-temperature conditions of hydrothermal circulation may be episodic and/or depend on competing rates at which some processes close off fracture permability while other processes create additional fracture permeability. The evidence from ore deposits is that multiple episodes of fracturing, hydrothermal alteration, and vein formation occur.

It is likely that a region will exist, between the heat source and the domain of fluid flow at hydrostatic pressure, where fractures in the rock will be filled with fluids at greater than hydrostatic pressure. In that region, the maximum fluid pressure that can be attained is equal to the minimum principal stress in the rock, because hydrofracturing will occur where fluid pressures exceed the minimum principal stress (Hubbert and Willis 1957). In some places the least principal stress may approach the total overburden pressure, so the fluid pressure may approach the lithostatic load, as is the situation in some deep sedimentary basin environments where "geopressured" fluids are found (Wallace and others 1978). Lithostatic fluid pressures might be attained where rocks are subjected to horizontal compressive stress, and where the deposition of minerals in veins (self sealing) prevents flow of fluid along fractures in response to pressures that are less than lithostatic. The evolution of hydrothermal fluid from crystallizing magma also might contribute to high pore fluid pressure and result in hydrofracturing (Phillips 1973; Henley and McNaab 1978).

In most current models, thermal energy is transferred mainly by conduction (and in part by flow if fluids evolved from crystallizing magma) from very hot rock or magma to hydrostatically pressured hydrothermal systems. Circulation of heated fluid is likely to be induced at the sides as well as the tops of intrusive magmatic bodies. Figure 1 shows a schematic depth-temperature profile for a situation in which the base of the convecting hydrothermal system (point C) is above the magmatic body, and the temperature gradient for conductive heat flow is shown by line DC. Upward flowing fluid in the convection system transfers heat relatively efficiently, and eventually temperatures along the flow path may approach an adiabatic profile that is nearly isothermal

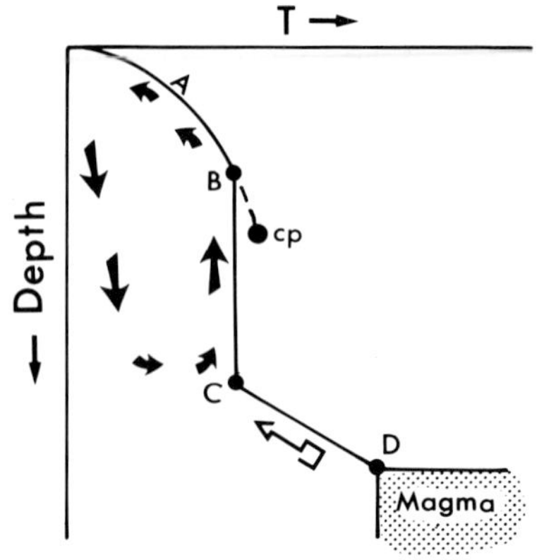


Fig. 1. Schematic depth-temperature diagram showing conductive transfer of heat from a magma (point *D*) to the basis of circulation *(arrows)* of a convecting hydrothermal system (point *C*) at hydrostatic pressure. Maximum temperatures at the upper part of the convecting system are controlled by the boiling point of the liquid, curve *A*. Point *B* shows where boiling of the ascending solution first occurs. See text for additional discussion

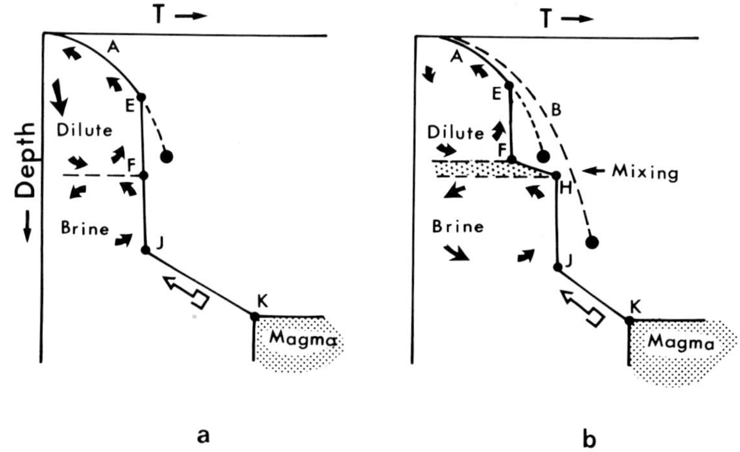


Fig. 2a,b. Schematic diagrams showing dilute, convecting hydrothermal systems that overlie convecting brines. Heat is transferred by conduction from magma to the base of circulation of the brines that are at hydrostatic pressure. In *a* the brine and dilute system are at the same temperature. In *b* the brine is at a higher temperature than the dilute system, which may promote physical mixing at the contact. See text for additional discussion

(line CB, Fig. 1). The upper limit of the nearly isothermal region is set by the boiling point curve (curve A, Fig. 1). The position of the boiling point curve, in turn, is determined mainly by the location of the water table and the composition of the particular fluid in the system. The boiling point curve might not be intersected where an ascending hydrothermal fluid cools entirely by conduction or where it discharges either at the bottom of the ocean or at the bottom of a deep lake.

It is also possible that highly saline brines might develop in and around bodies of hot, intrusive rock, with more dilute hydrothermal waters floating on top of those brines (Truesdell and Fournier 1976; Fournier 1983b). Cold, dilute water may surround the hydrothermal system in hydrostatic equilibrium with the hot brine. There may be no

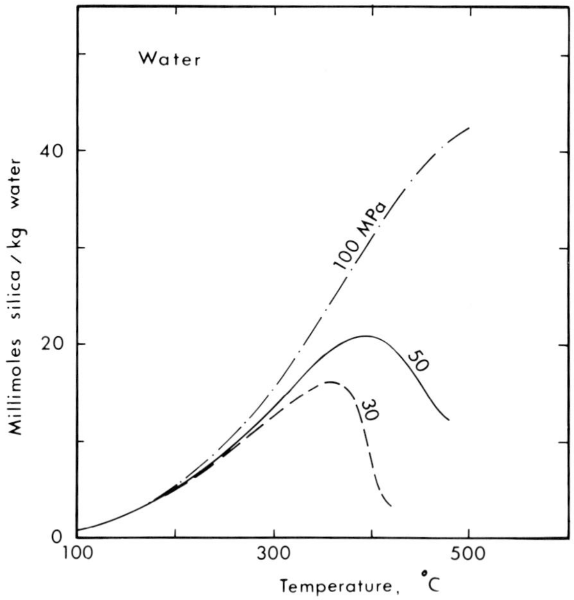


Fig. 3. Solubility of quartz in water as a function of temperature at the indicated pressure

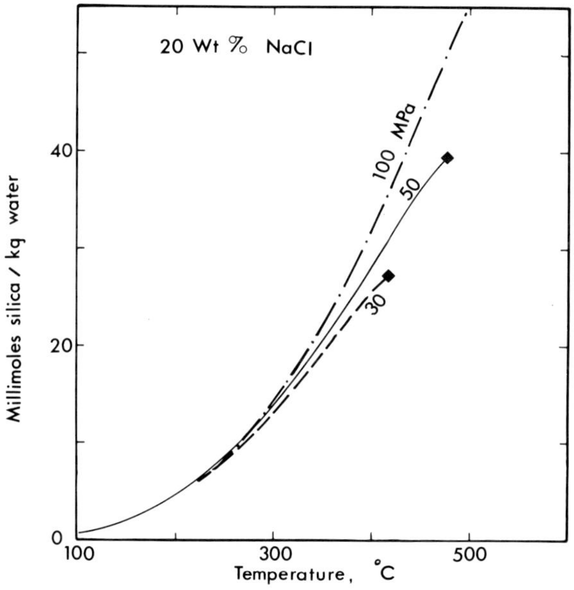


Fig. 4. Calculated solubility of quartz in 20 wt % aqueous NaCl as a function of temperature at the indicated pressures. (Fournier 1983c)

change in temperature across the upper interface of the brine and more dilute water (Fig. 2a), or a sharp temperature increase may occur (Fig. 2b). A higher temperature of the deeper circulating brine, compared to the cooler, more dilute overlying water, would diminish the density contrast between the two, so that directly mixing (in addition to diffusion) might occur at the interface.

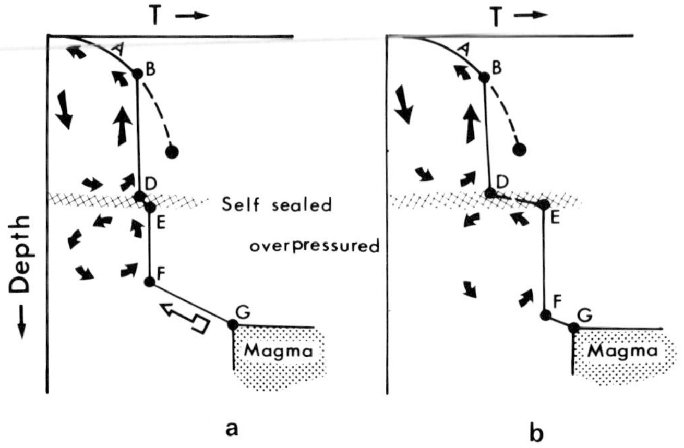


Fig. 5a,b. Schematic diagrams showing dilute, convecting hydrothermal systems that are separated from deeper convecting brines by impermeable zones that may have formed by self-sealing processes. Heat is transferred by conduction from magma to the base of circulation of the brines (point *F*) that are at pressures greater than normal hydrostatic for the given depths, but less than the least principal stress in the rock. In Fig. 2a the brine and dilute system are at nearly the same temperature. In Fig. 2b the brine is at a much higher temperature than the dilute system. See text for additional discussion

The possibility that chloride-rich brines may underlie dilute hydrothermal systems is of particular importance in formulating models of ore formation. Base metal sulfide deposition may occur entirely within the brine, or in a zone of mixing of brine and dilute water, while the main part of the dilute system remains barren of base metal sulfides.

At temperatures above about 360°C, the precipitation of minerals that decrease permeability may be less of a problem for sustaining deep circulation of brine than it is for sustaining deep circulation of dilute solutions. Figure 3 shows the effect of temperature and pressure on the solubility of quartz in pure water. At constant temperatures greater than about 360°C, pressure has a large effect on quartz solubility. A dilute hydrothermal solution flowing through the regions BC in Fig. 1 at a temperature greater than 360°C will deposit significant amounts of quartz, decreasing the permeability and rate of flow accordingly. Figure 4 shows calculated quartz solubilities (Fournier 1983c) in 20 wt % NaCl solutions. In comparison with quartz solubilities in pure water, pressure has relatively little effect upon the solubility of quartz in 20 wt % NaCl solutions. Thus, at temperatures above about 360°C, much less quartz will precipitate from an upward-flowing brine than from a dilute solution. Also, the densities of brines do not decrease as rapidly upon heating as do those of dilute solutions, so the buoyancy effects referred to by Truesdell and Fournier (1976) and Bischoff (1980) would be less upon heating to a given temperature.

A "self-sealed" zone deep in a hydrothermal system (Fig. 5) might separate an upper system with fluids at hydrostatic pressure from a lower convecting hydrothermal system in which fluid pressures exceed hydrostatic pressures for the given depths (an "overpressured" region). Because fluids are prevented from flowing across the self-sealed zone, fluid pressures may build up within the lower hydrothermal system as a result of various processes, including heating of a confined fluid, liberation of water by metamorphic reactions, influx of magmatic emanations that are evolved from crystallizing magma, and ductile flow of rock that slowly decreases the porosity. In the overpressured deeper hydrothermal system, the maximum fluid pressure is limited by the least principal stress in the rock, as discussed above, but in some systems that pressure may approach lithostatic. There may be little change in temperature across the self-sealed zone (Fig. 5a) or a significant temperature increase (Fig. 5b). The liquid in the part of the system where fluid pressure is greater than normal hydrostatic might be highly saline or relatively dilute, depending upon its origin (meteoric, ocean, saline lake, connate, metamorphic, magmatic, or a mixture of these waters) and subsequent history. Where thermally induced convection occurs in a part of the hydrothermal system where fluid overpressures are present, the fluid pressure will increase downward owing to the increased weight of the overlying liquid. If the least principal stress in the rock is less than lithostatic and independent of depth, the fluid pressure might become equal to that least principal stress at the bottom of the convecting system, and induce hydrofracturing downward, possibly toward the heat source, rather than upward, away from the heat source.

Deep drilling in active hydrothermal systems should be designed to gain information that can be used to set limits on the above models: namely, (1) detailed temperature and fluid pressure-depth profiles, (2) fluid composition-depth profiles, including chemical and isotopic measurements, (3) rock composition-depth profiles, including mineralogic, chemical and isotopic characterization, (4) the number, kinds, thicknesses, and attitudes of vein fillings as functions of depth and type of host rock, and (5) fluid inclusion compositions and filling temperatures as functions of depth and type of vein filling.

## Factors Affecting Salinities of Hydrothermal Fluids

Dissolved metal concentrations in hydrothermal solutions commonly increase with increasing dissolved chloride (Helgeson 1969; Barnes 1979; Bischoff and others 1981). Therefore, for models of transport and deposition of hydrothermal ore deposits it is important to determine the conditions that promote the formation of hydrothermal fluids that are rich in chloride. It is also important to determine the conditions that lead to the formation of brines in general, because total salinity has a major influence upon physical parameters that affect fluid flow, such as density and viscosity.

In the parts of presently active hydrothermal systems thus far explored by drilling, cation ratios are controlled by temperature-dependent water-rock reactions while total salinity and anions appear to be controlled by a combination of factors, including: (1) initial salinity of the recharge water, (2) salts leached from rocks, (3) diffusion of water and ions in response to chemical and thermal gradients, (4) physical mixing of dilute and concentrated waters, (5) adiabatic boiling in response to decompression of an ascending solution, and (6) precipitation of salts as solutions become supersaturated with respect to various components owing to heating (mainly affecting sulfates and carbonates, such as $CaSO_4$ and $CaCO_3$), to conductive cooling (possibly affecting NaCl), and to boiling (affecting all salts, but particularly $CaCO_3$ where $CO_2$ separates from the liquid phase).

Where dilute meteoric water recharges a hydrothermal system, and evaporite deposits are not present in the rock traversed by the circulating water, the type of host rock has a large influence on the amount of dissolved chloride found in the hot fluid. Representative chemical analysis of geothermal waters from various localities and geologic environments are given in Ellis and Mahon (1977). Data published by Arnorsson and others (1983) show that hydrothermal waters in basaltic rocks in Iceland far inland from the ocean generally contain about 20 to 50 mg $kg^{-1}$ chloride. In contrast, hot-spring waters issuing from silicic volcanic rocks in Iceland contain about 100 to 300 mg $kg^{-1}$ chloride (Arnorrson and others 1983). Hydrothermal waters in silicic volcanic rocks throughout the world, in general, contain a few hundred to a few thousand mg $kg^{-1}$ chloride. Hot-spring waters flowing through sedimentary rocks attain a wide variety of salinities, depending upon the type of sediment and the environment of deposition of that sediment. Highly concentrated brines thus far encountered in active hydrothermal systems all are in environments in which the waters appear to have interacted with evaporite deposits in the surrounding or nearby rock.

Fluid inclusions in many hydrothermal mineral deposits associated with shallow intrusive rocks contain both highly saline and moderately saline to dilute fluids (Roedder 1967, 1971, 1972, 1979). Fluids in some of these inclusions appear to be magmatic in origin; especially hypersaline, chloride-rich, high-temperature fluid inclusions found in copper and molybdenum porphyry systems. Experimental data (Sourirajan and Kennedy 1962; Ryabchikov and Hamilton 1971; Kilinc and Burnham 1972) show that highly saline as well as moderately concentrated brines may evolve from crystallizing magma, depending on the initial water and chloride concentrations in the magma and on the depth of crystallization. Steam and other gases are evolved from magma during eruptions of volcanoes. It is reasonable to assume that magmatic water (water that was once dissolved in magma, but not necessarily juvenile water) should be present in some presently active hydrothermal systems. However, to date there is no direct evidence for the incorporation of magmatic water into any presently active hot-spring system.

There may be various reasons why magmatic water has not been found in presently active hot-spring systems: (1) the release of magmatic water may be short-lived or episodic, and we may be observing presently active systems after or between magmatic hydrothermal episodes (continual influx of surficial recharge water and advection of the hydrothermal fluid may have caused most of earlier magmatic fluids to be flushed from the systems that we are now investigating; (2) much of the magmatic water may become incorporated in hydrous minerals during crystallization of the magma (particularly at greater depths); (3) the magmatic water may be released so slowly and in such small quantities that it becomes too diluted by other groundwaters to detect chemically or isotopically; and (4) the magmatic water may become trapped in heavy, deep brines that underlie presently active systems at greater depths than present drilling has attained. If magmatic water and gases are slowly being released from magmas into brines that underlie more dilute hydrothermal systems, sulfides should precipitate where the sulfur species in that gas come in contact with the presumably metal-rich brine. The distribution of sulfides in active hydrothermal systems relative to changing conditions of temperature, pressure, salinity, pH, and oxidation state is of major importance in theories of ore formation.

Theoretically, brines may also develop where dilute meteoric water interacts with very hot rock. However, in the parts of presently active systems that have been drilled, there is no direct evidence for significant quantities of brine formation by this mechanism.

## The Influence of Temperature and Pressure on the Evolution of Chloride-Rich Brains in Hydrothermal Systems

Chemical analysis of waters from geothermal wells and investigations of fluid inclusions trapped during hydrothermal ore deposition show that NaCl is the dominant dissolved constituent in most hydrothermal waters (Ellis and Mahon 1977; Roedder 1979). A large amount of information is available about the solubility of NaCl in $H_2O$ at high temperatures over a wide range of pressures (Keevil 1942; Sourirajan and Kennedy 1962), and some data are available for the system NaCl-KCl-$H_2O$ (Ravich and Barovia 1949).

Relatively few data are available for mixed salt systems in the compositonal range generally encountered in natural hydrothermal solutions. As a first approximation, the experimental work of Sourirajan and Kennedy (1962) can be used to determine the general influence of temperature and pressure on the evolution of brines. This has been done by others, such as Henley and McNabb (1978), Cunningham (1978), and Cloke and Kesler (1979), but there still seems to be widespread confusion about the attainment and likely significance of critical pehnomena in natural hydrothermal systems.

In applying the Sourirajan and Kennedy (1962) experimental results in this presentation, and with a geologic perspective in mind, pressure is plotted increasing downward corresponding to increasing depth. The depths corresponding to given fluid pressures in hydrostatically pressured hydrothermal systems were calculated with the assumptions that the pressure-controlling fluid has a uniform density of 1 and that the water table is at ground level. These assumptions yield a 10 MPa $km^{-1}$ (100 bar $km^{-1}$ increase in fluid pressure that greatly facilitates plotting the data, and avoids the problem of having to estimate densities of highly saline fluids over widely ranging temperatures and pressures. Even though the above assumptions were made expediently, they appear to be valid in some hydrothermal systems; measured fluid pressures in wells in the dilute hydrothermal system at Yellowstone National Park (White and others 1975; Fournier 1983b) and in the hypersaline hydrothermal system at the Salton Sea (Helgeson 1968) were found to increase at about 10 MPa $km^{-1}$.

Figure 6 shows the approximate positions of boiling point curves for water, and for 10 and 25 wt % NaCl solutions, extending to the respective critical points. Haas (1971) published calculated boiling point curves for water and various salt solutions ranging up to 25 wt percent with an upper temperature limit of 330°C. His results are slightly different from those shown in Figure 6 because he assumed that hydrostatic pressure is controlled by the weight of the overlying column of boiling solution, and corrected for changing density. Lithostatic depths in Figure 6 and subsequent figures were calculated using an average rock density of 2.5 kg $m^{-3}$.

The solubility of NaCl in water at atmospheric pressure is about 26 wt percent at 25°C and 28 wt percent at 100°C. Above 100°C, boiling point curves for NaCl solutions of salinity greater than 28 wt % terminate at a maximum temperature and depth (pressure) that corresponds to the critical condition, and a minimum temperature and depth given by the intersection of the boiling point curve with the stability field of gas plus solid salt. The approximate trace of the gas plus solid field for NaCl-$H_2O$ in temperature-depth space, at hydrostatic conditions with the density of the pressure-controlling fluid equal to unity, is shown in Figure 1. The gas plus solid region should contract to much

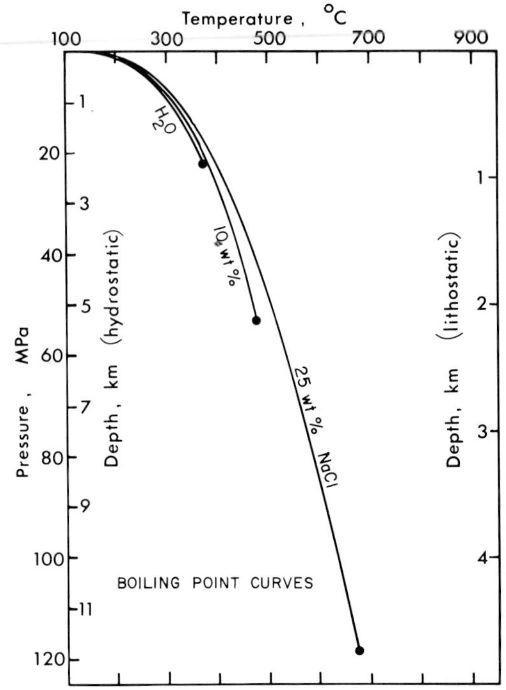


Fig. 6. Calculated boiling point curves for pure water and 10 and 25 wt % NaCl solutions, using data from Sourirajan and Kennedy (1962), and assuming pressure controlled by a fluid with a uniform density of 1, as discussed in the text

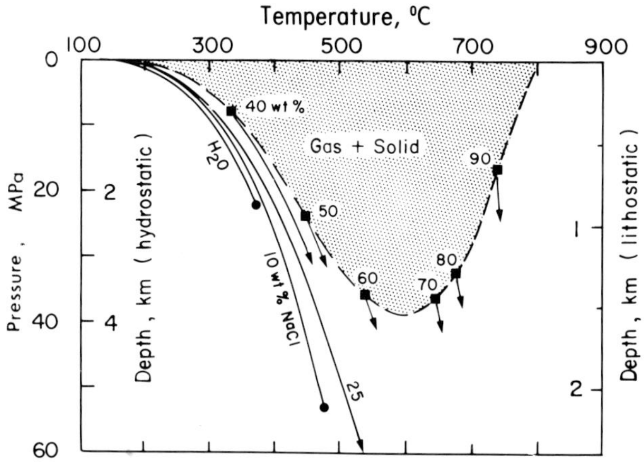


Fig. 7. Calculated boiling point curves and stability field of gas plus solid for the system $NaCl-H_2O$, using data from Sourirajan and Kennedy (1962), and assuming pressure controlled by a fluid with a uniform density of 1. The numbers beside the small squares indicate the wt % NaCl dissolved in the liquid phase along the three-phase curve gas plus liquid plus solid. *Arrows* indicate the general downward directions taken by boiling point curves of the indicated salinities. See text for additional discussion

shallower depths (lower pressures) when other salts are added to the system. However, confining pressures less than hydrostatic, as defined above, are likely in many volcanic systems in which gases vent directly to the atmosphere. Therefore, the gas plus solid salt region may extend to much greater depths than are shown in Figure 7. Underground gas pressures are likely to change step-wise where the gases expand through throttles.

Within the field of gas plus solid shown in Figure 7, NaCl reacts with water (hydrolizes) forming HCl and NaOH (Briner and Roth 1948; Martynova and Samoilov 1959; Galobardes and others 1981; Fournier 1983b). The yield of HCl is greater at lower pressures. The NaOH, in turn, will react with quartz, clays and micas to form sodic feldspars. The HCl that is generated at high temperatures is highly associated and relatively unreactive (Helgeson 1969; Frantz and Marshall 1984). However, where the gas is cooled, and particularly where steam condenses, the HCl dissociates and becomes very reactive.

For pure water, once the critical point is attained on heating, the system remains homogeneous upon further increase in temperature or pressure. Supercritical water changes back and forth from liquid-like to gas-like, without a perciptible change in phase, upon heating or cooling at pressures above the critical pressure or upon changing pressure at temperatures above the critical temperature. There is a tendency for those geologists who are only casually acquainted with phase equilibria involving saline solutions to think that these solutions also pass back and forth from liquid-like to gas-like, without a perceptible phase change, upon heating or cooling at pressures above the critical point. This idea is fostered by erroneous statements to this effect, published prior to publication of the work of Sourirajan and Kennedy (1962) (for example, F.G. Smith 1953). In the following discussion it will be shown that what is generally thought of as "supercritical" behavior of a fluid occurs only at pressures great enough to ensure that the gas phase may retain all the salt that was initially dissolved in the solution at some other temperature. Heating a saline solution at a pressure that is greater than the critical pressure will generally result in separation of that fluid into a less saline gas and a much more saline liquid (brine). For example, the solid line in Figure 8 shows the boiling point curve for a 10 wt % NaCl solution, ending at the critical point (C.P). The dashed line extending from the critical point shows the isopleth of 10 wt % NaCl dissolved in the gas phase that, in turn, will be in equilibrium with a more concentrated brine. Heating a 10 wt % NaCl solution at a confining pressure greater than the critical pressure (about 550 bars), and less than the pressure given by point H (Fig. 8), will cause an initial liquid to change gradually to a more gas-like phase, without a change in composition, until the solubility limit of 10 wt % NaCl in the gas phase is attained (the dashed line in Fig. 8).

Additional heating will cause the gas to disproportionate into a less concentrated gas and a brine containing more than 10 wt % NaCl. At a confining pressure less than the critical pressure, heating a 10 wt % NaCl solution will cause a liquid to disproportionate into gas plus a more concentrated brine when a temperature greater than the boiling point curve for that composition is attained. An indication of the approximate brine compositions that are required at different depths and temperatures can be obtained by projecting to greater depths the short curves with downward-pointing, arrow-shaped terminations shown in Figure 7. Isopleths showing NaCl solubilities in the gas phase that would be in equilibrium with those brines are shown by dashed lines in Figure 9.

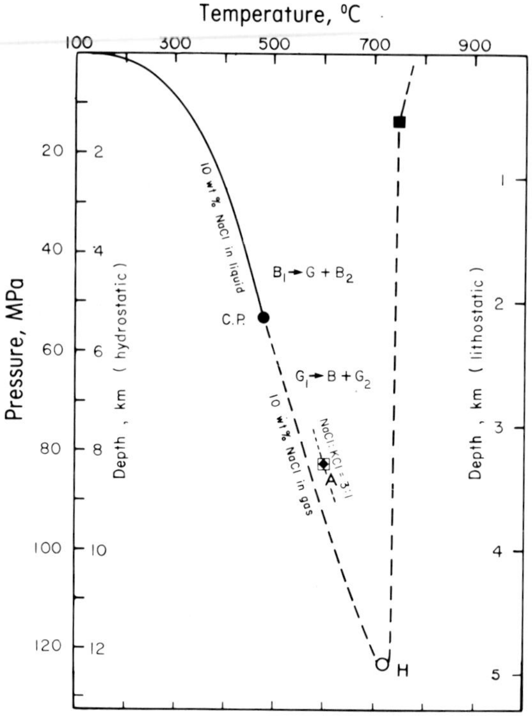


Fig. 8. Boiling point curve *(solid line)* and solubility in the gas phase of 10 wt % NaCl with water as solvent, using data from Sourirajan and Kennedy (1962), and assuming pressure controlled by a fluid with a uniform density of 1. The dot labeled *C.P.* is the critical point, the *square* marks the location of the three-phase boundary, gas plus liquid plus solid, and point *H* is informally called the hypercritical point. See text for additional discussion

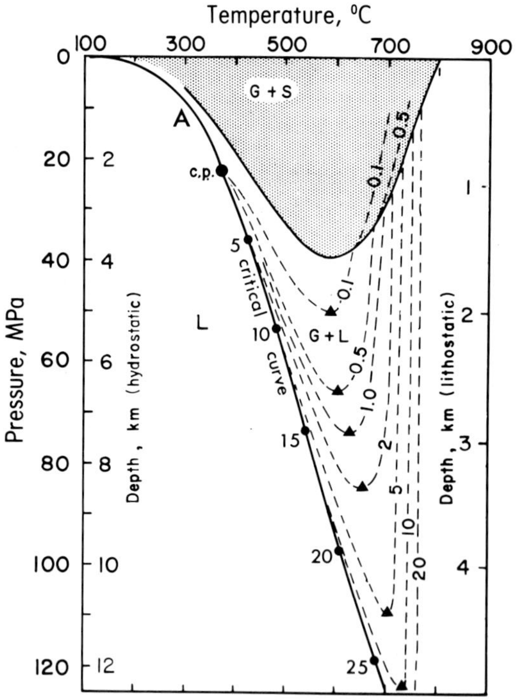


Fig. 9. Temperature-depth diagram for the system $NaCl-H_2O$, using data from Sourirajan and Kennedy (1962), and assuming pressure controlled by a fluid with a uniform density of 1. The stippled region shows the gas plus solid field. The line labeled *A* is the boiling point curve for pure water, ending at the critical point, *c.p.* Critical points for various aqueous NaCl solutions ranging from 5 to 25 wt % dissolved salt are shown by the *smaller dots*. *Dashed lines* show isopleths of wt % NaCl dissolved in the gas phase with water as solvent. *Triangles* are hypercritical points

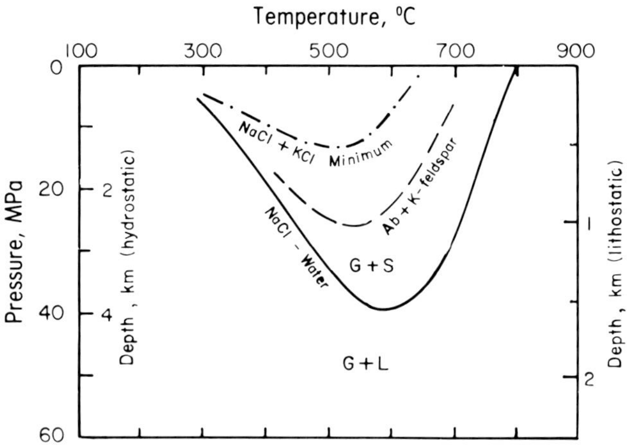


Fig. 10. Pressure vs. temperature diagram showing the three-phase boundary curves, gas plus liquid plus solid salt, for the system NaCl-$H_2O$ *(solid line)*, the system NaCl-KCl-$H_2O$ with Na/K controlled by the base exchange between albite and K-feldspar *(dashed line)*, and Na/K values appropriate for the ternary minimum *(dot-dashed line)*. See text for additional discussion

For geologic applications, point H, in Figure 8, may have more significance than the crictical point. At pressures greater than that given by point H, heating $H_2O$ containing 10 wt % dissolved NaCl to magmatic temperatures will result in the fluid changing from a liquid to a gas-like condition without a change in composition by disproportionation into less concentrated gas and more concentrated brine. For purposes of this discussion, point H will be called a "hypercritical" point. Approximate positions of hypercritical points for various amounts of NaCl dissolved in $H_2O$ are shown by the triangles in Figure 9. Addition of other salts to the system will cause the isopleths of salt solubility in the gas phase to be displaced to lower pressures (less depth), while addition of dissolved gases, such as $CO_2$, will have the opposite effect. These compositional effects may be relatively large. Point A in Figure 8 shows the position of the isopleth for 10 wt % dissolved salt in the gas phase determined in an experiment starting with a 3:1 by weight mixture of NaCl:KCl (unpublished results of the Author). Figure 10 shows the decrease in extent of the gas plus solid salt field that results when KCl is added to the system NaCl-$H_2O$.

Minimum depth-temperature conditions for interactions of magmas directly with hydrothermal solutions are given by the minimum melting curve for the respective magmas that are saturated with water (Burnham 1967, 1979). Figure 11 shows approximate depths and minimum melting temperatures for granites saturated with water with partial pressures of water in the melt controlled by hydrostatic (curve A) and lithostatic (curve B) conditions. Also shown are boiling point curves and isopleths of solubilities in the gas phase of 10 and 25 wt % NaCl (Sourirajan and Kennedy 1962) at hydrostatic and lithostatic fluid pressures. Note that for either hydrostatic or lithostatic fluid pressure conditions the gas solubility isopleths for 10 and 25 wt % NaCl lie to the high temperature side of the granite minimum melting curves. This indicates that the evolving magmatic fluid is likely to immediately disproportionate into a highly saline brine and a gas phase with less than 10 wt % dissolved NaCl. Studies of mineral phases found in the rhyolitic lavas and tuffs at Yellowstone show that most of the silicate melts in the magma reservoirs prior to eruption contained less than about 1 to 2 wt % water. Therefore, at depths greater than 1 to 2 km the magmas are unsaturated with water and at temperatures considerably higher than those given by the water-saturated minimum melting curves, A and B in Figure 11.

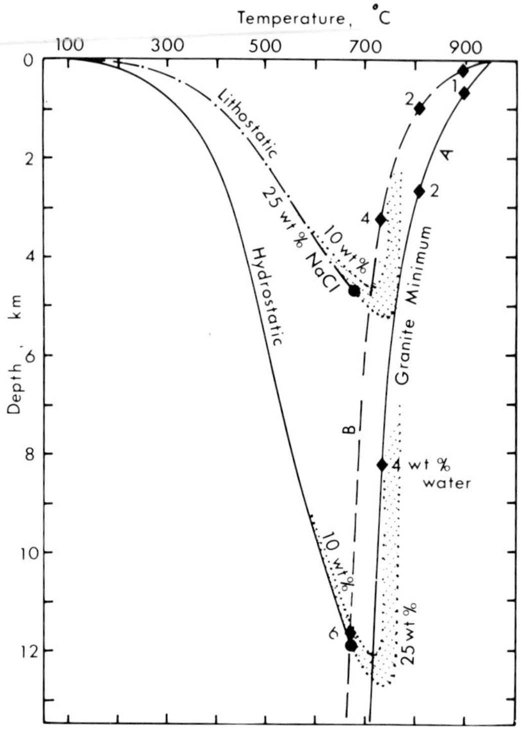


Fig. 11. Temperature-depth diagram showing boiling point curves for a 25 wt % NaCl solution hydrostatic pressure *(solid line)*, and lithostatic pressure *(dot-dashed line)* for the system NaCl-$H_2O$, using data from Sourirajan and Kennedy (1962). The *upper boundaries* of the *cross-hatched regions* show the solubilities of 10 wt % NaCl in the gas phase at hydrostatic and lithostatic pressures, and the *lower boundaries* show the solubilities of 25 wt % NaCl in the gas phase at those pressure. *Solid line A* shows the position of the minimum melting curve for granite, with the melt saturated with water at hydrostatic fluid pressure, and *dashed line B* shows the position of the minimum melting curve for granite with the melt saturated with water at lithostatic fluid pressure ($PH_2O$ = Lithostatic pressure). *Numbers next to diamonds* show the maximum wt's/% of water that can dissolve in magma at the given depths and hydrostatic and lithostatic fluid pressures

## Drilling Targets to Investigate Hydrothermal Magma Interactions

### General Considerations

In addition to being situated at drillable depths beneath physically and legally accessible locations, the igneous intrusive targets selected for drilling to investigate hydrothermal-magma interactions and related ore-forming processes should be young enough and/or large enough to have temperatures at or near magmatic temperatures, yet old enough and of sufficient size to have a well-established associated hydrothermal system. Ideally, heat sources at various depths should be drilled to investigate the effects of changing pressure and temperature on fluid circulation and brine evolution.

Magmatic temperatures at depths of 10 km or less may occur beneath active or recently active volcanoes in California, Oregon, Washington, Alaska, Hawaii, and probably elsewhere. There is geophysical evidence that magmatic material may be present beneath the hydrothermal systems at Yellowstone National Park, Wyoming (Eaton and others 1975; Iyer and others 1981; Schilly and others 1982), Long Valley, California (Steeples and Iyer 1976). The Geysers, California (Iyer and others 1979), and Roosevelt, Utah (Robinson and Iyer 1979). The largest of these magmatic-hydrothermal systems is at Yellowstone National Park. The recharge water at Yellowstone is dilute meteoric water (Allen and Day 1935; Truesdell and others 1977), but there is a possibility that magmatic water may be present at depth. The Yellowstone system is a huge thermal anomaly that offers an opportunity to study hydrothermal activity associated with magmas that may be only 3 to 5 km deep.

The Salton Sea hydrothermal system at the Imperial Valley presents a unique opportunity to study a landward extension of an oceanic spreading center (Elders and Biehler 1975; Elders and others 1982). Geophysical data suggest that the heat source results from repeated basaltic dike intrusions into tensional zones at depths generally ranging from about 10 to 14 km (Biehler and others 1964; Elders and others 1972; Fuis and others 1982; Johnson and Hill 1982). The recharge water is highly concentrated brine that attains its salinity by dissolving halite that is present in the sedimentary section (Rex 1983). The remainder of this paper will focus on the Salton Sea hydrothermal system in the Imperial Valley and the Yellowstone hydrothermal system.

## The Salton Sea Hydrothermal System

The hydrologic characteristics of the drilled portion of the central part of the Salton Sea geothermal field (to depths of about 2 km) have been described by Helgeson (1968), who showed that temperatures in the upper parts of wells, where relatively impermeable shales are present, increase almost linearly with depth. In contrast, in the bottom portions of wells, where permeabilities are relatively high, temperature-depth profiles are nearly isothermal, indicative of upward flow of hot brine and temperatures below the boiling point. However, isothermal temperatures in the bottom portions of different wells were shown to be different. The maximum depth and temperature of fluid circulation is unknown. The brine may have formed by dissolution of halite at a temperature near 25°C (Rex 1983). The maximum chloride concentration found at depth, corrected for loss of steam that occurs during production, is about 15 wt %. If all the dissolved chloride were balanced by dissolved sodium, the solution would be about 25 wt % NaCl. Even higher salinities at depths greater than those that have yet been drilled might result from dissolution of halite beds by waters at temperatures above 25°C, or by boiling where brines come into contact with very hot rock.

The boiling point curve and critical point for a 25 wt % NaCl solution are shown in Figures 6 and 11. The boiling point curve of the deep Salton Sea brine could be displaced several degrees toward higher temperatures because of high calcium and potassium concentration. However, because the density of the pressure-controlling fluid may be less than unity in its hotter and deeper regions, and because dissolved gases, such as $CO_2$, will displace the boiling point curve to lower temperatures, the net effect is likely to be a boiling point curve that is not greatly different than the one for the 25 wt % NaCl solution shown in Figures 6 and 11.

One conclusion that can be drawn from Figure 6 is that a well drilled to a depth of 5 or 6 km is not likely to encounter a brine near its

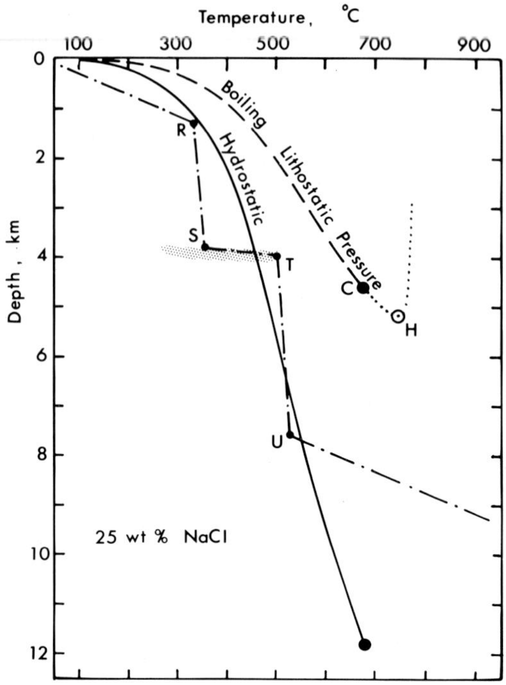


Fig. 12. Temperature-depth diagram showing a hypothetical situation in which an impermeable region (line segment *ST*) separates a 25 wt % NaCl brine convecting at hydrostatic pressure (line segment *RS*) from a similar brine convecting in a region where fluid pressure is greater than normal hydrostatic for the given depths, but less than the least principal stress in the rock (line segment *TU*). Also shown are boiling point curves for a 25 wt % NaCl solution at hydrostatic pressure *(solid line)*, and lithostatic pressure *(dashed line)* for the system NaCl-$H_2O$, using data from Sourirajan and Kennedy (1962). *Large dots* show the positions of the critical points appropriate for the respective depth-pressure conditions, and point *H* is the hypercritical point for lithostatic fluid pressure

critical point at hydrostatic pressure unless there is a dramatic decrease in salinity at greater depths than have been attained by present drilling. Another conclusion is that the maximum temperature attained by brine that flows up to within about 2 km of the surface at hydrostatic pressure is likely to be less than about 400°C. If a higher temperature were attained, the solution would boil during its upward flow, but there is no indication of boiling. Other convection cells with higher temperatures and, perhaps, higher salinities might be present at deeper levels, separated from each other by contrasting fluid denisities or by relatively impermeable strata. If the strata are truly impermeable because of rock type, deformation, or self-sealing, the fluid in one or more deep convection cells might be considerably above hydrostatic pressure.

Figure 12 shows boiling point and gas solubility curves for 25 wt % NaCl solutions at hydrostatic and lithostatic conditions. It also shows a hypothetical situation in which a deeply convecting brine, at a temperature of about 500°C and with properties similar to a 25 wt % NaCl solution, is at a pressure much greater than hydrostatic but less than lithostatic (line segment UT). That deep brine is separated from the upper hydrostatic system by an impermeable zone at about 4 km (line segment TS). The impermeable zone could be located at shallower or deeper levels. The brine is well below boiling conditions for fluid pressures significantly less than lithostatic, but above the boiling point curve for hydrostatic pressures at depths shallower than about 6 km. Rupturing of the impermeable zone (perhaps as a result of seismic activity or a hydrofracturing process) would result in an immediate drop in fluid pressure just below the rupture, in local boiling, in massive deposition of minerals, such as quartz and sulfides, and in formation of a more concentrated brine. The mineral deposition would tend to heal the rupture, so that fluid pressures could once again return to greater than hydrostatic. Deep drilling into the Salton Sea hydrothermal system might penetrate into a region where fluid pressures are considerably in excess of hydrostatic.

For lithostatic fluid pressures, critical and hypercritical conditions could be attained somewhere between 4 and 6 km deep. However, the likelihood of a convecting brine at lithostatic pressure with 25 wt % dissolved solids attaining the critical temperature at 4 to 6 km appears remote.

## The Hydrothermal System at Yellowstone National Park

The geologic, tectonic, and volcanic setting of Yellowstone National Park has been summarized by Smith and Christiansen (1980), Christiansen (1984), and Smith and Braile (1984). Volcanism in the Yellowstone Park region has been active for the last 2.2 m.y., and there is no compelling reason to think that this volcanism has ended (Christiansen 1984). Geologic and geophysical data suggest that a huge body of magma still underlies much of the Yellowstone caldera (Eaton and others 1975; Iyer and others 1981; Lehman and others 1982; Smith and Braile 1984). The lavas and tuffs that have been erupted from the Yellowstone volcanic system during the past 2.2 m.y. are predominantly rhyolitic and subordinately basaltic. Compositions of coexisting iron-titanium oxides in the rhyolite indicate that magmatic temperatures ranged from about 800° to 925°C prior to eruption, and the lack of water-bearing minerals coupled with the presence of fayalite suggest that the magma contained not more than 1 to 2 wt % dissolved water (Hildreth and others 1984).

Fournier and others (1976) estimated the total advective heat flow from the hydrothermal system by measuring the total discharge of water in the rivers flowing from the area and equating chloride concentrations in those waters with enthalpies of fluids in the deep hydrothermal system. The thermal power output by the advecting hot-spring water is estimated to be about $5.5 \times 10^9$ W. The area of Yellowstone caldera is about 2500 $km^2$, so the average thermal energy flux that results just from advection of the hot-spring water is about 1800 mW $m^{-2}$. If the assumption is made that thermal energy is transferred uniformly by conduction from magma across a 2500 $km^2$ surface, through hot, dry rock (with an assumed thermal conductivity ranging from 2.5 to 1.7 W $m^{-1}$°K) to the base of circulation of the hydrothermal system, a thermal gradient of about 700°C to 1000°C $km^{-1}$ would be required to sustain the average convective thermal output.

The water in the deep part of the Yellowstone hydrothermal system that advects to the surface is estimated to contain about 400 mg $kg^{-1}$Cl (Fournier and others 1976), and is about equivalent in salinity to a 0.07 wt % NaCl solution. Figure 13 shows the boiling point curve and isopleth of salt solubility in the gas phase for a 0.07 wt % NaCl solution. Substitution of KCl for about one third of the NaCl will move the isopleth of solubility in the gas phase to lower pressures (shallower depths), about to the dotted line labeled NaCl + KCl. Addition of $CaCl_2$ to the system may cause the field of gas plus solid salt to contract to slightly lower pressures. The stippled area in Figure 13 shows the approximate depth-temperature region from which dilute waters of the compositions thus far found in the Yellowstone hydrothermal system are excluded. However, in the parts of the hydrothermal system where the convective up-flow is relatively rapid, the temperature-depth limitation is probably given by the combined boiling point curve and isenthalpic decompression line shown in Figure 13. Solutions that are heated to temperatures to the right of the isenthalpic curve must

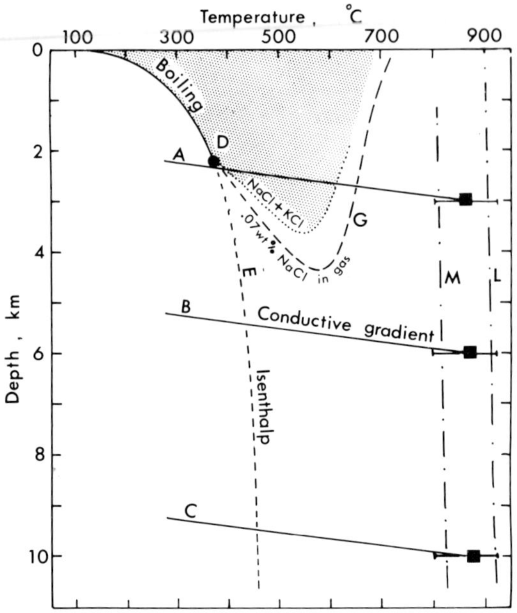


Fig. 13. Temperature-depth diagram showing the boiling point curve ending at the critical point, *D*, the isenthalpic expansion curve, *E*, extending from the critical point, and solubility in the gas phase of a 0.07 wt % NaCl solution, curve *G*. *Stippled region* shows the temperature-depth conditions at which a liquid of the composition found in the Yellowstone hot-spring system is likely to disproportionate to brine plus gas. *Squares* show likely magmatic temperatures at 3, 6, and 10 km depths, and the lines *A*, *B*, and *C* show minimum conductive thermal gradients that are required to account for the advective heat flux from magmas at the depths indicated by the squares. *Dot-dashed lines*, *L* and *M*, show respectively the melting curves for granite with 1 and 2 wt % dissolved water. See text for additional discussion

cool in part by conduction as they rise or disproportionate to steam and highly concentrated brine upon decompression.

Lines A, B, and C in Figure 13 show thermal gradients for conductive transfer of heat from magmas at 3, 6, and 10 km, respectively, to the convecting hydrothermal system. If the maximum temperature attained by the convecting hydrothermal water is just 350°C and the magma has a temperature of 850°C, the circulating water would have to come within about 0.5 to 0.7 km of the magma to maintain the calculated advective thermal flux, no matter how deep the magma was situated. Because the maximum temperature attained by the convecting hydrothermal system is likely to be greater than 350°C, and the conductive transfer of heat from magma to the hydrothermal system is not likely to be uniform, convecting water is likely to come much closer to the heat source than 0.5 km.

If the heat carried in the advective flow of water from the system were supplied entirely by the latent heat of crystallization of dry rhyolite magma, about $5.33 \times 10^{11}$ kg yr$^{-1}$ of magma would be required to furnish that heat (using an assumed latent heat of crystallization equal to $3.14 \times 10^5$ J $kg^{-1}$). This requires solidification of about 0.22 $km^3$ of rhyolite magma each year (using an assumed magma density of 2.37 g $cm^{-3}$). However, the rhyolitic magma is likely to contain about 1 to 2 wt % water, as previously noted. If the water that is released from the crystallizing magma mixes with and partly heats the dilute hydrothermal system, crystallization of a little less than $5.33 \times 10^{11}$ kg $yr^{-1}$ of magma would be required to supply the heat carried by the advecting hydrothermal fluid. When a very high value for the enthalpy of the water that would escape during crystallization of the magma ($3.5 \times 10^6$ J $kg^{-1}$) is assumed, about $4.8 \times 10^{11}$ kg $yr^{-1}$ of magma with 1 wt % dissolved water are required to furnish the advective heat flux, and $4.4 \times 10^{11}$ kg $yr^{-1}$ of magma with 2 wt % dissolved water. The corresponding volumes of magma that would be required to crystallize are 0.20 and 0.19 $km^3$, respectively. However, the uncertainties in the latent heat of solidification of the magma, the amount of water retained in hydrous minerals, and enthalpy of the evolved hydrothermal fluid are so large that the above calculated numbers may be in error by at least 20%.

The maximum quantities of water that could be liberated by crystallizing the above amounts of magma are $4.8 \times 10^9$ and $8.8 \times 10^9$ kg $yr^{-1}$, assuming 1 and 2 wt % water respectively in the magma. In comparison, Fournier and others (1976) estimate that about $10^{11}$ kg $yr^{-1}$ water is discharged by the hot-spring system (not including water incorporated by near-surface dilution). Thus, the water evolved from crystallizing magma could amount to about 5 to 9 % of the total flow from the deep part of the hydrothermal system. However, when the chloride that is likely to be evolved with the magmatic water also is considered, it seems likely that only a small fraction of the magmatic water can reach the dilute hydrothermal system.

According to Wes Hildreth (oral communication 1984), the maximum chloride concentration measured in obsidian glass erupted from the Yellowstone volcanic system is about 1500 to 1800 mg $kg^{-1}$. If this chloride were all balanced by sodium, there would be about 0.25 to 0.30 wt % NaCl in the magma. When water separates from a crystallizing magma, NaCl is strongly fractionated into the aqueous phase (Kilinc and Burnham 1972). If the above range in chloride found in obsidian glass at Yellowstone is indicative of the amount of NaCl that will follow 1 to 2 wt % water into the evolved magmatic fluid where magma crystallizes slowly at depth, the salinity of that magmatic fluid will be equivalent to about 23 to 11 wt % NaCl. The quantity of chloride evolved from a crystallizing magma that initially contains 1 wt % water and 1500 mg $kg^{-1}$ Cl is $7.2 \times 10^8$ kg $yr^{-1}$; for 2 wt % water and the same chloride it is $6.6 \times 10^8$ kg $yr^{-1}$. In contrast, the calculated amount of chloride discharged by the hot-spring activity is only about $0.4 \times 10^8$ kg $yr^{-1}$. Therefore, if all the heat supplied to the hydrothermal system is derived from the latent heat of crystallization of magma plus the heat carried in the evolved water, only about 5 to 6 wt % of the co-evolved chloride becomes incorporated in the dilute hot-spring system.

One explanation for the above is that all or most of the thermal energy carried by the advecting hot-spring water is derived from hot, but already crystalline rocks that are being cooled by the circulating meteoric waters. Simple cooling of hot crystalline rock, however, should lead to thermal contraction of the rocks beneath the caldera. The uplift reported by Pelton and Smith (1979, 1982) suggests just the reverse; the underlying rock apparently is heating up and/or magma is being added to the system from the lower crust or mantle. Another explanation for the apparent deficiency of chloride in the shallow part of the hydrothermal system is that brine is accumulating under the dilute, circulating hydrothermal water that reaches the surface (Figs. 2,

3). It is likely that the thermal energy carried to the surface by the advecting hot-spring water comes in part from cooling already crystalline rock and in part from crystallizing magma, and that brine is accumulating at depth.

Referring to Figures 9, 11, and 12, an aqueous fluid with 10 to 25 wt % dissolved salt that separates from crystallizing magma into a lithostatically pressured hydrothermal system should not disproportionate into more concentrated brine plus gas if the magma is at a depth greater than about 5 km (possibly 3 to 4 km because of the effects of other salts in addition to NaCl). Disproportionation is very likely at depths less than about 3 km. In contrast, where an aqueous fluid with 10 to 25 wt % dissolved salt separates from crystallizing magma into a hydrostatically pressured hydrothermal system, disproportionation into more concentrated brine plus gas should occur to depths of 10 or 12 km, depending on the effects of mixed salts upon the hypercritical points. However, the brines that form by disproportionation will be more concentrated at shallower than at greater depths.

## Conclusions

The depths of circulation and temperatures attained by fluids at hydrostatic pressure are controlled by combinations of circumstances and processes. Some processes decrease permeability, such as deposition of minerals in fractures, and ductile or plastic flow of rock in response to increasing lithostatic load and increasing temperature. The plastic deformation of rock may squeeze shut fractures and pore spaces until the fluids that remain in the rock attain very high pore pressures. Other processes increase permeability, such as thermal cracking in response to heating or cooling, dissolving of minerals in response to changing fluid temperatures and pressures, brittle fracturing that results from regional and local tectonically induced stresses, and hydrofracturing that may occur when fluids are squeezed upward from rocks that are deforming plastically.

At temperatures above about 350°C, solubilities of some minerals in water, such as quartz, change greatly as a function of pressure at constant temperature. However, pressure has little effect upon the solubility of quartz, and presumably many other minerals, in concentrated salt solutions. Therefore, deep circulation of hydrothermal solutions at temperatures above 350°C will be less limited by deposition of minerals in fractures when those solutions are brines.

The recharge water for hydrothermal systems at hydrostatic pressure may vary greatly in salinity, ranging from dilute meteoric water to ocean water and evaporated brines in saline lakes. Pore waters in deep sedimentary basins are generally highly saline, and may become incorporated into convecting hydrothermal systems. Brines may also form as a result of circulating waters dissolving halite and other evaporites that may be present in sedimentary rocks. Boiling point curves (Figs. 6, 7) and solubilities of salts in the gas phase (such as those shown for NaCl in Fig. 9) place limits on the salinities of hydrostatically pressured fluids as they become heated while flowing toward a heat source. Where the recharge water is dilute, and relatively little salt is dissolved from the surrounding rock, brines are likely to form only where temperatures in excess of about 500°C are attained at less than about 5 km (the heat source is relatively shallow).

The critical point is geologically of minor importance for natural waters that contain dissolved salts. Heating a solution of given composition at a pressure above its critical pressure may result in its disproportionation into a more concentrated liquid (brine) and a gas. What is generally thought of as "supercritical" behavior of a fluid occurs only at pressures great enough to ensure that the gas phase may retain all the salt that was initially dissolved in the solution at some other temperature. In this presentation that pressure is called the "hypercritical" pressure: the "hypercritical point" is at a temperature that is fixed by the maximum pressure on an isopleth of salt solubility in the gas phase (such as point H in Fig. 8).

The salinities of fluids that are evolved from a crystallizing magma depend on the initial salt and water concentrations in the magma and on the fractionation of salt between the magma and the separating aqueous phase (Ryabchikov and Hamilton 1971; Kilinc and Burnham 1972). Chloride is strongly fractionated into the aqueous phase, so the first fluid to separate is likely to be much more saline than the last (Kilinc and Burnham 1972). Whether or not the separating fluid retains its initial salinity or disproportionates into more saline brine and dilute gas will depend on the fluid pressure regime surrounding the magma. When an aqueous phase separates from cooling and crystallizing magma directly into a hydrostatically pressured environment, it is likely that slight cooling of that aqueous phase will cause it to disproportionate into a brine and a more dilute gas (Figs. 9 - 11), even at depths as great as 10 km. Magmatic fluids separated from magmas at hydrostatic pressure at depths less than 2 to 4 km are likely to undergo retrograde boiling to dryness (Fig. 10) and release large amounts of HCl to the gas phase. At depths below about 7 or 8 km the "more dilute" gas phase may retain enough dissolved salt to be considered a brine (Fig. 9).

The situation is very different when fluids separate from the magma and cool at lithostatic pressure. Only those fluids that separate and cool at depths less than 4 to 5 km are likely to disproportionate into more concentrated brine and more dilute gas. However, if a fluid that is initially at lithostatic pressure passes to a hydrostatically pressured condition, the temperature and depth of that transition become very important. If the temperature of the transition is to the high temperature side of either the hydrostatic boiling point curve or the solubility in the gas phase curve appropriate for the particular fluid composition, the magmatic fluid will disproportionate into more concentrated brine and gas, as previously discussed.

The general conclusion that can be drawn from the above discussion is that highly concentrated brines are much more likely to evolve around shallow magmatic heat sources than around deep ones. Where concentrated brines form by disproportionation of dilute to moderately saline fluids or magmatic gases, layered or stacked hydrothermal systems are likely to evolve in which dilute circulating waters float on top of the denser brines.

Deep scientific drilling into the Salton Sea hydrothermal system offers the opportunity to investigate the movement and geochemical evolution of hypersaline brines at very high temperatures above a deep heat source (perhaps 10 km deep). That heat source appears to contribute few, if any, components directly to the hydrothermal fluid. The Yellowstone caldera is one of the few places in the world where drilling might encounter brines that are genetically related to magmas that are crystallizing in the 5 to 7 km depth range, a depth range where a transition from hydrostatic to higher than hydrostatic fluid pressures is very likely, and brines may range widely in salinity.

*Acknowledgements.* The manuscript was reviewed by Susan W. Kieffer, Rosemary J. Vidale, and Donald E. White. I am grateful for their many helpful comments and suggestions.

## References


Allen ET, Day AL (1935) Hot springs of the Yellowstone National Park. Carnegie Inst Washington Publ 466:525

Arnorsson S, Gunnlaugsson E, Svavarsson H (1983) The chemistry of geothermal waters in Iceland. II. Mineral equilibria and independent variables controlling water compositions. Geochim Cosmochim Acta 47(3):547-566

Barnes HL (1979) Solubilities of ore minerals. In: Barnes HL (ed) Geochemistry of hydrothermal ore deposits, 2nd edn. Wiley, New York, pp 404-460

Biehler S, Kovach RL, Allen CR (1964) Geophysical framework of northern end of Gulf of California structural province. Am Assoc Petrol Geol Mem 3:126-143

Bischoff JL (1980) Geothermal system at 21°N, East Pacific Rise; physical limits on geothermal fluid and role of adiabatic expansion. Science (Wash DC) 207:1465-1469

Bischoff JL, Radtke AS, Rosenbauer RJ (1981) Hydrothermal alteration of graywacke by brine and seawater: Roles of alteration and chloride complexing on metal solubilization at 200° and 350°C. Econ Geol 76(3):659-676

Bowen NL (1933) The broader story of magmatic differentiation, briefly told. In: Ore deposits of the Western States (Lindgren vol). Am Inst Min Metall Engineers, New York, p 106-128

Briner E, Roth P (1948) Recherches sur l'hydrolyse par la vapeur d'eau de chlorures alcalins seuls on additionnes de divers adjuvants. Helv Chim Acta 31:1352-1360

Burnham CW (1967) Hydrothermal fluids at the magmatic stage. In: Barnes HL (ed) Geochemistry of hydrothermal ore deposits. Holt, Rinehart and Winston, New York, pp 34-76

Burnham CW (1979) Magmas and hydrothermal fluids. In: Barnes HL (ed) Geochemistry of hydrothermal ore deposits, 2nd ed. Wiley, New York, pp 17-136

Cathles LM (1977) An analysis of the cooling of intrusives by groundwater convection which includes boiling. Econ Geol 72(5):804-826

Cathles LM (1980) Fluid flow and genesis of hydrothermal ore deposits. Econ Geol 75th Anniv Vol, pp 424-457

Cathles LM (1983) An analysis of the hydrothermal system responsible for massive sulfide deposition in the Hokuroku Basin of Japan. In: Ohmoto H, Skinner BJ (eds) The Kuroko and related volcanogenic massive sulfide deposits. Econ Geol Monogr 5:439-487

Christiansen RL (1984) Yellowstone magmatic evolution: Its bearing on understanding large-volume explosive volcanism. In: Explosive volcanism: Inception, evolution, and hazards. Nat Acad (Wash DC), pp 84-95

Cloke PL, Kesler SE (1979) The halite trend in hydrothermal solutions. Econ Geol 75(8):1823-1831

Cunningham CG (1978) Pressure gradients and boiling as mechanisms for localizing ore in porphyry systems: J Res US Geol Surv 6(6):745-754

Dunn JC, Hardee HC (1981) Superconvecting geothermal zones. J Volcanol Geotherm Res 11:189-201

Eaton GP, Christiansen RL, Iyer HM, Pitt AM, Mabey HR, Blank HR Jr, Zeitz I, Gettings ME (1975) Magma beneath Yellowstone National Park. Science (Wash DC) 188:787-796

Elder JW (1965) Physical processes in geothermal areas. In: Terrestrial heat flow. Am Geophys Union Monogr 8:211-239
Elders WA, Biehler S (1975) Gulf of California rift system and its implication for the tectonic of Western North America. Geology (Boulder) 3:85-97
Elders WA, Rex R, Meidav T, Robinson PT, Biehler S (1972) Crustal spreading in Southern California. Science (Wash DC) 178:15-24
Elders WA, Bird DK, Williams AE, Schiffman P, Cox B (1982) A model for the heat source of the Cerro Prieto magma-hydrothermal system, Baja, California, Mexico. Symp Cerro Prieto Geothermal Field. 4th Proc, Guadalajara, Mexico, 1982, vol 1, pp 265-284
Ellis AJ, Mahon WAJ (1977) Chemistry and geothermal systems. Academic Press, New York, p 392
Emmons WH (1933) On the mechanism of the deposition of certain metalliferous lode systems associated with granitic batholiths. In: Ore deposits of the western states (Lindgren vol). Am Inst Min Metall Engineers, New York, pp 327-349
Fehn U, Cathles LM (1979) Hydrothermal convection at slow-spreading mid-ocean ringes: Tectonophysics 55:239-260
Fenner CN (1933) Pneumatolytic processes in the formation of minerals and ores. In: Ore deposits of the western states. AIME, Rocky Mountain Fund Series, pp 58-106
Fournier RO (1968) Depths of intrusion and conditions of hydrothermal alteration in porphyry copper deposits (abs). Geol Soc Am Program with Abstracts, 1968 Annu Meeting, p 101
Fournier RO (1972) The importance of depth of crystallization on the character of magmatic fluids (abs). 24th Int Geol Congr, Montreal, 1972, sec 10, p 214
Fournier RO (1977) Constraints on the circulation of meteoric water in hydrothermal systems imposed by the solubility of quartz. Geol Soc Am Programs with Abstracts, vol 9, p 979
Fournier RO (1983a) Self-sealing and brecciation resulting from quartz deposition within hydrothermal systems. 4th Int Symp Water-Rock Interaction, Misasa, Japan, 1983. Extended Abstracts, pp 137-140
Fournier RO (1983b) Active hydrothermal systems as analogues of fossil systems. In: The role of heat in the development of energy and mineral resources in the basin and range province. Geotherm Resources Spec Report No 13, pp 263-284
Fournier RO (1983c) A method of calculating quartz solubilities in aqueous sodium chloride solutions. Geochim Cosmochim Acta 47:579-586
Fournier RO, White DE, Truesdell AH (1976) Convective heat flow in Yellowstone National Park. 2nd UN Symp Development and Use of Geothermal Resources, San Francisco, 1975, Proc Wash DC, US Government Printing Office, vol 1, pp 731-739
Frantz JD, Marshall WL (1984) Electrical conductances and ionization constants of salts, acids and bases in supercritical aqueous fluids. I. Hydrochlorid acid from 100° to 700°C and at pressures to 4000 bars. Am J Sci 284:651-667
Fuis GS, Mooney WD, Healy JH, McMechan GA, Lutter WJ (1982) Crustal structure of the Imperial Valley. In: The Imperial Valley, California Earthquake of October 15, 1979. US Geol Surv Prof Pap 254: 25-49
Galobardes DR, Van Hare DR, Rogers LB (1981) Solubility of sodium chloride in dry steam. J Chem Eng Data 26:363-366
Haas JL (1971) Effect of salinity on the maximum thermal gradient of a hydrothermal system at hydrostatic pressure. Econ Geol 66:940-949
Hardee HC (1982) Permeable convection above magma bodies: Tectonophysics 84:179-195
Hartline BK, Lister CRB (1981) Topographic forcing of supercritical convection in a porous medium such as the ocean crust. Earth Planet Sci Lett 55:75-86

Helgeson HC (1968) Geologic and thermodynamic characteristics of the Salton Sea geothermal system. Am J Sci 266:129-166
Helgeson HC (1969) Thermodynamics of hydrothermal systems at elevated temperatures and pressures. Am J Sci 267:729-804
Henley RW, Ellis AJ (1983) Geothermal systems ancient and modern: A geochemical review. Earth-Sci Rev, vol 19, p 50
Henley RW, McNaab A (1978) Magmatic vapor plumes and ground-water interaction in porphyry copper emplacement. Econ Geol 73:19
Hildreth W, Christiansen RL, O'Neil JR (1984) Catastrophic isotopic modification of rhyolitic magma at times of caldera subsidence, Yellowstone Plateau volcanic field. J Geophys Res 89:8339-8369
Holland HD (1972) Granites, solutions and base metal deposits. Econ Geol 67:281-301
Hubbert MK, Willis DG (1957) Mechanics of hydraulic fracturing. J Petroleum Technol, Petroleum Trans 210:153-166
Iyer HM, Evans JR, Zandt G, Stewart RM, Cookley JM, Roloff JN (1981) A deep low velocity body under the Yellowstone caldery, Wyoming; Delineation using teleseismic P-wave residuals and tectonic interpretations. Geol Soc Am Bull 92:1471-1646
Iyer HM, Oppenheimer DH, Hitchcock T (1979) Abnormal P-wave delays in the Geysers-Clear Lake geothermal area, California. Science (Wash DC) 204:495-497
Johnson C, Hill D (1982) Seismicity in the Imperial Valley. In: The Imperial Valley, California Earthquake of October 15, 1979. US Geol Surv Prof Pap 1254:15-24
Keevil NB (1942) Vapor pressure of aqueous solutions at high temperatures. J Am Chem Soc 64:841-850
Kilinc IA, Burnham CW (1972) Partitioning of chloride between a silicate melt and coexisting aqueous phase from 2 to 8 kilobars. Econ Geol 67:231-235
Knapp RB, Norton D (1981) Preliminary numerical analysis of processes related to magma crystallization and stress evaluation in cooling pluton environments. Am J Sci 281:35-68
Lehman JA, Smith RB, Schilly MM, Braile LW (1982) Upper crustal structure of Yellowstone from seismic and gravity observations. J Geophys Res 87:2713-2730
Lindgren W (1937) Succession of minerals and temperatures of formation in ore deposits of magmatic affiliation. Am Inst Min Metall Engineers, Trans 126:356-376
Lister CRB (1974) On the penetration of water into hot rock. R Astron Soc Gophys J 39:465-509
Lister CRB (1980) Heat flow and hydrothermal circulation: Earth Planet. Sci Ann Rev 8:95-117
Lister CRB (1983) The basic physics of water penetration into hot rock. In: Hydrothermal processes at seafloor spreading centers. Plenum, New York, pp 141-168
Martynova OI, Samoilov YuF (1959) Dissolution of sodium chloride in an atmosphere of water vapor of high parameters (Trans). Zh Neorganisheskoi Khimii 2(12):2829-2833
Neuman H (1948) On hydrothermal differentiation. Econ Geol 43:77-83
Norton D (1978) Sourcelines, sourceregions, and pathlines for fluid flow in hydrothermal systems related to cooling plutons. Econ Geol 73:21-28
Norton D, Cathles LM (1973) Breccia pipes – products of exsolved vapor from magma. Econ Geol 68:540-546
Norton D, Cathles LM (1979) Thermal aspects of hydrothermal ore deposition. In: Geochemistry of hydrothermal ore deposits (2nd ed). Wiley, New York, pp 611-631
Norton D, Knapp R (1977) Transport phenomena in hydrothermal systems: The nature of porosity. Am J Sci 277:913-936

Norton D, Knight J (1977) Transport phenomena in hydrothermal systems: Cooling plutons. Am J Sci 277:937-981
Norton D, Taylor HP (1979) Quantitative simulation of the hydrothermal systems of crystallizing magmas on the basis of transport and oxygen isotope data: An analysis of the Skaergaard intrusion. J Petrol 20: 421-486
Pelton JR, Smith RB (1979) Recent crustal uplift in Yellowstone National Park. Science (Wash DC) 206:1179-1182
Pelton JR, Smith RB (1982) Contemporary vertical displacement in Yellowstone National Park. J Geophys Res 87:2745-2761
Phillips WJ (1973) Mechanical effects of retrograde boiling and its probable importance in the formation of some porphyry ore deposits. Am Inst Min Metall Engineers, Trans, Sec B 82:B90-98
Ravich MI, Borovia FE (194) Phase equilibria in ternary water-salt systems at elevated temperatures (in Russian) Akademiia Nauk SSSR. Institut obshchei i neorganicheskoi. Izvestiia. Sektor fiziko khimicheskogo analiza 19:69-81
Rex RW (1983) The origin of the brines of the Imperial Valley, California. Geotherm Res Council Trans 7:321-324
Ribando RJ, Torrance KE, Turcotte DL (1976) Numerical models for hydrothermal circulation in the Ocean Crust. J Geophys Res 81:3007-3012
Robinson R, Iyer HM (1979) Evidence from teleseismic P-wave observations for a low velocity body under the Roosevelt hot springs geothermal area, Utah. Geotherm Res Council Trans 3:585
Roedder E (1967) Fluid inclusions as samples of ore fluids. In: Barnes HL (ed) Geochemistry of hydrothermal ore deposits. Holt, Rinehart and Winston, New York, pp 515-574
Roedder E (1971) Fluid inclusion studies on the porphyry-type or deposits at Bingham, Utah, Butte, Montana, and Climax, Colorado. Econ Geol, 66(1)98-120
Roedder E (1972) Composition of fluid inclusions. In: Data of geochemistry. US Geol Surv Prof Pap 400-JJ, 164 p
Roedder E (1979) Fluid inclusions as samples of ore fluid. In: Barnes HL (ed) Geochemistry of hydrothermal ore deposits (2nd ed). Wiley, New York, pp 684-737
Ryabchikov ID, Hamilton DL (1971) Possible separation of concentrated chloride solutions during crystallization of felsic magma. Akad Nau SSSR Dokalady 197:219-220
Schilly MM, Smith RB, Braile LW, Ansorge J (1982) The 1978 Yellowstone-eastern Snake River Plain seismic profiling experiment. Data and crustal structure of the Yellowstone region. J Geophys Res 87: 2692-2704
Sleep NH (1983) Hydrothermal convection at ridge axes. In: Hydrothermal processes at seafloor spreading centers. Plenum, New York, pp 71-82
Smith FG (1953) Review of physico-chemical data on the state of supercritical fluids. Econ Geol 48:14-38
Smith RB, Braile LW (1984) Crustal structure and evolution of an explosive silicic volcanic system at Yellowstone National Park. In: Explosive volcanism: Inception, evolution, and hazards. National Academy, Wash DC, pp 96-109
Smith RB, Christiansen RL (1980) Yellowstone Park as a window on the earth's interior. Sci Am 242:104-117
Sourirajan S, Kennedy GC (1962) The system $H_2O$-NaCl at élevated temperatures and pressures. Am J Sci 260:115-141
Steeples DW, Iyer HM (1976) Low-velocity zone under Long Valley as determined from teleseismic events. J Geophys Res 81:849-860
Torrance KE, Sheu JP (1978) Heat transfer from plutons undergoing hydrothermal cooling and thermal cracking. Numerical Heat Transfer, vol 1, pp 147-161
Truesdell AH, Fournier RO (1976) Conditions in the deeper parts of Yellowstone National Park, Wyoming. US Geol Surv Open-File Report 76-428, 29 p

Truesdell AH, Nathenson M, Rye RO (1977) The effects of subsurface boiling and dilution on the isotopic compositions of Yellowstone thermal waters. J Geophys Res 82:3694-3704
Wallace RH, Kraemer TF, Taylor RE, Wesselman JB (1978) Assessment of geopressured geothermal resources in the northern Gulf of Mexico basin. In: Muffler LJP (ed) Assessment of geothermal resources of the United States--1978. US Geol Surv Circ 790:132-155
White DE, Fournier RO, Muffler LJP, Truesdell AH (1975) Physical results of research driling in thermal areas of Yellowstone National Park, Wyoming, US Geol Surv Prof Paper 892, 70 p
Whitney JA (1975) Vapor generation in a quartz monzonite magma: A synthetic model with application to porphyry copper deposits. Econ Geol 70:346-358

# SCIENTIFIC DRILLING TO STUDY THE ROOTS AND MARGINS OF HYDROTHERMAL MINERAL SYSTEMS

J.J. Eidel[1] and C. Meyer[2]

Geological interpretations generally require projections from direct observations at the earth's surface. Crustal projections include mineral deposits as well as the geochemistry of the crust and its fluids, structure and tectonics. Drilling offers opportunities for verification of the required projections, and is thus a required step in furthering crustal understanding. Like any new science, the direct observations that drilling provides commonly generate major surprises and stimulate totally new ideas.

Drilling is costly. There are many *critically* important scientific problems with social and potential economic benefits in mineral deposits which have not been drilled because the holes are difficult to justify as economically sound risks for private capital. Such holes, however, *are* highly justifiable research goals for government because of the significant new opportunity for the insight into ore deposit and exploration that they afford. The earth science community is prepared with excellent drilling projects for research in mineral deposits.

Mineral deposits have been formed by circulation of hydrothermal solutions through the continental crust. Most of the energy for hydrothermal systems is supplied by cooling igneous rocks at variable crustal levels (Fig. 1). Magma invasion into cooler crustal rocks initiates fluid convection. The near surface expression of such convection is observed in

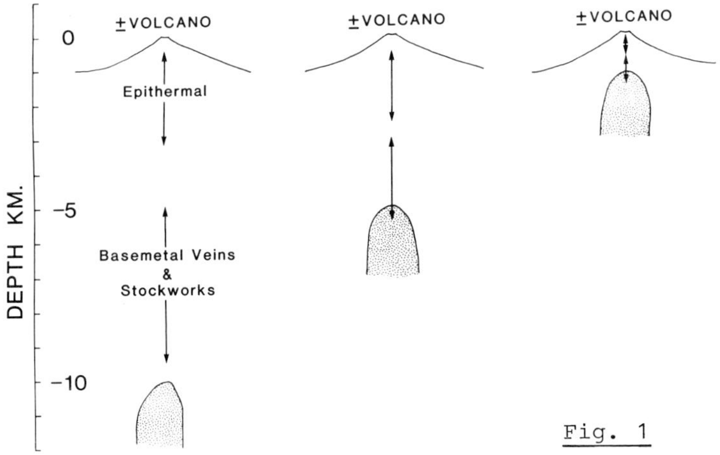


Fig. 1

[1]Illinois State Geological Survey, 615 E. Peabody Drive, Champaign, IL 61820, USA
[2]380 Smith Rd., Sedona, AZ 86336, USA

Observation of the Continental Crust through Drilling I (ed. by C.B. Raleigh)

currently active hot springs and steam vents. A few such active geothermal areas have ore minerals directly associated with them. Some important gold and mercury deposits were associated with ancient hot springs. Interesting but currently sub-economic gold, silver and lead concentrations are known in hot spring areas, and sub-aqueous brine emanations contain many metals.

In western North America there are hundreds of shallow veins and disseminations of gold and silver in volcanic and sub-volcanic environments which have the appearance of "fossil" hot springs or the roots of hot springs. The direct evidence for surface sinter is only rarely present, presumably having been eroded away. These "epithermal" deposits are again making a significant contribution of gold and silver in the United States, and their restudy is currently a major effort in mineral exploration in Western U.S.

The upper half of Figure 2 demonstrates a hypothetical relationship between a near surface magmatic heat source (crosshatched), the resultant isotherms, paths of hydrothermal fluid circulation and epithermal Au-Ag-Hg-As-Sb-Pb-Zn mineralization. Surficial expulsion of silica-laden fluids results in silicious sinter that may contain or be underlain by precious metals, Hg, As and Sb. Advanced argillic (alunite-kaolinite) alteration may be peripheral to the silicified core of the system. Increasing zinc and lead values at depth may be indicative of porphyry-type mineralization as illustrated in the bottom half of Figure 2. Here similar surficial environments overlie porphyry copper and/or molybdenum mineralization high in a sub-volcanic intrusive.

Historically, most "epithermal" gold-silver veins were discovered and mined in the late 19th and 20th centuries, when bonanzas like the Tonopah, Goldfield, Comstock, and Cripple Creek contributed immensely to the westward expansion of the United States. The precious metal ores in these deposits gave out at relatively shallow depths – few contained economic ore below about 700 m beneath the surface; however, there were a few cases such as Butte, Montona, where lateral development of the initial silver mines encountered base metals nearer the core of the district. The recognition of these zoning patterns encouraged speculation that silver originally projected over the top of base metals for the district as a whole, prior to being removed by erosion.

Consideration of the possible transition in depth between "epithermal" veins and deeper porphyry copper or molybdenum stockwork ores is not new. European geologists, including Stelzner, de Beaumont, Posepny and others began postulating such a relationship late in the 19th century. In the early 1900's American geologists including Waldemar Lindgren and S.F. Emmons argued that it was possible that base metal and related ore types underlay "epithermal" or "volcanic" precious metal veins. But then, as evidence accumulated during the extensive mining of the "epi-

---

Fig. 2. *(Upper)* Epithermal mineralized zone at the top of a hydrothermal system typical of silicic volcanic terranes. In this model, there is no relation to porphyry type mineralization. *(Lower)* Epithermal mineralized zone above a porphyry mineralized zone, separated by a barren zone, in a geothermal system typical of andesitic volcanic terranes. *Hachures* pre-volcanic basement; *cross-hachures* intrusive volcanics; *horizontal lines* low permeability stratum, e.g., mudstone; *dots* steam heated acid $SO4^{=}$ +/- $HCO3^{-}$ waters; *vertical lines* $SO4^{=}$ - $CI^{-}$ waters; *blank* near neutral chloride waters (within 200° d C isotherm - approx.); *circles* two phase region - water liquid + steam (+ gas) (after Henley and Ellis 1983)

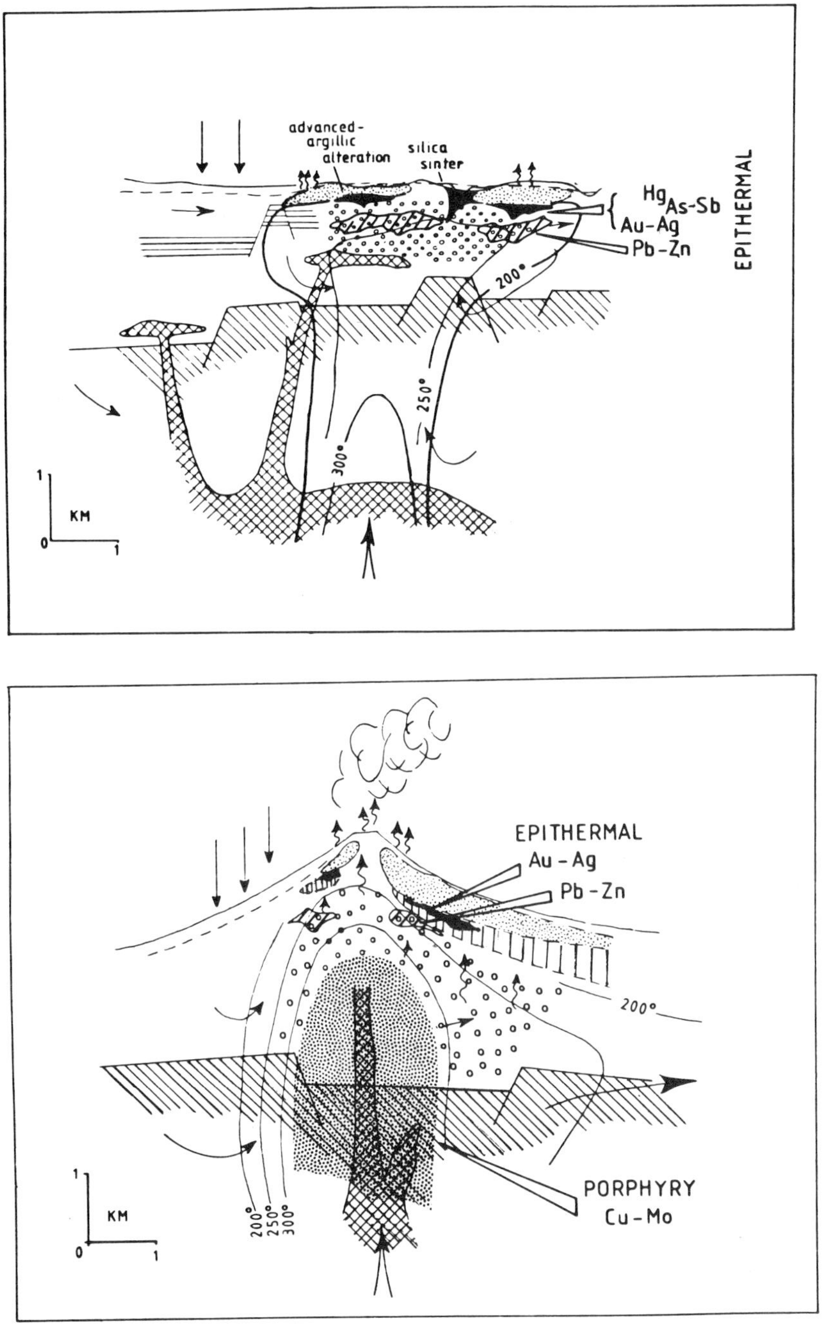

advanced-argillic alteration
silica sinter
Hg As-Sb
Au-Ag
Pb-Zn
EPITHERMAL
200°
250°
300°
1
KM
0
1
EPITHERMAL
Au - Ag
Pb - Zn
200°
PORPHYRY
Cu-Mo
200°
250°
300°
1
KM
0
1


Fig. 2

thermal" ores in the U.S., many geologists including Lindgren himself, began to doubt the "stacking" hypothesis as applied generally between "epithermal" gold-silver deposits and subjacent porphyry-related subvolcanic base metal ore systems. This critical relationship cannot be established until there is deep drilling under selected "epithermal" gold-silver districts aided by deep drilling of active hydrothermal systems.

"Epithermal" targets for deep drilling have the widest range of wallrock alteration of any class of ore deposit, i.e., from adularia representing weak to moderate $K^+$ ion metasomatism, through sericitization to argillic assemblages containing kaolinite and alunite. Cripple Creek, Colorado, has predominant K-metasomatic alteration at the depth of mining, but another gold district, Goldfield, Nevada, shows strong $H^+$ ion metasomatism with its ore. How does each alteration type equate with known classes of active hot springs and with possible deeper base metal deposits? Research has been restricted to small accessible parts of hydrothermal systems which may or may not be genetically related to each other. Although we currently have advanced knowledge of mineralization processes in each case, we cannot answer the question of whether these mineralization/alteration types are: (1) genetically unrelated products of hydrothermal systems that operated independently at different crustal depths (Fig. 1), or (2) vertically separated parts of a single hydrothermal system which we have not examined through its full vertical dimension (Fig. 2). From the viewpoint of understanding the hydrodynamics and chemical transitions in such systems, we need continuous drill hole intersections of complete ore deposit systems. We have only incomplete mass balance or heat budget data from such fragmentary portions of the systems that they cannot be fit together with reasonable confidence.

In 1978 a workshop was held in Los Alamos on Continental Drilling for Scienctific Purposes. The results of the workshop were published by the National Academy of Sciences/National Research Council, in a volume entitled *Continental Scientific Drilling Program*. Appendix C of this publication listed the mineral resource objectives potentially solvable by deep drilling and add-on experimentation.

The Panel on Mineral Resources was subsequently formed under the Continental Drilling Committee in 1981. The Panel constructed a set of criteria for evaluating the Los Alamos and subsequent mineral resource objectives:

1. The research objectives must address fundamental scientific problems with potentially broad impact on the understanding of mineral deposits.
2. The objectives should be sufficiently broad to address research relevant to national needs, e.g., enhanced exploration techniques based on the relationship of ore-forming systems; hydrothermal systems relevant to geothermal energy development; and strategic mineral resource distribution.
3. The targets should represent broad classes of mineral deposits so that the scientific results may be applied to exploration for mineral deposits in other areas.
4. Previous studies of the area should be in an advanced stage, allowing selection of specific drill site(s).
5. Targets should be as specific as possible to maximize the value of data obtained and to ensure achievement of the scientific objectives with minimum expenditure of time and funds.

| Location of Fossil System | Major Resource Objective | Anticipated | |
|---|---|---|---|
| | | No.of holes | Depth (km) |
| Creede, CO | Test Creede Fm. as source of fluid and sulfur. | 2 | 1 |
| | Quantify source rock system, rock-hydro T reactions + heat source. | 2 | 5 |
| Tonopah, NV. | Refine epi-T/bx geometry, detect low dipping faults. | 1-3 | 1-2 |
| | Similar to Creede in intermediate complex. | 1-2 | 2-5 |
| | Investigate Mo source + intrusive complex. | | |
| Red Mtn. AZ. | Examine relationship epi-T + P.Cu mineralization to bottom potash alteration zone. | 1 | 3-5 |
| | Examine relationship bx. to mineralization and alteration | 1 | 3-5 |
| Butte, Mt. | Determine extent of bx and relationship to mineralization. Examine Roots of Cu-Mo deposit. | 2 | 2-3 |

Fig. 3

6. The target should be unique in the sense that data sought are rarely available from continuous, uninterrupted rock sequences in the United States.
7. Geological models should be well defined, utilizing near-surface data to derive substantiated geometric relationships.
8. Results of geophysical field studies should be integrated with surface and near-surface geological and geochemical data.
9. The location should be accessible to scientists and suitable for surface geological, geochemical, and geophysical studies. If the site is in a mining district, working should be accessible and required subsurface data should be available.
10. The site should be on property providing to insurmountable environmental difficulties, and studies on the site should be compatible with the interests of the owners of land and mineral rights.

Based on these criteria the *Panel on Mineral Resources* proposes a program (Fig. 3) first to investigate deep hydrothermal fluid circulation in precious metal epithermal and base metal mesothermal mineralized systems. (Depths are indicated in kilometers.) Research will focus on heat and metal sources, fluid evolution and compositions, rock permeability and metasomatism, and on relating features observed at various depths in the systems. By applying the criteria defined to establish priorities, the Panel selected the mining districts of Creede and Tonopah as containing epithermal systems most likely to produce new scientific information through deep drilling.

These systems may be near-surface parts of hydrothermal systems which evolved from disseminated and stockwork mineralization of copper and/or molybdenum at depth. Evaluation of this hypothesis by deep drilling is

| Location<br>Active System | Major Resource Objective | Anticipated<br>No.of | <br>Depth |
|---|---|---|---|
| Yellowstone, WY. | For all mineralized or potentially mineralized geothermal systems: | 2 | 3-6 |
| Valles, Caldera, NM | 1. Temp., composition + source of fluids as function of depth | | |
| Salton Sea, CA | 2. Source of high salinities | | |
| | 3. Role of boiling in mineral deposition and alteration | | |
| The Geysers/Clear Lake, CA | 4. Permeability variation + extent of self-sealing | | |

Fig. 4

critical to the evaluation of partially exposed epithermal mineralized systems of similar type elsewhere.

On the other hand, the mining districts of Red Mountain, Arizona and Butte, Montana contain zones of disseminated and stockwork copper-molybdenum mineralization at depth. Vein systems superimposed on parts of the stockworks contain increasing proportions of silver outward and perhaps upward from a central zone richer in base metals. Red Mountain may offer an opportuniy to investigate the entire range of mineralization in a porphyry copper system of moderate size and vertical extent. Butte offers a potential opportunity to drill the root zone of a well-defined copper-molybdenum stockwork in a much larger granodiorite-base metal system.

*The Panel on Mineral Resources has selected the Creede mining district as its highest priority target and recommends that a research drilling project be undertaken at Creede in order to examine in depth a caldera-related silver and base metal-bearing hydrothermal mineral system.*
Selection of this target is based, in part, on the economic importance of the epithermal precious metal deposits, the availablity of an extensive data base that includes industry drill hole data, and the existence of a comprehensive and well-documented geological-geochemical model. The Creede program will test the hypothesis of subjacent copper and/or molybdenum-bearing porphyry systems beneath epithermal precious metal systems. The proposed program will include two (each approximately 5 km deep drill holes) supplemented with information from active, caldera-related geothermal systems. New data gained from this program would be integrated into existing models for enhanced interpretations of the Creede system.

Tonopah, Red Mountain, and Butte are lower priority targets that offer opportunities for future research. Much like Creede, the Tonopah and Red Mountain targets would test the vertical and possible genetic relationships between epithermal and subjacent porphyry copper-molybdenum deposits, whereas drilling at Butte would address the separate question of what may lie at the roots of a very large hydrothermal system known to have formed initially over depths of at least 4 km. Each of these targets are fossil hydrothermal systems that require additional data in order to define specific drilling requirements and sites.

*The Panel on Mineral Resources recommends that geological and related studies at Tonopah, Red Mountain, and Butte be continued in order to better define future research drilling targets in these important mineral systems.*

Available data indicate that the Creede system represents a mineralized fossil analog of certain active geothermal systems associated with silicic calderas, such as the caldera complexes at Yellowstone, Wyoming, and Valles, New Mexico. A major research objective that requires drilling of active geothermal systems (Fig. 4 – depths in kilometers) is the ability to interpret properly the individual stages of hydrothermal development of fossil mineral systems which are rarely well preserved in ore deposits. Active hydrothermal systems represent an instant of geologic time whereas the ore deposit represents the accured effects of an evolved hydrothermal system. Typical hydrothermal mineral deposits exhibit features that formed during later stage thermal collapse and fracture that produced major overprinting of early zonations and obscures critical field relationships of earlier mineralization. Recognizing whether diverse types of mineral deposition in subjacent parts of the system occurred simultaneously in separate convection cells, or separately in a single cell, is made difficult by this overprinting. Because of the similarities between active and fossil systems, information obtainable from the deeper portions of a presently active system should be directly applicable to the Creede system and to the construction of accurate models for epithermal mineralization in volcanic settings. *The Panel on Mineral Resources therefore recommends that a research program, complementary to that proposed at Creede, be initiated in an active hydrothermal system in a silicic caldera complex. Although Yellowstone is the most attractive target from a mineral deposits viewpoint, the Valles geothermal system in New Mexico may be the best available target.*

Perhaps the most important mineral resource aspect of the Continental Scientific Drilling Program is that of add-on experiments. Many critical aspects of mineral deposits and related geothermal activity can be addressed by investigations added on to existing or planned government and industry drill holes made available to scientific experiment and/or deepening. *The Panel recommends that experiments addressing significant problems of mineral deposits be designed for possible add-on investigations.* Such investigations involve sampling and measurements in host rocks which, through removed from mineral concentrations, have played an essential role in mineral emplacement.

Mining and drilling in mining areas normally do not provide information on unproductive, distal portions of mineral systems. High priority problems dependent for solutions on such peripheral studies have been defined by the Panel.

The CSDC Panel on Mineral Resources report outlines required add-on investigations for six or deposit systems.

What is required is a dialogue between researchers and industry, the former defining critical ore deposit problems for industry, and the latter providing a catalogue of available holes that could be made available to researchers. The CSDC and its Panel on Mineral Resources is available to facilitate such a dialogue, in part, via the Drill Hole Early Warning (DEW) Newsletter.

## Reference

Henley RW, Ellis AJ (1983) Geothermal systems ancient and modern: a geochemical review. Earth-Science Reviews 19, 1-50. Elsevier Sci. Publ. Co, Amsterdam

# IN SITU BOREHOLE MEASUREMENTS

# SCIENTIFIC DRILLING TO STUDY PHYSICAL PROPERTIES AND THE STATE OF STRESS IN THE EARTH'S CRUST: AN OPPORTUNITY FOR MID-CRUSTAL MEASUREMENTS NEAR THE SAN ANDREAS FAULT

Mark D. Zoback[1]

## Introduction

In recent years we have conducted studies of the state of stress, distribution of natural fractures, hydraulic conductivity, and P- and S-wave seismic velocity to depths of about 1.5 km in wells drilled primarily in crystalline rock. Similar studies are needed at greater depth and in boreholes drilled at critical locations to provide data to test a variety of hypotheses which affect a broad range of problems in the earth sciences. In this paper, some of these problems are briefly reviewed to indicate the important need for additional data and direct in situ experimentation. One specific site for downhole measurements is presented where in situ stress and temperature measurements would have the opportunity to resolve the order-of-magnitude uncertainty about the level of shear stress on the San Andreas fault.

## In Situ Stress

As first suggested by studies at the Rangely oil field (Raleigh et al. 1972, 1976) and the Rocky Mountain arsenal in Denver (Healy et al. 1968), in situ stress data at depth suggest that the state of stress near active faults is controlled by the frictional strength of the faults. Zoback and Healy (1984) present a number of cases where simple frictional theory (Coulomb 1773) and laboratory determined coefficients of friction (Byerlee 1978) accurately predict measured values of the principal stress in areas of active faulting.

Figure 1 illustrates a simple normal faulting case in which the greatest horizontal stress is parallel to the fault strike and the dip is about 60°. As shown in the figure, if the frictional strength of the fault is controlling the magnitude of the difference between the maximum and minimum principal stresses (after Jaeger and Cook 1969, p.89), the value of the least principal horizontal stress, $S_h$, can be predicted from the vertical stress, $S_v$, the pore pressure P, and the coefficient of friction, $\mu$, Fig. 2 shows measurements of $S_h$ made near active normal faults at the Nevada Test Site (Healy et al. 1984; Stock et al. 1984). The data show that for expected $\mu$ values (between 0.6 and 1.0), the expected overburden stress, and the observed pore pressure, the measured values of $S_h$ and those predicted by the equation in Fig. 1 are quite similar.

---

[1]Department of Geophysics, Stanford University, Stanford, CA 94305, USA

Observation of the Continental Crust
through Drilling I (ed. by C.B. Raleigh)

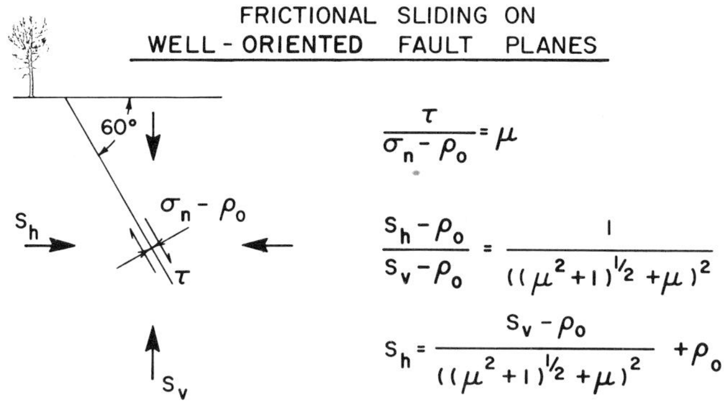


Fig. 1. Cartoon illustrating a normal fault that strikes perpendicular to the least horizontal principal stress, $S_h$. The simple equation shown (derived from Jaeger and Cook 1969) can be used to estimate the magnitude of $S_h$ in terms of the overburden stress, $S_v$, the coefficient of friction, $\mu$, and pore pressure $P_o$

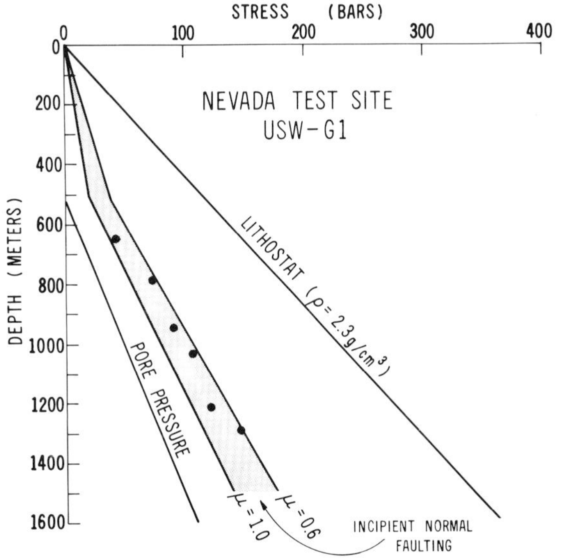


Fig. 2. Least principal stress, $S_h$, as a function of depth in hole USW-G1 at the Yucca Mountain on the Nevada Test Site (after Healy et al. 1984). The lithostat is taken from calibrated density logs and pore pressure is determined from the observed water table. The two lines indicated "incipient normal faulting" are derived from the equation shown in Fig.1

One outstanding paradox concerning the frictional strength of faults and stresses in the earth's crust concerns the San Andreas fault. Stress measurements to about 1 km depth indicate an increase in shear stress with depth of about 90 bar $km^{-1}$ (Zoback et al. 1980; McGarr et al. 1982). This rate of increase is basically consistent with expectations from frictional models as discussed above, and implies that to the base of the seismogenic zone (about 15 km), the average shear stress on the San Andreas is about 1 kbar. Extensive heat flow data, however, do not show any anomaly near the fault as would be expected at such stress levels. The heat flow data imply that the mean shear stress on the fault throughout the seismogenic layer is less than a few hundred bars (Lachenbruch and Sass 1980). The gross mechanical differences between these models have profound implications for the forces that drive lithospheric plates, the nature of deformation at transform plate boundaries, and the mechanics and prediction of major earthquakes. Lachenbruch and Sass (1980) for example, argue that the primary resistance to plate motion is due to tractions along subhorizontal detachment faults adjacent to the San Andreas fault.

A 2 km deep well near the San Andreas fault at Cajon Pass (Fig. 3) has recently become available which presents an opportunity to resolve this very important question. The well currently extends about 1 km into granodiorite (Fig. 4) and can be deepened to about 5 km. This would enable (1) stress measurements to be made at seismogenic depths where the validity of the heat flow constraint can be tested directly, (2) temperature and conductive heat flow to be determined over a sufficiently great depth range to confirm, or deny, the possibility of appreciable convective heat transport, and (3) seismic and strain sensitive instruments to be emplaced at seismogenic depth near the fault. This latter point is especially important because the Cajon Pass area was the southern terminus of the great 1857 earthquake. A repeat of this earthquake has about a 50% chance for recurrence in the next 30 years. Monitoring at this site may yield data close to the nucleation point of a great earthquake.

## Fractures in the Crust

The presence of natural fractures and joints have a profound effect on the physical properties and seismic velocity of crustal rock. Although fractures are ubiquitous in surface exposures, very little data is available on their distribution with depth and their mode of origin is basically not understood. In boreholes studies of macroscopic fractures in crystalline rock to depths of 1 km, Seeburger and Zoback (1982 and Moos and Zoback (1983) pointed out several unexpected findings. First, fractures can be quite prevalent to depths of 1 km and do not decrease with depth as rapidly as theorized in earlier studies (e.g. Snow 1968). Figure 5 shows the frequency of fractures encountered with apparent apertures greater than a few mm in two wells only 5 km apart in the Piedmont province of South Carolina (Mont-1 and Mont-2), and two wells in the western Mojave desert in southern California (XTLR and HI VISTA). All four wells are in granitic rock. While there is a modest decrease in fracture frequency with depth in all but the HI VISTA well, numerous fractures are observed at depth in each well. How fractures decrease with still greater depth, and the depth at which their effect on seismic velocity and hydraulic conductivity ceases to be important is unknown. Surprisingly, reports from the ultra-deep hole on the Kola Peninsula in the Soviet Union seem to indicate the presence of hydraulically conductive fractures at depths of about 10 km

Another unusual aspect of the distribution of fractures is their spatial variability. For example, Fig. 5 shows the marked difference in fracture frequency in the Mont-1 and Mont-2 wells which are only 5 km apart. Figures 6 and 7 show lower hemisphere stereographic plots of fracture poles for these two wells. In Mont-2 (Fig.6), a set of fractures with a north-south strike and steep west dip is observed in the 34-305 m and 610-1203 m intervals. However, from 305-610 m this fracture set is not observed although a set of fractures are present with a northwest-southeast strike and southwesterly dip. The fractures in Mont-1 (Fig.7) are fairly randomly distributed and show only minor concentrations of fractures of specific orientation. Similar disparities in fracture orientations were found in a set of shallow (~ 250 m deep) wells drilled in the Gabilan range of central California and in the western Mojave desert (Seeburger and Zoback 1982). The causes of such diverse fracture sets in wells must be a combination of both deep-seated tectonic events and near surface causes. Data from more holes, holes at greater depth, and holes sampling more environments are needed to better understand the distribution and origin of fractures in the crust.

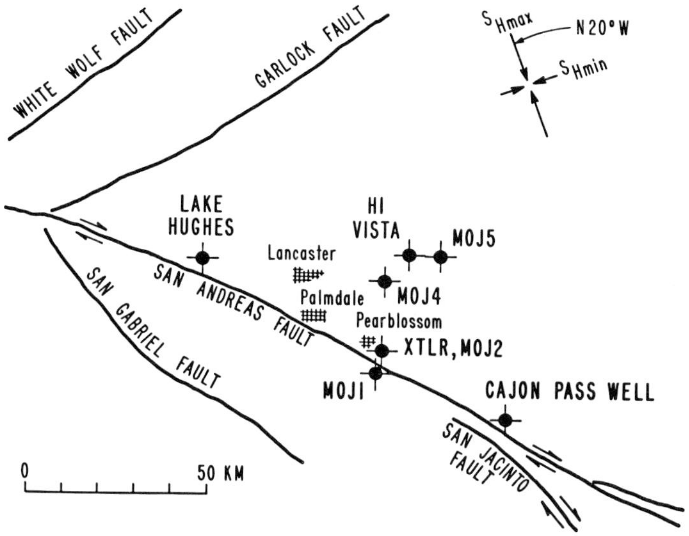


Fig. 3. Location of the Cajon Pass well near the San Andreas fault in the western Mojave desert, California. Shallow wells (less than 1 km deep) in the area previously used for stress measurements are also shown

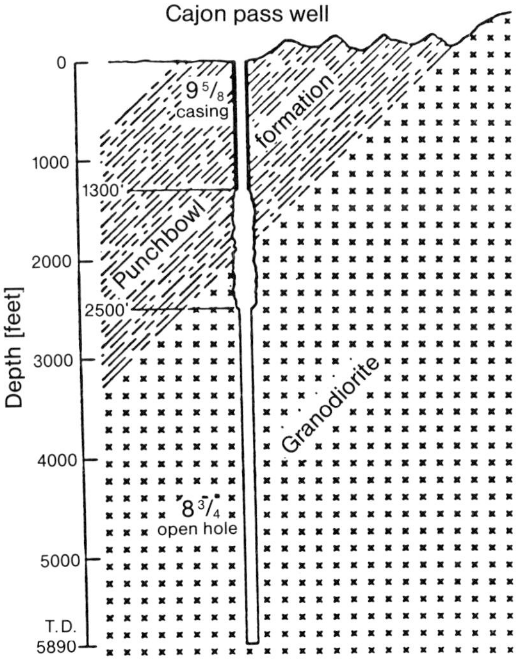


Fig. 4. Schematic cross-section of the Cajon Pass well. Total depth is 1.8 km, the bottom 1.1 km is in granodiorite. Nearly all the Punchbowl formation (a Miocene terrestrial sandstone) is cased. The well is 4 km from the San Andreas fault

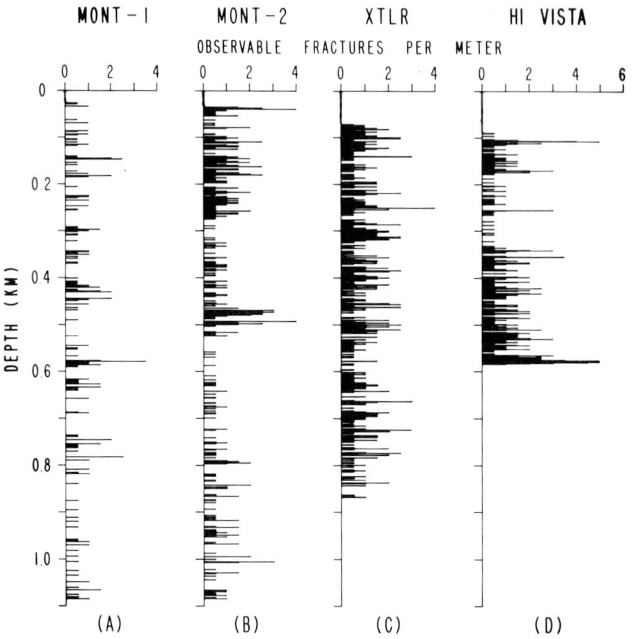


Fig. 5A-D. The frequency of observable macroscopic fractures in granitic rock observed in two wells drilled in the Piedmont province in South Carolina (Mont-1 and Mont-2) and two wells in the western Mojave desert. (Moos and Zoback 1983)

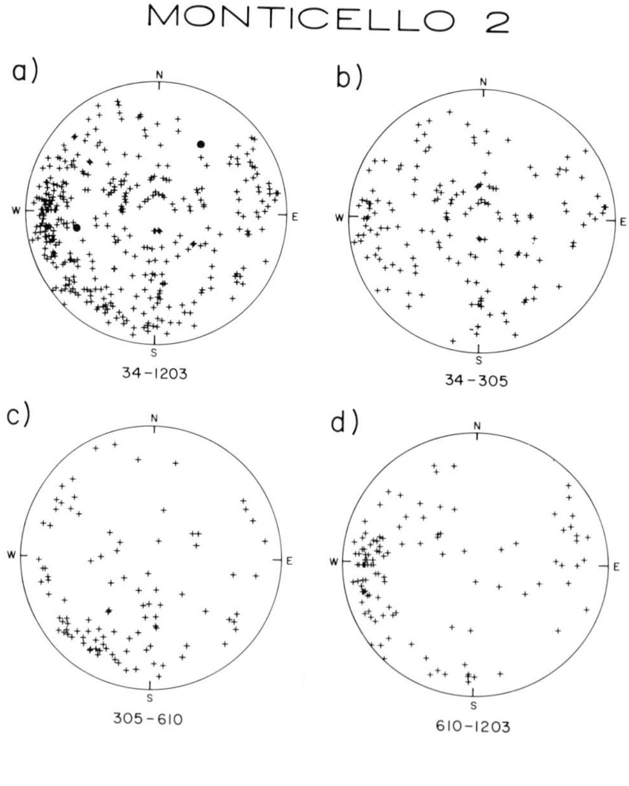


Fig. 6. a-d. Lower-hemisphere stereographic projection of fracture poles in Mont-2. Depth interval for each plot is indicated. (Zoback and Hickman 1982)

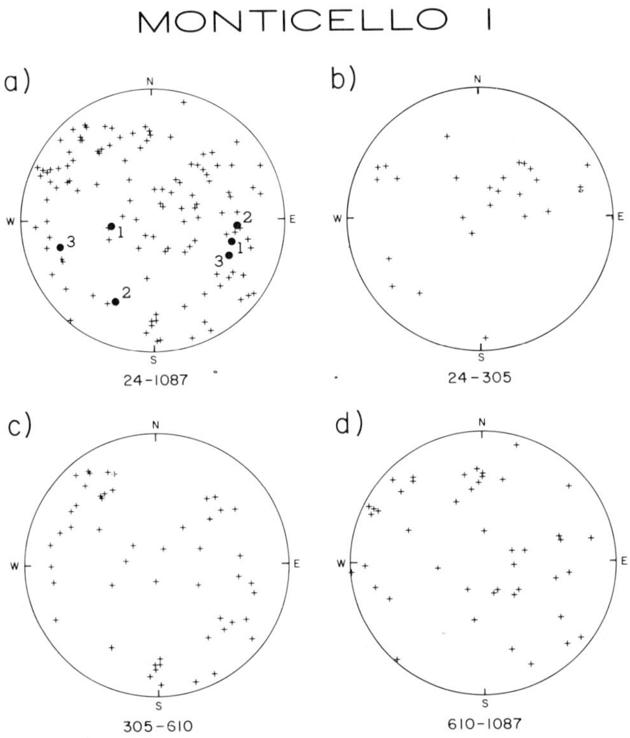


Fig. 7. a-d. Lower-hemisphere stereographic projection of fracture poles in Mont-1. Depth interval for each plot is indicated. (Zoback and Hickman 1982)

## Permeability and Fluid Flow

The permeability of in situ crystalline rock has important consequences for many geological problems related to fluid transport. A number of workers have made measurements in wells and boreholes to depths of about 1-2 km (Brace 1980, reviews many available measurements). For example, permeability measurements made in the Mont-1 and Mont-2 wells are quite dissimilar (Fig. 8). Mont-1 shows that the bulk permeability of the fractured granite is about 1 md at depths of 700-1100 m. The Mont-2 data, however, show values about 1 md only at much shallower depth, and much lower permeabilities (about 0.01 md) at depths greater than 500 m. This result is especially surprising since Mont-2 has many more fractures at depth than Mont-1 (Fig. 5). The fractures in Mont-2 are largely filled with alteration minerals which apparently severely lowers their hydraulic conductivity. Like the status of our knowledge about the distribution of fractures in the crust, the extremely wide range of reported in situ permeability values ($10^{5} - 10^{-9}$ md), and it's high spatial variability, make it nearly impossible to estimate what reasonable values might be in the mid-crust.

## Seismic Velocity

A final subject which has received considerable attention in recent years is the origin of deep seismic reflectors in the crust (Smithson et al. 1978; Lynn 1979; Schilt et al. 1979; Jones and Nur 1982, 1983;

Hale and Thompson 1982). While such reflections are fairly ubiquitous, no adequate explanation has been put forth to explain their existence. For example, Jones and Nur (1982) proposed that high pore pressure in fracture zones could be responsible for the presence of the reflectors. Hale and Thompson (1982) point to layered sequences of high-and low-velocity rock. On the basis of synthetic seismograms in some of the wells discussed above (Fig. 9) Moos and Zoback (1983) point to the importance of geochemical alteration of the country rock adjacent to fracture zones. None of these hypothesis have been tested at mid-crustal depth and the proper interpretation of deep events on seismic reflection profiles will remain unclear until these (and other) hypotheses can be tested with in situ data.

## Conclusions

Several examples are presented in which our lack of understanding of important geologic phenomena is directly impeded by the lack of critical data. These data can only be obtained through drilling and downhole experimentation. Such experimentation should be an integral part of scientific drilling programs.

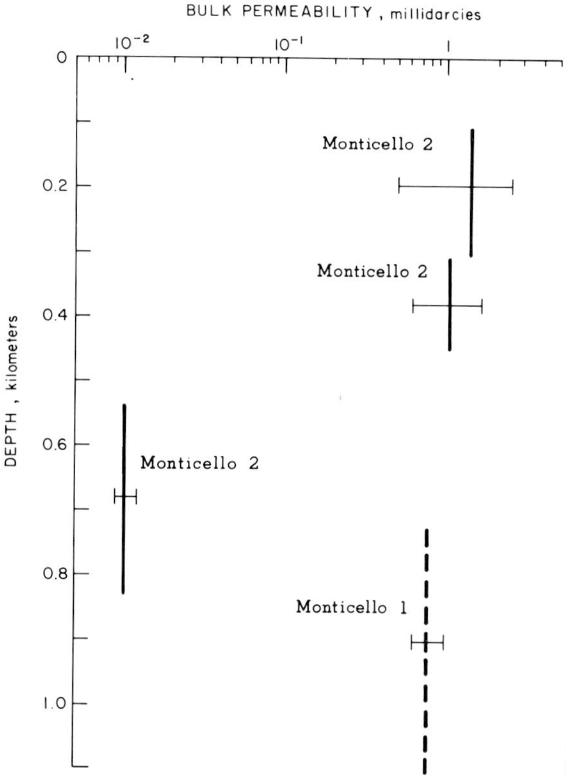


Fig. 8. Bulk permeabilitymeasurements in wells Mont-1 and Mont-2 (Zoback and Hickman 1982). The large uncertainty in permeability is due to accumulated error associated with differencing transmissivity tests

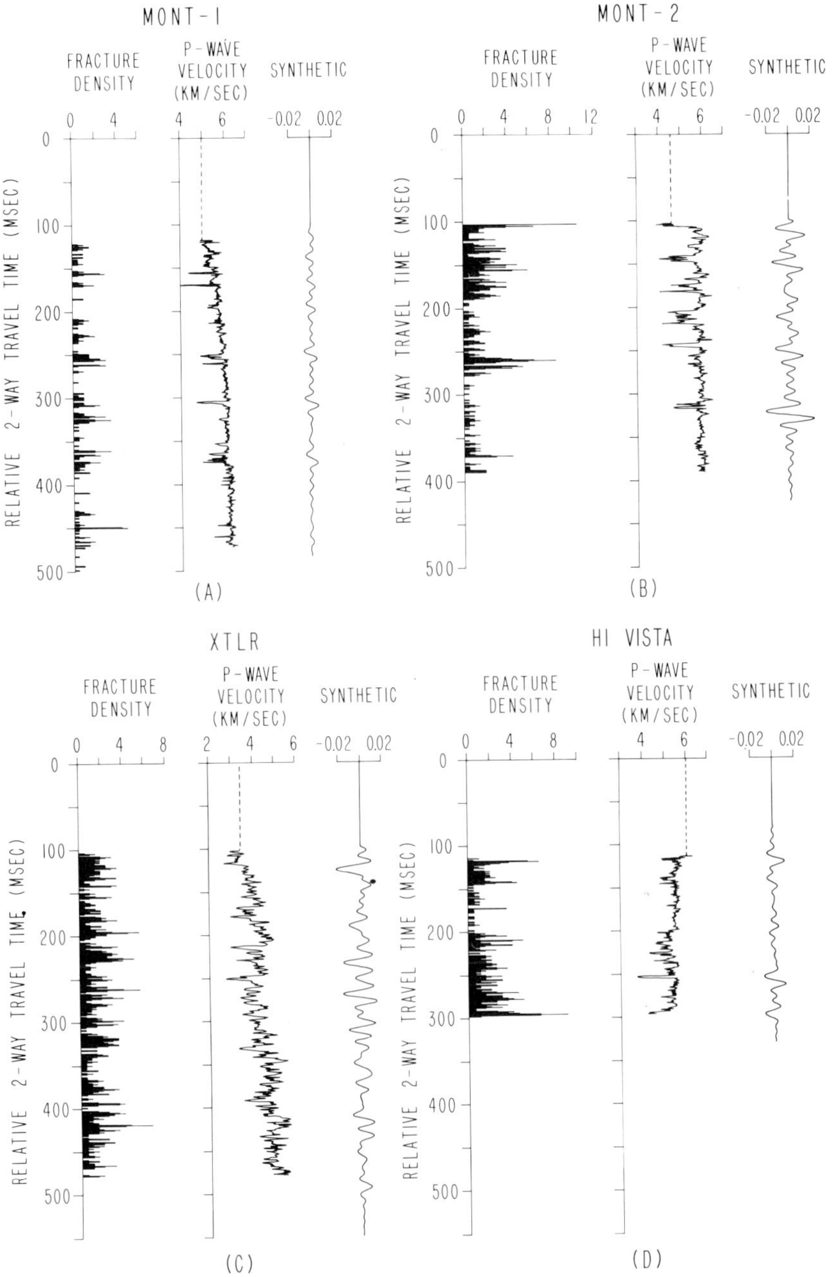


Fig. 9A-D. Synthetic reflection seismograms for the four wells shown in Fig. 5, with the corresponding macroscopic fracture and P-wave velocity data. (Moos and Zoback 1983)

## References

Brace WF (1980) Permeability of crystalline and argillaceous rocks. Int J Rock Mech Min Sci Geomech Abstr 17: 241-251
Byerlee JD (1978) Friction of rock. Pure Appl Geophys 116: 615-626
Coulomb CA (1773) Sur une application des regles de maximus et minimus a quelques problèmes de stitique relatifs à l'architecture. Acad R Sci Mem Mech Min Sci 7: 343-382
Hale LD, Thompson GA (1982) The seismic reflection character of the continental Mohorovicic discontinuity. J Geophys Res 87: 4625-4635
Healy JH, Griggs DT, Rubey WW, Raleigh CB (1968) The Denver earthquakes. Science, (Wash DC) 161: 1301-1310
Healy JH, Hickman SH, Zoback MD, Ellis WL (1984) Report on televiewer log and stress measurements in core hole VSW-G1. Nevada Test Site, December 13-22, 1981. US Geol Surv Open-File Report, pp 84-15
Jaeger JC, Cook NGW (1969) Fundamentals of Rock Mechanics. Methuen London, p 515
Jones TD, Nur A (1982) Seismic velocity and anisotropy and the reflectivity of deep crustal fault zones. Geology (Boulder) 10: 160-263
Jones TD, Nur A (1983) Comment and Reply on "Seismic velocity and anisotropy in mylonites and the reflectivity of deep crustal fault zones". Geology (Boulder) 11: 488-489
Lachenbruch AH, Sass JH (1980) Heat flow and energetics of the San Andreas fault zone. J Geophys Res 85: 6185-6222
Lynn HB (1979) Migration and Interpretation of Deep Crustal Seismic Reflection Data. Ph D Thesis, Stanford Univ, Stanford, CA, p 153
McGarr A, Zoback MD, Hanks TC (1982) Implications of an elastic analysis of in situ stress measurements near the San Andreas fault. J Geophys Res 87: 7797-7806
Moos D, Zoback MD (1983) In situ studies of velocity in fractured crystalline rocks. J Geophys Res 88: 2345-2358
Raleigh CB, Healy JH, Bredehoeft JD (1972) Faulting and crustal stress at Rangeley, Colorado. In: Flow and Fractures of Rocks. Geophys Monogr Ser 16. Head HC et al. Am Geophys Union, Wash DC
Raleigh CB, Healy JH, Bredehoeft JD (1976) An experiment in earthquake control at Langeley, Colorado. Science (Wash DC) 191: 1230-1237
Schilt FS, Oliver J, Brown L, Kaufman S, Albaugh D, Brewer J, Cook F, Jensen L, Krumhens P, Long G, Steiner D (1979) The heterogeneity of the continental crust: Results from deep crustal seismic reflection profiling using VIBROSEIS technique. Rev Geophys Space Phys 13: 354-368
Seeburger DA, Zoback MD (1982) The distribution of natural fractures and joints at depth in crystalline rocks. J Geophys Res 87: 5517-5534
Smithson SB, Brewer J, Kaufman S, Oliver J, Hunch C (1978) Nature of the Wind River thrust, Wyoming, from COCORP deep reflection data and from gravity data. Geology (Boulder) 6: 648-652
Snow DT (1968) Rock fracture spacings, openings and porosities. J Soil Mech Found Div Am L Soc Civ Enr 94 (SM 1): 73-91
Stock JM, Healy JH, Hickman SH, Zoback MD (1984, in press) Hydraulic fracturing stress measurements at Yucca Mountain, Nevada and relationship to the regional stress field. J Geophys Res
Zoback MD, Hickman S (1982) In situ study of the physical mechanisms controlling induced seismicity at Montiello reservoir, South Carolina. J Geophys Res 87: 6959-6974
Zoback MD, Healy JH (1984, in press) Friction, faulting and in situ stress. Ann Geophys
Zoback MD, Tsukahara H, Hickman S (1980) Stress measurements at depth in the vicinity of the San Andreas fault; implications for the magnitude of shear stress at depth. J Geophys Res 85: 6157-6173

# DEEP-BOREHOLE TECHNIQUES FOR CONTINUOUS MONITORING OF CRUSTAL ACTIVITY AND IN-SITU STRESS MEASUREMENT IN JAPAN

Hiroaki Tsukahara[1]

## Introduction

Deep borehole measurements have been carried out in two fields in Japan; the geothermal energy extraction project in the Tohoku (Sengan) geothermal area, and the earthquake prediction research in the Kanto-Tokai area (as "an area of specific observation" (e.g., Mogi 1981) for prediction of the next hazardous earthquake). I introduce deep borehole observatories for continuous observation of microearthquakes and ground tilt, and in-situ stress measurements in the Kanto-Tokai area.

Continuous observation of microearthquakes and ground tilt in the Tokyo metropolitan region by use of deep boreholes was planned by the National Research Center for Disaster Prevention (NRCDP) as a part of the earthquake prediction program of the 1960's. Accurate observations of microearthquakes and ground tilt around Tokyo require deep boreholes because of heavy background noise caused by advanced industries and transportation and also because of the thick, soft layers on hard basement rocks. The first observatory 3510 m deep was constructed at the end of 1972. Furthermore, two observatories 2300 m and 2750 m deep were made in 1978 and in 1980, respectively. At present, ten microearthquake observatories 450 m deep and over (see Table 1), and about 60 observatories shallower than 200 m are being operated by the NRCDP in the Kanto-Tokai area. Borehole-type tiltmeters are installed with microseismometers at about one third of the observatories.

On the other hand, the in-situ stress measurements have been carried out since 1978 as a part of the earthquake prediction research. As shown in Table 1, eleven boreholes 450 m deep and over have been drilled and used for stress measurement successfully for the last 7 years. Some of the boreholes have been used for microearthquake observation and/ or for groundwater observation after measurement.

## Deep Borehole Observatories

### 1. Introduction to the Observatories

The three specially deep observatories IWT, FCH, and SHM, are introduced in this section. Some of the observatories have already been described by Takahashi and Hamada (1975) and Hamada et al. (1979), and detailed description has been made by Takahashi (1982).

[1]National Research Center for Disaster Prevention, Tennodai 3-1, Sakura-mura, Ibaraki-ken 305, Japan

Observation of the Continental Crust through Drilling I (ed. by C.B. Raleigh)

Table 1. In-situ measurements in boreholes 450 m deep and over in the Kanto-Tokai area

| DEPTH, m | 3510 | 2750 | 2300 | 800[a] | 800 | 650 | 600 | 550 | 500 | 450 | 450 |
|---|---|---|---|---|---|---|---|---|---|---|---|
| Seis | 0 | 0 | 0 | 0 | 0 | 0 | – | 0 | 0 | 0 | – |
| Tilt | 0 | 0 | 0 | – | – | – | – | – | 0 | – | – |
| Strain | – | – | – | – | – | 0 | – | – | 0 | – | – |
| Stress | – | – | – | – | 0 | – | 0 | – | – | 0 | 0 |
| Number | 1 | 1 | 1 | 1 | 1 | 1 | 1 | 1 | 1 | 2 | 7 |

[a]Operated by the Geographical Survey Institute; The other observatories are being operated by NRCDP. Seis: Seismometer, Tilt: Tiltmeter, Strain: Three-component strainmeter (Sakata 1981), and Stress: Hydraulic fracturing stress measurement.

The locations of the three boreholes are given in Fig. 1, IWT: lat. 35.93°N, long. 139.74°E, FCH: 35.65°N, 139.47°E, and SHM: 35.79°N, 140.02°E, and the depths are 3510 m, 2750 m, and 2300 m, respectively. Pre-Tertiary basement appears at depths of 2900 m (IWT), 2020 m (FCH), and 1510 m (SHM), that is, the boreholes have been drilled into the basement by about 600 m to 800 m.

Velocity, resistivity, spontaneous potential, density, and temperature logs have been carried out in the three boreholes (Takahashi et al. 1983; Suzuki and Takahashi 1983). S-wave velocity and attenuation in

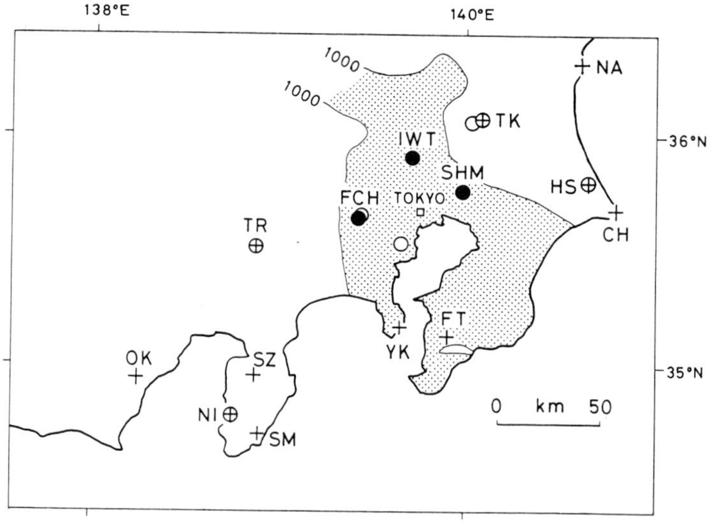


Fig. 1. Locations of boreholes 450 m deep and over which are used as observatories of crustal activity or have been used for in-situ stress measurement. Solid circle observatories 3510 m, 2750 m, and 2300 m deep; open circle and open circle with cross observatories 450 m to 800 m deep; cross and open circle with a cross hydraulic fracturing stress measurements; dotted area area overlain by soft sediments of post-Paleogene age which are thicker than 1000 m

deep soil deposit is very important especially in engineering seismology. Deep S-wave velocity measurements have been made using these boreholes (Ohta et al. 1980; Yamamizu et al. 1981, 1983). S-waves were produced on the surface by means of ordinary small explosions and a specially designed SH-wave generator, the S-gun. A set of three component seismometers was fixed at various depths in the deep boreholes for detecting the records of the S-wave. The S-wave velocities range from 0.4 to 0.9 km $s^{-1}$ in soft sedimentary Pleistocene rocks, 1.1 to 1.6 km $s^{-1}$ in Miocene rocks, and 2.5 to 2.6 km $s^{-1}$ in pre-Tertiary basement rocks. The corresponding P-wave velocities are 1.8 to 2.2 km $s^{-1}$, 2.8 to 3.2 km $s^{-1}$, and 4.7 to 5.0 km $s^{-1}$, respectively. The largest wave velocity difference is at the boundary above the basement rocks.

The inner diameter of the casing pipe of the borehole is 16 cm. The cased boreholes are filled with water containing rust-proofing chemicals. A high-pressure-proof vessel of 14 cm in diameter and about 11 m in length has been made for housing a three-component seismometer, a three-component accelerometer, a two-component tiltmeter, and electric parts such as signal transmitters and amplifiers. The vessel is hung down into the borehole by an armored cable with 19 conductors, and is rigidly fixed at the bottom of the borehole by devices driven by electric motors. This is resettable if necessary. Rigid contact between the vessel and the surrounding rocks is essential for accurate ground tilt observation. As the bottom temperatures of the three boreholes are rather high, 61° to 86°C, the heat resistance of the electric parts was the most difficult problem which had to be resolved for long-term continuous observation. Each electric part has been selected so that the value of reliability of the electric system is 0.908 at 90°C as a whole, that is, the mean time between failures (MTBF) is 10 years (Takahashi 1976).

Continuous observations have been made since May, 1973 (IWT), Apr., 1978 (SHM), and Apr., 1980 (FCH).

## 2. Microearthquake Observation

As expected, background noise at the bottom of the deep boreholes is extremely small. However, traffic noise from the surface carried by water within the borehole was found to be considerable at the bottom. Specially devised shelters have been made and have been set around the signal cable midway in the borehole. They reject noise from the surface effectively (Yamamoto et al. 1975; Yamamoto and Takahashi 1976). The background noises at the bottom of the boreholes are one part per several tens to one part per several hundred those at the surface for the period range 1 to 25 Hz. Figure 2 shows an example of reduction of the background noise with respect to depth (Yamamoto and Takahashi 1976) in the 3510 m deep well (IWT). This figure also shows frequency-dependent reduction of background noise. The reduction is more effective for high frequency waves than low frequency ones.

Any one of the three observatories can detect all the shallow microearthquakes (depth $<$ 30 km) with a magnitude of 1.5 or over in the area roughly 70 km (east-west) $\times$ 100 km (north-south) including Tokyo (Takahashi 1982).

The discovery of precursory microearthquake swarms is one of the noteworthy results obtained from deep borehole observation. The swarms occur repeatedly in a small area beneath Tokyo Bay at depths of 30 to 50 km (Takahashi 1977; NRCDP 1984). The earthquakes preceded by

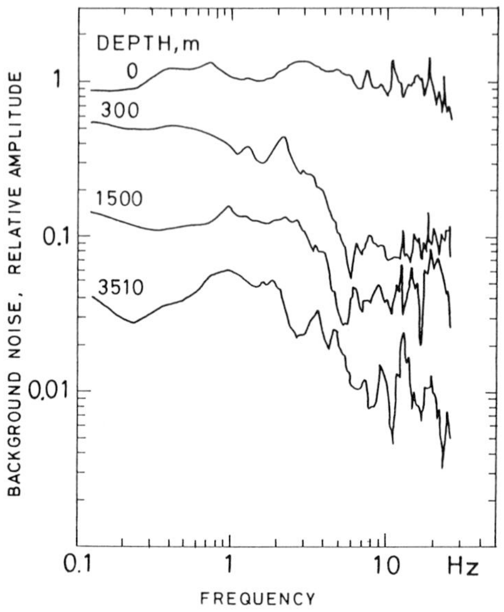


Fig. 2. Reduction of background noise with respect to depth at the IWT observatory. Vertical components of velocity-type seismometer. (After Yamamoto and Takahashi 1976)

the swarms have magnitude ranging from 3.7 to 7.0, and are found in the southwest quadrant of a circle with a radius of 150 km centering from the swarms. Within a period ranging from 5 to 40 days after a swarm, a major earthquake almost always occurs in the quadrant. Most of the microearthquakes in the swarm are less than 2 in magnitude so that they have been recorded only at the deep observatories. The region beneath Tokyo Bay will be particularly sensitive to the stress condition change around this area.

At present NRCDP has about 70 earthquake observatories including the three deep borehole observatories. About 6000 to 8000 locations of earthquakes, mostly larger than 1.5 in magnitude, are determined every year by use of the data from these observatories. The three-dimensional focal-distribution and focal mechanism solutions reveal the outline of three-dimensional structure of the three interacting plates in this area; the Pacific, the Eurasian, and the Philippine Sea plates (Kasahara 1980).

## 3. Ground Tilt Observation

A force-balanced pendulum tiltmeter has been developed for down-hole observation (Sato et al. 1980). The length of the pendulum is about 10 cm. The resolution is 0.01 microradian, and clear records of the tidal tilting, the amplitude of which is about 0.1 microradian, can be detected. The variation of recordings of tilt is smaller than a few microradian per year. However, as the greater part of the long-term variation is expected to be caused by the effect of aging of instruments and the boreholes themselves, continuous measurement of the net crustal-tilt during such a long-term as several years is still a problem. At present the recordings are very useful for detecting short-term unusual crustal-movements during one or two months.

## Hydraulic Fracturing Stress Measurements

### 1. Procedure for Measurement

The sites of the hydraulic fracturing stress measurements in the boreholes 450 m deep and over are shown in Fig. 1 (Tsukahara et al. 1978, 1980, 1981, 1983; Tsukahara 1983; Ikeda and Takahashi 1981, 1983). They have been made in various types of rocks, granite, tuff, tuff breccia and Miocene to Cretacious sediments.

Hydraulic fracturing was carried out by pumping water into a straddled interval 1.5 to 2.2 m in length between two 1.7 m long inflatable packers at suitable rates ranging from 5 to 100 l $min^{-1}$. The borehole diameter was 16 cm. Drill pipes of 3.5 inches in diameter were used for injection of water into the packers and the straddled interval. After fracturing, operations (pumping water, halting, and venting water) were repeated several times to obtain the reopening pressures of the fracture and the instantaneous shut-in pressures, the pressure necessary merely to keep the fracture open. The in-situ stresses were calculated from the reopening pressure and the instantaneous shut-in pressure. The pressure was measured simultaneously in the hydraulic line on the surface by using a strain-gauge-type pressure transducer located at the well head, and for some experiments a down-hole pressure transducer with logging cable was also used. The flow rate of the water was measured in an intake pipe of the pump. All the data was recorded by both a multi-pen chart recorder and a magnetic tape recorder.

The following relationships are used in calculating the maximum and minimum horizontal compressive stresses: $S_{Hmax}$ and $S_{Hmin}$ (Bredehoeft et al. 1976; Zoback et al. 1977; Haimson 1978).

$$S_{Hmin} = P_s \tag{1}$$

$$S_{Hmax} = 3P_s - P_b^{\,r} - P_p \tag{2}$$

where $P_b^{\,r}$, $P_s$ and $P_p$ are the reopening pressure of the hydraulic fracture, the instantaneous shut-in pressure, and the normally existing pore pressure in the rock, respectively. We take the value of $P_p$ as hydrostatic pressure because the water level of the wells was near the surface at all the sites.

An example of pressure vs. time and flow rate vs. time recordings obtained from a 450 m deep borehole is given in Fig. 3. These are data obtained from the Shimoda borehole (SM; see Fig. 1) which has been drilled into tuff breccia of Miocene age. The pressures $P_s$ and $P_b^{\,r}$ are simply determined from inflection points in pressure-time recordings after consideration of the reproducibility of the values in cyclic pressurization. In those cases where multiple pressurizations of a zone produce multiple values of $P_s$ and/or $P_b^{\,r}$, an average value is taken after omitting values which deviate greatly from other data. The down-hole pressures of $P_b^{\,r}$ and $P_s$ are calculated from the surface pressure by adding the hydrostatic pressure in the drill pipe.

The lithostatic pressure (i.e., vertical compressive stress) is calculated from the rock density which is taken from logging data and/or core samples of each borehole.

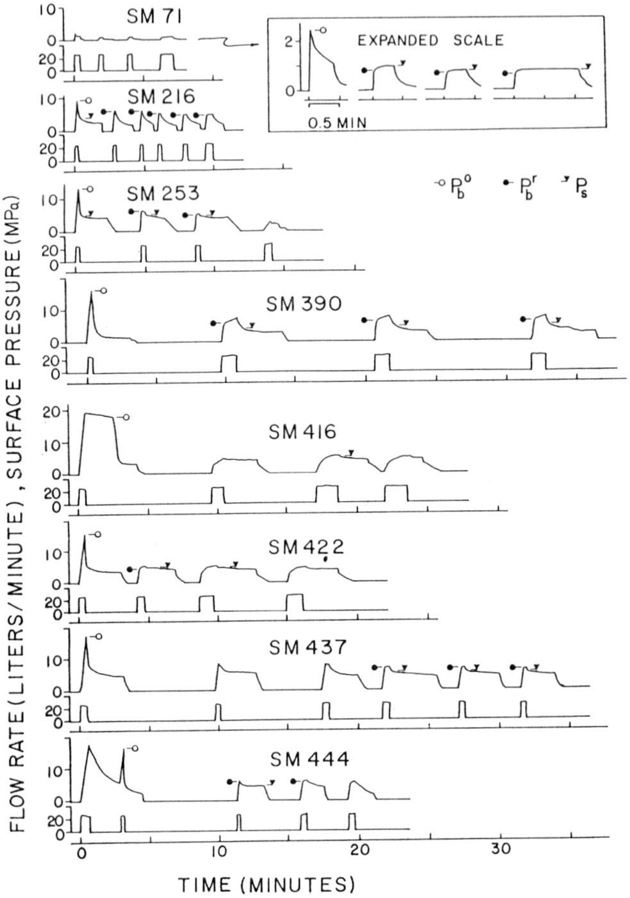


Fig. 3. Examples of recordings of pressure vs. time, and flow rate vs. time. Data from the 450 m deep Shimoda borehole. The upper and the lower recordings indicate the water pressure and the flow rate, respectively. The number following SM indicates depth in meters. $P_b^o$ initial breakdown pressure; $P_b^r$ reopening pressure; $P_s$ instantaneous shut-in pressure. (After Tsukahara et al. 1983)

The fracture azimuth of each experiment which indicates the azimuth of the maximum horizontal compressive stress (Hubbert and Willis 1957) was observed by an ultra sonic borehole televiewer and/or an impression packer. The azimuths of the fractures made by the hydraulic fracturing in a borehole scatter within a range of about ±20 degrees. No systematic variation of the azimuth with depth is observed at any measurement site. It follows that the amount of ±20 degrees are scattering mostly due to non-homogeneous nature of the rocks.

## 2. Stress Magnitude

More than 70 sets of in-situ stresses have been obtained from boreholes of various depths; nine of 450 m, one of 600 m and another of 800 m. Figure 4 gives an example of a stress-depth relation obtained from the Nakaminato borehole, NA. The solid line represents the best fit obtained by a least squares method. The dashed line represents the vertical stress calculated from density 2.3 g $cm^{-3}$. It is seen on the figure that (1) both stresses $S_{Hmax}$ and $S_{Hmin}$ increase linearly with depth, (2) difference between $S_{Hmax}$ and $S_{Hmin}$ also increases with depth, and (3) both the horizontal stresses are larger than the vertical stress at any depth. Condition (1) holds for all in-situ stress data measured in the Kanto-Tokai area. Condition (2) is also observed in almost all of our measurements with the exception of only one set of data from a single borehole. I will discuss the increase of the differential stress in the following paragraphs. Condition (3) does not hold generally, and relative magnitude among $S_{Hmax}$, $S_V$ and $S_{Hmin}$ depends on a measurement site. It is well known that the relative magnitude controls the features of faults, that is, in the case of $S_V > S_{Hmax} > S_{Hmin}$, normal faults appear, in the case of $S_{Hmax} > S_V > S_{Hmin}$, strike-slip faults appear and in the case of $S_{Hmax} > S_{Hmin} > S_V$, reverse faults appear. Therefore, we can examine the relative magnitude among the three components of in-situ stress comparing them with the features of the faults near the measurement site. It is found that the relative magnitude of in-situ stress is consistent with that estimated from the features of faults near measurement sites.

I will now discuss the increase of the differential stress with depth. The increase is not due to topographic effects because it can be observed in most measurements in spite of the different topography around each measurement site. Furthermore, as local upheaval or subsidence has not been observed in most measurement sites, rocks are expected to be strained almost uniformly from the surface to at least a depth of 1 km. If the elastic constant increases with depth under conditions of uniform strain, the differential stress increases with depth; but the rates of increase of most of the measurements are too large to be explained in terms of the elastic constant increase. Consequently, it is reasonable to assume that the anelastic deformation plays an important role in the stress variation with depth. That is, differential stress should be reduced in shallow parts due to the anelastic deformation of rocks. The relaxation of stress differences caused by the anelastic deformation in shallower parts is greater than that in deeper parts due to the mechanical properties of rocks and low confining pressure in shallower parts.

Let us consider the stress relaxation. Mogi (1971) proposed a criterion for the yielding of rocks, which corresponds to the generalized von Mises criterion,

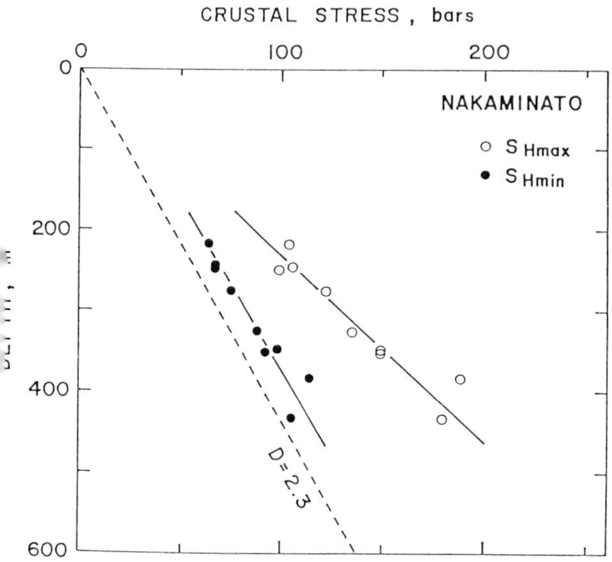


Fig. 4. An example of a plot of in-situ stress vs. depth. Data from the 450 m deep Nakaminato borehole. $S_{Hmax}$, maximum horizontal compressive stress; $S_{Hmin}$ minimum horizontal compressive stress. Dashed line indicates the vertical stress $S_V$ calculated from the rock density 2.3 g $cm^{-3}$ which is given by the density log data. (Ikeda and Takahashi 1983)

$$T_{oct} = f(S_1 + S_2 + S_3) \tag{3}$$

where $T_{oct}$ is the octahedral shear stress which is defined by

$$T_{oct} = (1/3)\sqrt{(S_1-S_2)^2 + (S_2-S_3)^2 + (S_3-S_1)^2},$$

and the right side of equation (3) is the monotonic increase function of sum of the principal stresses $S_1$, $S_2$ and $S_3$. The actual formula of $f(S_1+S_2+S_3)$ depends on the rock's properties and physical condition. In Fig. 5 the octahedral shear stresses calculated from all our in-situ stress data, that is, $(1/3)\sqrt{(S_{Hmax}-S_{Hmin})^2+(S_{Hmin}-S_V)^2+(S_V-S_{Hmax})^2}$ are plotted against mean effective stress $(S_{Hmax}+S_{Hmin}+S_V)/3-P_p$. It is worthy of note that the data are distributed in a rather narrow zone. This zonal distribution suggests that (1) the upper boundary of the zonal area, which is illustrated by a dash-dot line in the figure, will indicate the approximate upper limit of the criteria for in-situ yieldings of the rocks in all of the measurement sites, (2) the actual formulae for the yielding criterion of each rock are similar to one another, and the stress states at all the measurement sites are close to the stress condition for yielding, and (3) the yielding is going on in a steady state. On the other hand, detailed observation of the distribution of the data in Fig. 5 reveals that there seems to be a difference between the criteria for soft (solid circles) and hard (open circles) rock. Therefore, we need more detailed discussion in the future considering the properties of rocks.

The maximum differential stress at the depth of 400 m ranges from 2 MPa to 16 MPa depending on the site. The largest value of differential stresses is obtained at the TR borehole (90 km west from Tokyo, see Fig. 1) in Feb., 1983, which is only 15 km from the magnitude 6.0 earthquake of Aug. 8, 1983. However, it is not always the case that the regions of small differential stresses are inactive in microseismicity and crustal movement.

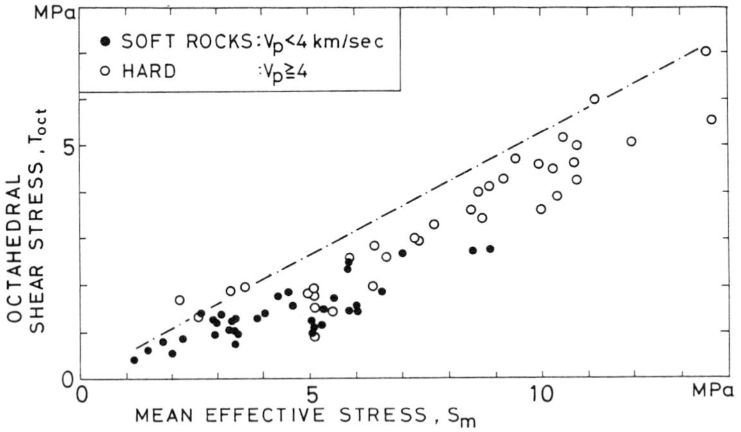


Fig. 5. Plot of octahedral shear stress $T_{oct}$ vs. mean effective stress $S_m$.

$$T_{oct} = (1/3)\sqrt{(S_{Hmax}-S_{Hmin})^2 + (S_{Hmin}-S_V)^2 + (S_V-S_{Hmax})^2}$$

$$S_m = (S_{Hmax}+S_{Hmin}+S_V)/3 - P_p$$

where $P_p$ is the pore pressure in the rocks

## 3. Stress Orientation

Figure 6 shows the maximum horizontal compressive stress direction measured by the hydraulic fracturing method in boreholes of a depth of 450 m and over. As there is not enough of the necessary data to discuss the regional distribution of the stress direction in this area, stress directions estimated from geologic and seismic data have been compiled. The segments in Fig. 7 show the maximum compressive stress directions estimated from various methods (Tsukahara and Ikeda 1983): hydraulic fracturing stress measurements, overcoring stress measurements, focal mechanism solutions of shallow (depth < 20 km) and more than moderate sized (M > 6.5) earthquakes, Quaternary dike trends, Quaternary cinder cone alignments, and movements of active faults of class A (average slip rate more than 1/1000 m/year) and of certainty ranking I and II defined by the Research Group for Active Faults in Japan (1980). It can be seen that the segments which lie close to one another give a similar orientation. The lines in the figure indicate the average direction of $S_{Hmax}$ in the provinces where the stress directions appear almost uniform. The classification of the province is made on the basis of the $S_{Hmax}$ direction considering the plate interaction.

The average stress direction distribution shown in Fig. 7 is consistent with the relative movements of two plates: the Eurasian and the Philippine Sea plates. The Philippine Sea plate, the northern region of which is bordered by the Suruga and the Sagami troughs, is estimated to be moving relatively northwestward against the Eurasian plate, and is subducting at the two troughs. The stress directions in areas SZ, HI and MB are attributed to the northwest movement of the Philippine Sea plate against the Eurasian plate. The stress direction in area NI is due to downward bending of the Philippine Sea plate at the Suruga trough (Tsukahara and Ikeda 1983).

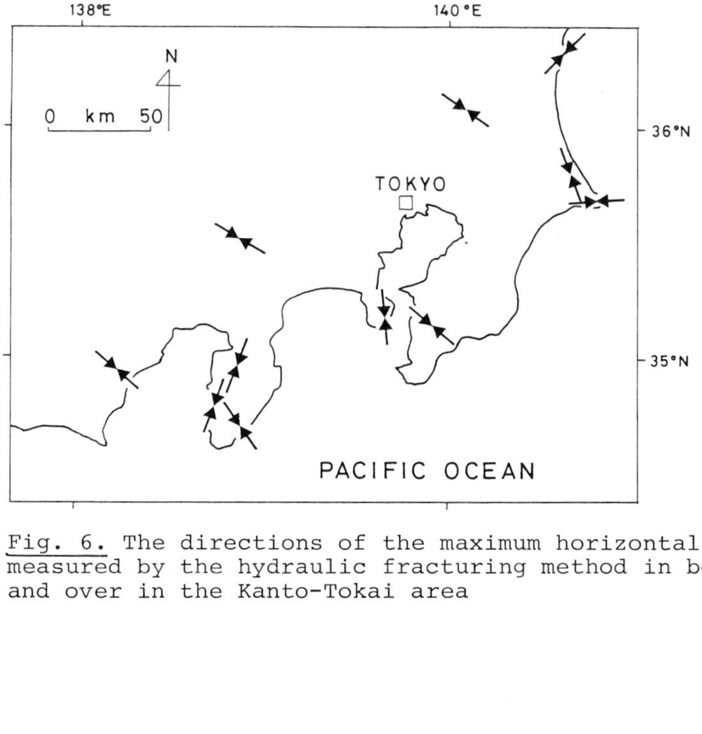


Fig. 6. The directions of the maximum horizontal compressive stress measured by the hydraulic fracturing method in boreholes 450 m deep and over in the Kanto-Tokai area

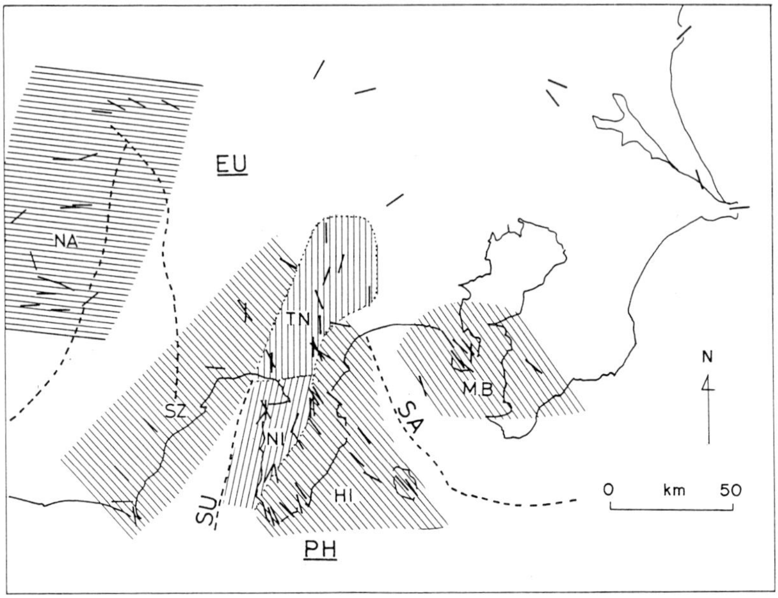


Fig. 7. Map of the maximum compressive stress direction. The azimuth of lines indicates the average direction of the maximum horizontal compressive stresses. Each segment shows the maximum stress direction obtained from various methods: hydraulic fracturing method, overcoring method, dike trend measurements, cinder cone alignments, active fault movements, and focal mechanisms of shallow earthquakes. SU and SA Suruga and the Sagami troughs; EU and PH Eurasian and the Philippine Sea plates; NA, SZ, NI, TN, MB names of the stress provinces. (After Tsukahara and Ikeda 1983)

## Summary

1. Three deep boreholes 3510 m, 2750 m, and 2300 m have been drilled for continuous and accurate observations of microearthquakes and ground tilt in a region overlaid with soft, thick sediments and dominated by heavy noise from human activity.
2. The amplitude of background microtremors at the borehole bottoms decrease to about 1/150 to 1/300 of those at the surface (at 10 Hz). Any one of the borehole observatories detects all shallow microearthquakes (depth 30 km) of magnitude 1.5 or larger within an area of roughly 70 km (east-west) $\times$ 100 km (north-south) including Tokyo.
3. Precursory microearthquake swarms are observed several times by the observatories, which occur repeatedly in a small area beneath Tokyo Bay preceding small to moderate earthquakes near Tokyo.
4. A force-balanced pendulum tiltmeter has been developed for downhole observation, and has been installed at the borehole bottom. It has a resolution of about 0.01 microradian which is enough for detecting the tidal tilting the amplitude of which is around 0.1 microradian. The data are expected to be useful for detecting unusual crustal-movement within as short a time as 1 or 2 months.

5. Hydraulic fracturing stress measurements have been made successfully in boreholes of various depths; nine of 450 m, one of 600 m, and another of 800 m.
6. The extent of the difference between the maximum and the minimum principal stresses increase with depth. This phenomenon is interpreted in terms of large stress relaxation in the shallows. A plot of the octahedral shear stress of the in-situ stress against the mean effective stresses suggests that a steady-state relaxation of stress in rocks is going on at depths of at least several hundred meters.
7. The measured stress directions tally with those estimated from geologic and seismic methods. The regional distribution of the directions are understood in terms of the relative movement of the Philippine Sea plate against the Eurasian plate.

## References


Bredehoeft JD, Wolff RG, Keys WS, Shuter E (1976) Hydraulic fracturing to determine the regional in situ stress field, Piceance Basin, Colorado. Geol Soc Am Bull 87: 250-258

Haimson BC (1978) Near-surface and deep hydrofracturing stress measurements in the Waterloo quartzite. Proc 19th US Symp Rock Mech, pp 345-361

Hamada K, Takahashi H, Takahashi M, Suzuki H (1979) Deep borehole measurements of crustal activities around Tokyo. Terrestrial and Space Techniques in Earthquake Prediction Research. In: Andreas Vogel (ed) Wieweg, Germany, pp 475-500

Hubbert MK, Willis DG (1957) Mechanics of hydraulic fracturing. Petroleum Trans AIME 210: 153-168

Ikeda R, Takahashi H (1981) Stress measurement by hydrofracturing at Futtsu City, Chiba Prefecture (in Japanese with English abstract). Zisin 34: 565-576

Ikeda R, Takahashi H (1983) Stress measurement by hydrofracturing at Nakaminato City, Ibaraki Prefecture (in Japanese with English abstract). Zisin 36: 213-223

Kasahara K (1980) On the pattern of earthquake occurence in the Kanto District (in Japanese), Abstr Seismol Soc Jpn 1980 (2): 66

Mogi K (1971) Fracture and flow of rocks under high triaxial compression. J Geophys Res 76: 1255-1269

Mogi K (1981) Earthquake prediction program in Japan. In: Simpson DW, Richards PG (eds) Earthquake Prediction. An International Review. Am Geophys Union, pp 635-666

NRCDP (1984) Microearthquake swarms of the northern Tokyo Bay area at June 1983 as possible precursor to felt earthquake. Rep Coord Committ Earthq Prediction 31: 165-168

Ohta Y, Goto N, Yamamizu F, Takahashi H (1980) S-wave velocity measurements in deep soil deposit and bedrock by means of an elaborated down-hole method. Bull Seismol Soc Am 70: 363-377

Research Group for Active Faults of Japan (1980) Active faults in and around Japan: The distribution and the degree of activity. J Nat Disaster Sci 2: 61-99

Sakata S (1981) On the concepts of some newly-invented borehole three-component strainmeters (in Japanese with English abstract). Rep Nat Res Center Disaster Prev 25: 95-126

Sato H, Takahashi H, Yamamoto E, Fukuo N, Uehara M (1980) Development of the crustal tilt observation method using borehole-type tiltmeters (in Japanese with English abstract). Zisin 33: 343-368

Suzuki H, Takahashi H (1983) Contraction and geology of the Shimohsa deep borehole observatory (in Japanese with English abstract). Res Note Nat Res Center Disaster Prev 48: 1-61

Takahashi M (1976) A system for telemetering the deep-well data at Iwatsuki and its reliability (in Japanese with English abstract). Rep Nat Res Center Disaster Prev 13: 35-57

Takahashi M (1977) Microearthquake swarms of the northern Tokyo Bay area (in Japanese), Abstr Seismol Soc Jpn 1977 (2) : 16

Takahashi H (1982) The deep borehole observatories and their contribution for revealing the characteristics of microearthquake activity in the Kanto district (in Japanese with English abstract). Rep Nat Res Center Disaster Prev 28: 1-104

Takahashi H, Hamada K (1975) Deep borehole observation of the earth's crust activities around Tokyo - Introduction of the Iwatsuki observatory. Pure Appl Geophys 113: 311-320

Takahashi H, Fukuda O, Suzuki H, Tanaka K (1983) Contraction and geology of the Iwatsuki deep borehole observatory (in Japanese with English abstract). Res Note Nat Res Center Disaster Prev 47: 1-113

Tsukahara H (1983) Stress measurements utilizing the hydraulic fracturing technique in the Kanto-Tokai area, Japan. In: Zoback MD, Haimson BC (eds) Hydraulic Fracturing Stress Measurements. Proc Workshop, December 2-5, 1981. Academic, Washington DC, pp 18-27

Tsukahara H, Ikeda R (1983) State of stress in the Kanto-Tokai area (in Japanese with English abstract). Zisin 36: 571-586

Tsukahara H, Ikeda R, Satake H, Ohtake M, Takahashi H (1978) Hydrofracturing stress measurements at Okabe Town, Shizuoka Prefecture (in Japanese with English abstract). Zisin 31: 415-433

Tsukahara H, Ikeda R, Satake H, Takahashi H (1980) Stress measurements by hydrofracturing at Nishiizu Town, Shizuoka Prefecture (in Japanese with English abstract). Zisin 33: 317-327

Tsukahara H, Ikeda R, Satake H, Takahashi H (1981) Stress measurements by hydrofracturing at Choshi City, Chiba Prefecture (in Japanese with English abstract). Zisin 34: 13-20

Tsukahara H, Ikeda R, Takahashi H (1983) Hydraulic fracturing stress measurements: Results at Okabe Town, Shuzenji Town and Shimoda City in Shizuoka Prefecture and Yokosuka City in Kanagawa Prefecture (in Japanese with English abstract). Zisin 36: 551-569

Yamamoto E, Takahashi M (1976) Depth distribution of background noise in the deep borehole of the Iwatsuki observatory (in Japanese), Abstr Seismol Soc Jpn 1976 (2): 95

Yamamoto E, Hamada K, Kasahara K (1975) Background seismic-wave noise and elimination of noise transmitted through water in a borehole at the Iwatsuki observatory (in Japanese with English abstract). Zisin 28: 171-180

Yamamizu F, Takahashi H, Goto N, Ohta Y (1981) Shear wave velocities in deep soil deposits III (in Japanese with English abstract). Zisin 34: 465-479

Yamamizu F, Goto N, Ohta Y, Takahashi H (1983) Attenuation of shear waves in deep soil deposits as revealed by down-hole measurements in the 2300 meter-borehole of the Shimohsa observatory, Japan. J Phys Earth 31: 139-157

Zoback MD, Healy JD, Roller JC (1977) Preliminary stress measurements in central California using the hydraulic fracturing technique. Pure Appl Geophys 115: 135-152

# ADVANCES IN DRILLING AND LOGGING TECHNOLOGY

# DEEP DRILLING FOR SCIENCE CONSTRAINTS/CHALLENGE

Frank J. Schuh[1]

It is generally accepted that deep drilling offers a significant window to the secrets of the continent. Although the oil and gas industry and geothermal and mining/drilling areas have developed a large array of highly sophisticated drilling tools and techniques, the unique needs of science will undoubtedly require new extensions and applications of this technology.

## Conclusions and Recommendations

1. Even with the industry's major advances in tools, technologies and equipment, the unique requirements of the scientific holes will require unique well designs and extensions and new applications of technology.
2. Success in these new efforts will require significant well design and planning efforts. Based on typical oil field extensions of technology, we would estimate efforts of 10 to 25 man years for the major technological steps being considered. Minimum lead times are probably in the one to 3-year range depending on the existence of feasible designs for the major well and drilling components.
3. Equipment developments for the industry's major steps have been performed by the drilling engineering units of major producing companies. Although much support is available from consultant engineering firms, the critical trade-off studies and the tough engineering choices have seldom been relegated to third party efforts. Success of the science program will require assembling an equivalent operating staff of highly skilled experienced drilling experts. One pattern for such an organization might be Shell's deep water drilling company that was organized to plan and conduct drilling operations in 7500' of water off the East Coast of the U.S. Other models might be obtained from ARCO's C.O.S.T. well program in the Bering Sea, Mobil's Mobile Bay deep exploratory operations, Gulf's Beaufort Sea drilling program, etc.
4. It will take a strong organization with an excellent balance between scientific objectives, operational constraints and costs to handle the diverse objectives of the program. The optimum organization from this standpoint would be to have a drilling manager and a science manager, each reporting to the overall project management authority. The science manager will be developing the scientific goals for each drilling project. The science manager will balance

[1]ARCO Oil and Gas Company, Exploration and Production Research, Post Office Box 2819, Dallas, TX 75221, USA

Observation of the Continental Crust
through Drilling I (ed. by C.B. Raleigh)

the mutually exclusive data requirements and prioritize the data and goals and prepare the well prognosis. The drilling manager will be responsible for the design, execution and cost control of the project. He will need to have reporting to him the following positions: A drilling supervisor who will handle the supervision of all field operations; a drilling engineering supervisor who will be responsible for well designs and well plans; the special projects manager who will be assigned technology development issues; a support manager to handle business functions such as purchasing, logistics, etc; and an engineering manager will develop well designs, well plans, and cost estimates for achieving the scientific prognosis. The overall project manager will be responsible for selecting the trade-offs between project cost and specific scientific objectives until the final plan has been accepted. The organization will need to include enough people so that the in house group will have sufficient expertise to make all of the trade-off technical decisions.

## Discussion

The major constraints on oil field drilling operations are well pressures, fracture gradients, temperatures, and formation stabilities. Of these, temperature has been the most difficult and persistent problem, with major changes requiring significant advances in high temperature electronics. For my analyses of the proposed deep North Carolina well, it has been assumed that pressure and fracture gradients will permit using a water-base unweighted drilling mud and most importantly, that temperatures be less than 400°F even if the depth is extended to 50,000'. Drilling such a hole will pose two major related problems. One is the drill string design required for a 50,000' long drill string and the second is the hole deflection and dogleg limits required to protect the drill string and prevent excessive casing wear. A number of serious trade-offs and well design decisions await the drilling management team. With conventional surface rotation, it would appear to be feasible to take wireline cores clear to the 50,000' target. However, the use of turbines offers the potential of improved penetration rates and much less drill string fatigue and casing wear problems. Use of conventional turbines would of course preclude wireline coring. There is also the possibility of developing a new turbine that would permit wireline core recovery through a hollow shaft. Additionally, one might duplicate the reported Russian concept of double wall replaceable protection casing strings to facilitate replacement of worn casing.

# GEOSCIENCE EXPERIMENTS IN BOREHOLES: INSTRUMENTATION

Richard K. Traeger[1]

## Introduction

> Geoscientist: "Provide a list of all the logging tools available and obtain a complete suite on all holes."
>
> Instrumentation Engineer: "Tell me what you want to measure - its accuracy, precision and environment, and I'll tell you what is available."

The above interchange has occurred frequently, in many forms, over the past several years during discussions of research in deep drill holes. This paper attempts to identify specific scientific and technological needs and proposes a priority list for development. The summary is based on the results of many conferences and individual discussions, and should be modified to meet specific needs for specific drill holes. Thus this is expected to serve as a guide; all critiques are welcome.

Direct, unique measurements of in situ properties and processes can only be made in a drill hole. Carel Otte, President of Union Geothermal, has stated many times that the only reliable exploration technique is drilling. Information from drilling is obtained in the form of resource production rates, commercial logs, cores and fluid samples. Because of high field costs, commercial borehole measurements are minimized to obtain only critical parameters needed to identify that resource. Also the current logging techniques do not directly measure the properties of interest (e.g., porosity) but infer the properties from other measurements (e.g., sonic, density and neutron logs). Two significant points to be made are that

1. Commercial logging and sampling technology is not designed to provide research information, and
2. Calculation of the property of interest requires calibration and/or knowledge of the specific lithology; this is currently oriented to oil and gas exploration.

Thus the researcher cannot rely on commercial logging to meet his needs. However an extensive suite of commercial logging tools and the extensive background available on sedimentary logging should both be exploited in any research activity.

Research sampling and diagnostic needs for deep hole geoscience have been reviewed in a number of documents (Deep Hole Drilling Feasibility Study 1969; Continental Drilling 1975; Varnado and Colp 1978; Conti-

---

[1]Sandia National Laboratories, Albuquerque, NM 87185, USA

Observation of the Continental Crust through Drilling I (ed. by C.B. Raleigh)

nental Scientific Drilling Program 1979; Hardee and Luth 1980; Hardee 1980; Traeger et al. 1981; Programme Geologie Profunde de la France 1982; Proposed Scientific Activities for the Salton Sea Drilling Program 1983; Haenel 1983; Cooper and Traeger 1984; Eidel 1984; Hermance 1984; Oliver 1984, Zoback 1984). The diversity of research goals in these reports leads to a large variation in instrumentation needs. Also the detail is not adequate to estimate needed requirements, to evaluate existing equipment or to determine if development of new tools is possible. This paper attempts to cross cut these past documents and identify common needs to focus initial technology development. However serious design can be initiated only after a specific research plan and drill site is defined.

The following discussion identifies research needs for instrumentation and sampling that are cited in the majority of the studies noted above. This is followed by a review of existing capabilities and current development activities. Tentative priorities are then set on technology needs to provide a focus for further discussion.

## Geoscience Instrumentation and Sampling Needs

Many conferences, workshops, papers and discussions have identified borehole instrumentation needs (Appendix A). These needs have been further condensed into a priority instrumentation list (Table 1) which includes the measurements identified as critical by all studies and, for a first cut, includes downhole measurements that cannot rely on core studies.

Table 1. Geoscience borehole research interests

| | |
|---|---|
| *Sampling* | *Formation properties* |
| Oriented core<br>Fluid samples | Porosity and permeability<br>In situ stress, magnitude and direction<br>Fracture density, geometry and composition<br>Chemistry<br>Resistivity |
| *Thermodynamics* | *In situ experiments* |
| T&P = f(depth)<br>S&P Wave velocities | Seismic<br>EM measurements<br>Gravity<br>Stress and strain = f(time) |

Sampling is the highest priority. The ultimate sampler would be a 100 %, oriented pressure core. Operational cost eliminates the ultimate but the pressure core should remain a goal for future development. The standard, oriented core must be made available for the zones of research interest. Oriented cores are needed for defining the directions of in situ stress and fractures, of the dip of bedding planes, and for paleomagnetic studies.

Samples of pristine fluids are also important. However the value of fluid samples is currently clouded by a number of issues as:

a) Contamination by drilling fluids.
b) Uncertain fluid source.
c) Mixing of several formation fluids due to effects of drilling.
d) Alterations due to sampling, temperature changes during recovery, and changes during sample extraction for analyses.

Thermodynamic properties as temperature, pressure and acoustic wave velocities are needed to aid interpretation of core analyses, determine heat flows, and evaluate bulk formation mechanical properties to compare with core data.

Porosity and permeability can be measured on cores although the correlation between in situ data and core data is not clear, particularly if the permeability is fracture dominated. Borehole measurements are usually needed for evaluation of fracture densities. Current measurements of in situ stress include minifracturing experiments and borehole deformation measurements. However recent work by Teufel (1983) has been successful in determining in situ stress and direction on core from sediments and may be applicable to hard rock cores. Downhole resistivity measurements are needed for true in situ property evaluation and subsequent interpretation of surface EM studies. The chemical composition of precipitates in fractures and knowledge of active chemical reactions are important to understanding in situ processes.

In situ experiments include long-term measurements of stress and strain. Also experiments are needed to extend the laboratory into the formation. Such experiments would include borehole to borehole and borehole to surface, seismic and EM measurements.

## Borehole Measurement Capabilities

Development of fossil energy resources has resulted in an extensive array of logging tools to provide borehole measurements. Some of the slim hole tools are listed in Appendix B. The German CSDP working group has also provided an extensive listing of tools (also in Appendix B) and discussed them in their report (Haenel 1983). As stated many times, and restated in the Albuquerque workshop (Cooper and Traeger 1984):

1. Existing sampling and logging technologies can meet 90% of the CSDP needs in all but the thermal regimes;
2. Meeting the additional 10% will be expensive but will have significant potential for broad application;
3. The existing technology is designed for petroleum and gas exploration in sedimentary fields thereby requiring calibration for research use in metamorphic and igneous systems;
4. Temperature limitations of existing logging tools are 200°-300°C. Electronics are restricted to temperatures below 300°C because of physics limitations of silicon devices.

Also of great importance is recognition of Timur's statement (1982); "Well logging methods usually do not measure directly the primary rock properties needed for engineering purposes: porosity, permeability, fluid saturation and fluid mobility. These parameters are inferred from borehole measurements of electromagnetic, acoustic and nuclear properties." Consequently theoretical developments are needed to obtain research quality data from existing tools.

The need for expanding energy resources and the "hi tech" revolution have spurred extensive development in the logging industry where new capabilities are continually evolving. Limited developments are also occurring in the various federally funded agencies in the U.S.

The European scientific community has been looking at borehole instruments for a number of years. Haenel's report (1983) lists several developments including induction, radar, flowmeter, thermal conductivity and magnetic susceptibility. The ultradeep Russian holes are also requiring new or modified logging techniques (Rezanov 1981) including spectral gamma, gamma-gamma, seismic, magnetic susceptibility, natural gamma, pulsed neutron and other tools. There exist two major technical limitations to the development of CSDP-related instrumentation:

1. Scientific experiments, precisions, accuracies, etc. are insufficiently defined to evaluate what can and cannot be done.
2. Downhole temperatures exceeding 200°-300°C will require significant technological developments.

The first limitation will be resolved as the drill sites are specified, geological models developed and questions identified. However such questions may be asked too late in time to allow any reasonable instrumentation development.

Resolution of the temperature limitation will initially have to be an engineering solution involving a cooled borehole, individually cooled instruments and cables, dewars and the high-temperature components. A discussion of some of these concepts is included in the paper by Traeger et al. 1981.

## Borehole Instrumentation and Sampling Priorities

Any scientifically oriented drilling program will include logging for drilling control, sampling, conventional wireline logging and downhole experimentation. The priorities for scientific research, as defined by many discussions and workshops are:

1. Sampling.
2. Application of conventional logging systems to non-sedimentary formations and to research (not exploration).
3. Far-field geophysics (i.e., looking 10's to 1000's M away from borehole).
4. High-temperature (500°C) instrumentation capabilities.
5. Unique equipment for downhole experiments – this is not a low priority area but can only be defined by individual researchers for specific studies.
6. Logging for drilling – logs run for drilling operations may be of scientific interest.

Each one of these areas will be briefly discussed.

## Sampling

Solid cores, liquids and gases, obtained in situ and representative of downhole conditions, provide the most definitive information on the crust's structure and processes at one point in the crust. The pressure core, core with entrapped gases and liquids, is the ideal sampler. Recent advances in pressure coring (Hyland 1983) have provided a powerful evaluation tool for soft to medium hard formations. Investigations for use in hard, highly fractured formations are needed and orientation capabilities developed. However, all current pressure coring operations require extensive planning, a specialized coring crew and a detailed plan of post-coring procedures.

Coring is usually done: (1) by adding a core bit and barrel to an existing drill string, coring 10-30 m, and tripping the entire string to recover the core (used in the oil and gas industry) or (2) using a wireline coring system wherein the core, held in an inner core barrel, is removed through the drill string without tripping out of the hole (used in mineral exploration). Coring can also be accomplished with downhole motors. Options have been reviewed in recent papers (Filshtinsky et al. 1983; Radford and Carroll, p. 67 in Cooper and Traeger 1984).

The capability of sidewall coring also exists. A recent workshop (Radford, p. 25 in Cooper and Traeger 1984) identified a high-priority need for a large (~20 mm diam.) sidewall core that could be obtained out of the mud invasion zone (~1 m from borewall). With this technology, logging could identify areas of scientific interest and cores then obtained at those specific locations.

Developments needed in coring include:

1. an improved orientation capability;
2. sidewall coring capability to penetrate 1-2 m into the formation (addition of orientation and pressure coring would be desirable);
3. higher temperature capabilities (from 100° to 300°C) (note: coring is done in geothermal systems to temperatures of 300°C using extensive fluid circulation for cooling but technology improvements are needed);
4. development of a wireline-downhole motor system — implied in the Russian work (Rezanov 1981) and proposed by Filshtinsky et al. (1983).

Fluid samplers exist, some of which have been used in 300°C systems (Weres, p. 62 and Dennis, p. 111 in Cooper and Traeger 1984). However extensive formation fluid movement is needed for sampling and interpretation of analyses is clouded by:

1. contamination by the drilling fluids;
2. uncertainty of source of the fluid due to borehole or formation effects;
3. effect of the sampler itself, i.e., flow patterns, flashing during sampling, etc., and
4. chemical and physical processes occurring during sample removal from the wellbore.

A complete systems analysis of fluid sampling is needed to determine if further development is justified and if so, to identify the areas requiring development.

## Conventional Logging

An extensive suite of logging tools, measurements and interpretations has been developed for oil and gas exploration. Any borehole research must take advantage of the tremendous investment that has been made in these systems. In addition, logging will be required because coring costs will dictate less expensive drilling-logging operations and core recovery is always incomplete. Also the majority of logging capabilities are designed for exploration – not research – and for sedimentary formations.

Therefore the first priority must be to identify calibration facilities and test pits for igneous, metamorphic and other systems of research interest. The facilities must be adequately characterized to understand limitations to and modifications needed for borehole research. One significant effort related to geothermal systems has been initiated by Matthews (Hallenburg 1982). Granite test pits with defined fractures are available at the Denver Federal Center (Matthews, p. IV-3 in Hallenburg 1983 and Matthews, p. 103 in Cooper and Traeger 1984). Existing test and calibration facilities need to be identified and adequately characterized. Any significant gaps must be identified and appropriate facilities developed.

The European instrumentation panel prioritized the use of commercial tools in CSDP activities by evaluating their individual contributions in seven research areas (i.e. hydrology, stress, geochemistry, physical properties, etc.). The top ten tools of the 22 considered were (Haenel 1984, personal communication):

1. Production logging tool
2. Temperature
3. Spectral Gamma
4. Natural Gamma
5. Pressure
6. Acoustic
7. Seismic
8. Fluid Sampler
9. Dipmeter
10. Density

I suggest that top priority should be placed on those measurements that cannot be made on core. This would include temperature and pressure. Other critical measurements that may not be obtained from core are fracture density and in situ stress.

Actual prioritization of commercial tool use will obviously depend on the chosen drill site and the scientific questions to be answered.

## Far Field Geophysics

Major limitations of any research centered on a borehole include the uncertainty that the borehole is representative of the region being explored and the need to obtain a more global view of the crustal structure. Also information is needed to evaluate interpretations of surface geophysical data. A high priority should be placed on the geosciences and instrumentation developments to look 10's to 1000's M away from boreholes. Some of this work is under way for VSP and hole-to-hole tomography technologies.

Instrumentation priorities (Hermance and Cheng, p. 26 in Cooper and Traeger 1984) identified are

1. Wide band, three-component, lockable geophones;
2. Three-component, time recording, vector magnetometer;
3. Borehole gravity;
4. HF-VHF electromagnetic probes.

A first look is needed to determine feasibility of actually building such tools. Then further developments must be accompanied by extensive, focused theoretical investigations of signal propagation through complex geological media.

## High Temperature Capabilities

High temperatures anticipated in CSDP wells (300°-350°C at 10 km for basement structures and mineral resources and 300°-800°C for 3-10 km for thermal regimes) are seen as a major barrier for logging or in situ research. Conventional oil and gas logging capabilities are good to 100°C with many usable at 220°C and some at 300°C. Instrumentation limitations have been discussed by Moran and Attali (1971) and Traeger et al. (1981). In summary, commercially available high-temperature cables (340°C), high-temperature electronics (250°-300°C) and dewar-heat sink systems can be used if the sensors can tolerate the temperatures or be insulated from the environment. Details on high-temperature instrumentation capabilities have been reviewed in two symposia (Veneruso et al. 1980; Hudson and Kelsey 1982) and a catalogue of high-temperature components has been developed (Veneruso 1981). Other options, as the use of slick line, Kuster-like tools must be considered.

Limited developments in higher temperature components are occurring. Prototype semiconductor devices, using aluminium gallium phosphide/ gallium phosphide heterojunction technology, have been built that functioned at 550°C (Zipperian and Dawson, p. 118 in Cooper and Traeger 1984). CISE in Italy is looking at optical transducers and cables that can function at temperatures over 400°C (Adorni et al. 1983). Sheathed thermocouples that survive 1000°C are available in kilometer lengths. Metal sheathed cables insulated with MgO for use to 600°C (BICC Pyrotenax 1981) are commercially available. However, development of ~300°C system capabilities will be expensive and require long times; particularly since commercial interest is low due to no market.

Therefore development of generic instrumentation that will operate over 300°C is considered to have a low priority. Emphasis should be placed on assessing engineering solutions as combinations of high-temperature electronics with dewared housings. Maintaining low bottom hole temperatures after drilling by continued fluid circulation would allow measurements to be made, but the effect of the cooling on the data obtained would need to be considered. Slim tools could be lowered through drill string while circulating water; but logging times would be extended due to tripping times. The trade-off results however are very dependent on the bottom hole temperature, temperature-depth profile and formation composition. These tradeoffs must be evaluated on a case-by-case basis. Priority should be placed on a set of trade-off evaluations for a deep, continental hole with nominal gradients, and a thermal regimes hole as now being considered in the Salton Sea.

## Unique Experiments

The borehole laboratory will provide an opportunity for extended in situ experiments as strain measurement, long-term teluric or seismic studies, flow mechanics, etc. Experiment and equipment design would be the responsibility of the scientist. Priorities cannot be defined until specific experiments are described.

## Logging for Drilling

A number of logs will be run to anticipate drilling needs, assure drilling safety and meet permitting requirements. These may require some developments and should incorporate scientific needs if desired.

Specific measurements will include:

1. Temperature - but hole will be cooled by drilling fluids.
2. Pressure.
3. Caliper - hole distortion or spalling may be useful for in situ stress measurement.
4. Hole direction/location.
5. Cuttings analysis - more science could be done using cuttings. However dispersion during circulation prevents accurate location of the cutting source and core bits make a very fine cutting that is difficult to analyze.
6. Packers - packer technology is needed for isolating sections of the wellbore, and also needed for minifrac measurements of in situ stress. Developments are needed on open hole packer concepts; conventional designs will not be useful in highly fractured zones.
7. Cement Bond Log - an acoustic log that could also be run at lower frequencies to look into the formations.
8. Natural Gamma - used for location in the hole.
9. Directional Surveys - identification of the bottom hole location would be important for interpretation of core and geophysics data.

## Summary

The following is a priority summary for future instrumentation developments needed for research in deep boreholes. The needs and priorities will obviously change somewhat from site to site and between difference scientific objectives. This list hopefully will catalyze discussions, provide some guidance and result in better experimental plans.

| | | |
|---|---|---|
| Priority 1: | Cores | Improved Orientation<br>Extended Side Wall core<br>Improved Pressure Core |
| Priority 2: | In Situ Properties | Temperature, Pressure, Permeability, In Situ Stress |
| Priority 3: | Conventional Logs | Calibration Facilities and Test Pits<br>Fracture Mapping<br>In Situ Stress |

| | | |
|---|---|---|
| Priority 4: | Far Field Geophysics | Downhole Geophones, Magnetometer, Gravity and EM Probes<br>Theoretical Modeling |
| Priority 5: | High Temperature | Engineering Alternative Studies |

References

Adorni N, Ceppatelli L, Ferrera GC, Mezzetti S, Pizzolati PL (1983) Design and testing of Downhole probes for operations in deep and hot environments. Proc 3rd Int Update. Eur Geotherm Update, November 29 - December 1, Munich/Germany, pp 552-563

BICC Pyrotenax (1981) Slickensender, Advertising Publ No 935. BICC LTD/Houston, TX

Continental Drilling (1975) Results of the Ghost Ranch Workshop, June 10-13, 1974. Carnegie Inst Wash, p 56

Continental Scientific Drilling Program (1979) Results of a Los Alamos Workshop, July 17-21, 1978. Nat Acad Sci, p 192

Cooper DL, Traeger RK (eds; 1984) Workshop on borehole measurements and interpretation in scientific drilling: identification of problems and proposals for their solutions, SAND 84-0288. Proc Albuquerque, NM Workshop November 15-17, 1983. Sandia Nat Lab Albuquerque, NM

Deep Hole Drilling Feasibility Study (1969) Prepared for the US AEC [Contract AT(26-1)-38] by Fenix and Scisson/Tulsa, OK, p 101

Eidel JJ (ed; 1984, in press) Nat Res Council report on drilling for research of mineral resources

Filshtinsky M, Aumann JT, Quinn J (1983) New tools improve the economics of coring. SPE 58th Annu Technic Conf/San Francisco, CA, October 5. Paper SPE 12092, p 7

Haenel R (Project Leader; 1983) Niedersächsisches Landesamt für Bodenforschung. Tätigkeitsbericht der Arbeitsgemeinschaft Bohrlochmessungen (ARGE 2; in German), September 15

Hallenburg JK (ed; 1982) Geothermal log interpretation handbook. Society of Professional Well Log Analysts/Tulsa, OK

Hardee HC, Luth WC (1980) Comparative assessment of five potential sites for hydrothermal-magma systems. Summary, US DOE report DOE/TIC 11303, p 11

Hardee HC (1980) Comparative assessment of five potential sites for hydrothermal-magma systems: energy transport. SAND 80-1671, Sandia Nat Lab/Albuquerque, NM, p 9

Hermance JF (ed; 1984, in press) Nat Res Council report on drilling for research in thermal regimes

Hudson SR, Kelsey JR (eds; 1982) Proc High Temperature Electronics and Instrumentation Conf, December 1981. SAND 82-0425, Sandia Nat Lab/Albuquerque, NM, p 201

Hyland CR (1983) Pressure coring - an oilfield tool. SPE 58th Annu Technic Conf/San Francisco, CA, October 5. Paper SPE 12093, p 8

Moran JH, Attali G (1971) Wireline operations to 50,000 feet. 8th World Petroleum Congr, June, pp 149-158

Oliver JE (ed; 1984, in press) Nat Res Council report on drilling for research on basement structures and continental basins

Programme Geologie Profonde de la France (1982) Report of a national colloquium in Paris, June 15, 1982. Bureau de recherches geologiques et minieres (in French)

Proposed Scientific Activities for the Salton Sea Scientific Drilling Project (1983) A report of the experiments panel of the Salton Sea scientific drilling project, December. Inst Geophys Planetary Phys, Univ California, Rough Draft, p 134

Rezanov IA (1981) Ultra-deep drilling. DOE-TR-252. Nauka (Science), Moscow, p 160

Steinman DK (1976) Future research in borehole assaying technology, vol 1. Technology assessment of borehole logging techniques, IRT 4611-001, prepared by IRT for US Bureau of Mines (March 18), p 325 (note: for orebody evaluation)

Teufel LW (1983) Determination of the principal horizontal in situ stress directions from anelastic strain recovery measurements of oriented cores from deep wells: application to the Cotton Valley Formation of East Texas. Geomechanics, vol 57. Book No G00229/ASME, pp 55-63

Timur A (1982) Advances in well logging, J Petroleum Technol 6: 1181-1185

Traeger RK, Varnado SG, Veneruso AF, Behr VL, Ortega A (1981) Drilling, instrumentation and sampling considerations for geoscience studies of magma-hydrothermal regimes, SAND 81-0800. Sandia Nat Lab/Albuquerque, NM, (May), p 53

Varnado SG, Colp JL (1979) Report of the workshop on magma/hydrothermal drilling and instrumentation, SAND 78-1365C. Sandia Nat Lab/Albuquerque, NM, (June), p 67

Veneruso AF (1981) Sourcebook on high temperature electronics and instrumentation, SAND 81-2112. Sandia Nat Lab/Albuquerque, NM, (October), p 227

Veneruso AF, Simpson RS, Arnold C (eds; 1980) High temperature electronics and instrumentation. Seminar Proc, December 3-4, 1979. SAND 80-0834C. Sandia Nat Lab/Albuquerque, NM, (May), p 258

Zoback MD (ed; 1984, in press) Nat Res Council report on drilling for research in fault zones

## Appendix A

### Identified Borehole Measurement Needs

#### Steinman (1976)

A. Neutron activation
B. X-Ray fluorescence
   X-Ray diffraction
   Photoluminescence
   Infrared spectroscopy

#### Haenel (1983; p. 28)

A. Temperature, pressure and density = f(depth)
   Chemistry and phases
B. Porosity, permeability,
   Fluid phases
C. Electrical, thermal cond., magnetics
D. Elastic and anelastic properties

Deep Hole Drilling Feasibility Study (1969; p. 27)

Acoustic properties
Elastic moduli
Density
Natural radioactivity
Reaction to neutron bombardment

Borehole workshop (Cooper and Traeger 1984)

| | |
|---|---|
| In situ stress | In situ pore pressure<br>Packers<br>Caliper |
| Fr. mapping | BHTV<br>Dipmeter |
| Coring and sampling | Oriented core<br>(Pressure-sidewall) |
| Far field | Wide band, lock geophones<br>3-comp. magnetometer<br>borehole gravity |
| Logging | Calibration<br>interpretation |

Proposed Scientific Activities for the Salton Sea Scientific Drilling Project (1983)

| | |
|---|---|
| Thermodynamic | T, P, thermal conductivity; T&P profile |
| Rock | Permeability and porosity; drill, stem, well-to-well |
| Fluids | Chemistry |
| Fracture | Faults structure and direction; BHTV, dipmeter, radar |

U.S. Geodynamics Committee (1979)

| | |
|---|---|
| Basement structures | Density, mag. suscept., porosity, perm.<br>Radioactivity<br>Stress orientation and magnitude |
| Thermal regimes | P, T = f(depth)<br>Porosity, permeability<br>Fluid compositions |
| Minerals | T = f(depth)<br>Permeability<br>Fracture density and orientation<br>Flow direction and rate<br>In situ chem. |
| Earthquakes | Stress magnitude and orientation<br>Temp. and heat flo.<br>Pore pressure<br>Seismic - velocity<br>Strain - long term |

Workshop on Magma/Hydrothermal Drilling and Instrumentation (Varnado and Colp 1978)

High resolution temp.
Coring and fluid sampling
Chemistry
Cables                                    Needs to 500°C

Continental drilling, Results of Ghost Ranch Workshop (1975)

Rock and fluid samples
S&P wave velocities
Densities
Strain for extended times
Stress
Porosity
Permeability
Heat production (radioactive)
In situ thermal conductivity

## Appendix B

### Commercially Available Logging Tools

Logging tools, primarily used for oil and gas exploration, are designed to provide the minimum information needed to assess resources in sedimentary formations at the lowest cost. Logs provided by service companies should be considered as information, not scientific data. This is particularly true for use of conventional logging tools in igneous or metamorphic formations where tool use or log interpretation may be incorrect (Sanyal and Meidav 1976).

Table 2 lists slim hole tools that are noted in catalogues from the service companies. No operating data or insight is included or implied by this listing. Tools for use in 4.5 inch wellbores should have diameters less than 3.5 inches to prevent sticking with possible loss of hole.

### Special Logging Tools

Commercial tools can be modified or new tools developed to provide desired, downhole information. The majority of this type of work has been done at the USGS in Denver and the tools are listed in Table 3. These tools have been used with the USGS truck to provide data from 3-inch wellbores; the "Illinois Hole" (Daniels et al. 1982) and Utah (Daniels et al. 1980).

### Geophysics Instrumentation

Given a question and a set of boundary conditions, special downhole instrumentation can be developed to answer that question. Such development has been accomplished at many laboratories, but, to my knowledge, has not been summarized in any available documentation.

Table 2. Commercially available slim hole logging tools

| Tool type | Wellbore | O.D. (in) | Max. Press. (ksi) | Max. Temp. °F/°C |
|---|---|---|---|---|
| Schlumberger (Schlumberger Services Catalog 1978) | | | | |
| Resistivity | | | | |
| Induction | Open | 2-3/4, 3-7/8 | 20 | 350/175 |
| Electrical | Open | 3-3/8 | 25 | 500/260 |
| Induction-spherically focused | Open | 3-1/2 | 20 | 350/175 |
| Dual induction laterolog | Open | 3-3/8, 3-7/8 | 20 | 350/175 |
| Dual laterolog | Open | 3-5/8 | 20 | 350/175 |
| Ultralong spaced electrical | Open | 3-5/8 | 20 | 350/175 |
| Porosity | | | | |
| Formation density | Open | 2-3/4 | 25 | 500/260 |
| | Open | 3-3/8 | 20 | 400/205 |
| Compensated sonic | Open | 1-11/16 | 16.5 | 300/150 |
| | Open | 3-3/8, 3-5/8 | 20 | 350/175 |
| | Open | 2-3/4, 3-3/8 | 25 | 500/260 |
| Long spaced sonic | Open | 3-5/8 | 20 | 350/175 |
| Compensated neutron | Open | 2-3/4 | 25 | 500/260 |
| Natural gamma | Open | 3-5/8 | 20 | 350/175 |
| Temperature | | | | |
| Temperature | Open | 1-11/16 | 15 | 350/175 |
| Flowmeter-temperature | Open | 1-11/16 | 20 | 500/260 |
| Drill String | | | | |
| Electrical | Through | 1-1/2 | 20 | 350-500/175-200 |
| Induction | Drill stem | 2-3/4 | 20 | 350-400/175-205 |
| Sonic | Drill stem | 1-11/16 | 16 | 300/150 |
| Neutron | Drill stem | 2-3/4 | 25 | 500/260 |
| Formation density | Drill stem | 2-3/4 | 25 | 500/260 |
| Gamma ray | Drill stem | 2-3/4 | 25 | 500/260 |
| Gamma neutron | Drill stem | 2-5/8 | 25 | 500/260 |
| Thermal decay | Drill stem | 1-11/16 | 16.5 | 300/150 |
| Temperature | Drill stem | 1-11/16 | 15 | 350/175 |

Table 2 (continued)

| Tool type | Wellbore | O.D. (in) | Max. Press. (ksi) | Max. Temp. °F/°C |
|---|---|---|---|---|
| Production logging | | | | |
| Continuous flowmeter | Cased | 1-11/16 | 15 | 350-600/175-315 |
| Gradiometer | Cased | 1-11/16 | 15 | 350/175 |
| High resolution thermometer | Cased | 1-11/16 | 15 | 350/175 |
| Fluid sampler (650 and 836 cc) | Cased | 1-11/16, 2-1/2 | 10 | 350/175 |
| Radioactive tracer | Cased | 1-11/16 | 20 | 275/135 |
| Logging in casing | | | | |
| Gamma ray | Cased | 1-11/16, 2, 2-3/8, 3-3/8 | 20 | 350-500/175-260 |
| Neutron | Cased | 1-11/16, 2, 2-3/8, 3-3/8 | 12-25 | 350-500/175-260 |
| Thermal neutron decay | Cased | 1-11/16 | 16.5 | 300/150 |
| Dresser Atlas, 1980 | | | | |
| Electrical | | | | |
| Induction-electrolog | Open | 2.0 | 17 | 350/175 |
| Dual induction focused | Open | 3-5/8 | 18 | 350/175 |
| | | 3-3/8 | 25 | 400/204 |
| Dual laterlog | Open | 3-5/8 | 20 | 400/204 |
| Radioactive | | | | |
| Compensated neutron | Open | 2-3/4, 3-5/8 | 20 | 300/150 |
| Gamma-neutron | Open | 1-11/16, 3-3/8 | 17 | 300/150 |
| | | 2-3/4, 3-3/8 | 20 | 300/150 |
| Compensated densilog | Open | 3 | 20 | 300/150 |
| Epithermal neutron | Open | 3 | 20 | 300/150 |
| Gamma spectra | Open | 3-5/8 | 20 | 400/204 |

Table 2 (continued)

| Tool type | Wellbore | O.D. (in) | Max. Press. (ksi) | Max. Temp. °F/°C |
|---|---|---|---|---|
| Acoustic | | | | |
| Acoustilog | Open | 2-3/4 | 20 | 450/ |
| | | 3-3/8, 3-7/8 | 20 | 350/175 |
| Production | | | | |
| Nuclear flolog | Cased | 1-1/2 | 12 | 350/175 |
| Tracerlog | Cased | 1-1/2 | 12 | 350/175 |
| Fluid density | Open or cased | 1-3/4 | 15 | 400/204 |
| Temperature | Open or cased | 1-1/2, 1-11/16 | 18 | 325/163 |
| | | 1-11/16 | 17 | 400/204 |
| Flowmeter | Open or cased | 1-11/16, 2-1/8 | 18 | 300/150 |
| Fluid sampler | Open | 1-11/16 | 10 | 300/150 |

Veneruso 1982

Gearhart Industries is set to market a complete line of logging tools, 2-3/8 inches diameter available for service to 225°C. Catalogs and data are currently in process of being printed.

Table 3. USGS logging tools, January 1982 (Daniels, 1982, personal communication)

| Temperature range (°C) | Log type | Required hole diameter |
|---|---|---|
| 100 | Sonic ratio (far/near) | 4-1/2" hole; 3" if smooth hole |
| 100 | Sonic delta-t | 4-1/2" hole; 3" if smooth hole |
| 100 | Sonic waveforms | 4-1/2" hole; 3" if smooth hole |
| 100 | Sonic velocity | 4-1/2" hole; 3" if smooth hole |
| 100 | sonic amplitude | 4-1/2" hole; 3" if smooth hole |
| 300 | 8" resis., normal, uncomp. | 3" |
| 300 | 16" resis., normal, uncomp. | 3" |
| 300 | 32" resis., normal, uncomp. | 3" |
| 300 | 64" resis., normal, uncomp. | 3" |
| 300 | 8" resis., normal, comp. | 3" |
| 300 | 16" resis., normal, comp. | 3" |
| 300 | 32" resis., normal, comp. | 3" |
| 300 | 64" resis., normal, comp. | 3" |
| 300 | 8" wenner resis., uncomp. | 3" |
| 300 | 16" wenner resis., uncomp. | 3" |
| 300 | 8" wenner resis., comp. | 3" |
| 300 | 16" wenner resis., comp. | 3" |
| 300 | 8" wenner IP | 3" |
| 300 | 16" wenner IP | 3" |
| 300 | 10 cm Dakhnov IP | 3" |
| 300 | 40 cm Dakhnov IP | 3" |
| 100 | Compensated dual density | 4-1/2"; 3" if smooth |
| 100 | Density computer from near detector | 4-1/2"; 3" if smooth |
| 100 | Density computed from far detector | 4-1/2"; 3" if smooth |
| 100 | Near detector count rate | 4-1/2"; 3" if smooth |
| 100 | Far detector count rate | 4-1/2"; 3" if smooth |
| 100 | Caliper from density probe | 4-1/2"; 3" if smooth |
| 300 | 3-arm caliper | 3" |

Table 3 (continued)

| Temperature range (°C) | Log type | Required hole diameter |
|---|---|---|
| 100 | Neutron count rate | 3" |
| 100 | Neutron porosity | 3" |
| 100 | Neutron API units | 3" |
| 100 | Gamma ray count rate | 3" |
| 100 | Gamma ray API units | 3" |
| 100 | Equivalent U from gamma ray | 3" |
| 80 | Magnetic susceptibility (volts) | 4-1/2"; 3" if smooth |
| 80 | Magnetic susceptibility (micro CGS) | 4-1/2"; 3" if smooth |
| 80 | Magnetic susceptibility (micro SIO) | 4-1/2"; 3" if smooth |
| 300 | Self potential (single point) | 3" |
| 300 | Self potential (8" differential) | 3" |
| 300 | Self potential (16" differential) | 3" |
| 300 | Single point resistance | 3" |
| 300 | 8" differential resistance | 3" |
| 300 | 16" differential resistance | 3" |
| 300 | Temperature | 3" |
| 100 | Vertical component magnetometer | 3" |
| 100 | Directional survey | 3" |

Work (Hardee 1982) on unique, slim hole diagnostics at Sandia includes:

Temperature:

1. Purchase of sheathed thermocouple 500 ft lengths spliced to 1-2 km lengths with temperature capabilities to 1100°C.
2. Use of sheathed, MgO insulated multiconductor cable (BICC 1981) for tying to resistance thermometers or other temperature sensors. Cable diameters 0.09-0.15 in/2.4-3.8 mm. Temperature to 1500°F/800°C, lengths to 300 kft/9km.

In situ Periodic Source:

Downhole oscillator providing either Por Swaves into the formation at frequencies of 10-100 Hz, to temperatures of 300°C; diameter = 2-3/8 inch.

## Summary

At the request of CSDC panels, a preliminary list of slim hole diagnostic capabilities has been gathered. Many commercial logging tools exist that are oriented toward solving problems in petroleum fields. The USGS has developed a number of slim hole tools for geoscience research. Finally, speciality diagnostics have been or are being developed by individual laboratories.

The purpose of this memo is to provide a "tool menu" for the scientific community to have for CSDP considerations. Any comments or additions are solicited so this memo can be updated to be more useful.

## References

BICC Pyrotenax (1981) Slicksender mineral insulated logging cable (November) - Manufacturers information
Continental Drilling (1975) Carnegie Institution of Washington (Ghost-Ranch Meeting (June), yellow book), pp 37-41
Daniels JJ, Hite RJ, Scott JH (1980) Geophysical well-log measurements in three drill holes at Salt Valley, Utah. USGS Open File Report, pp 81-86
Daniels JJ (1982) Communication (January)
Daniels JJ, Olhoeft GR, Scott JH (1982) Interpretation of core and well log physical property data from UPH-3 (January). Paper in process
Dresser Atlas (1980) Wireline Services Catalog (August). Dresser Industries, Houston, Texas, with notes from JF Petko, Dresser in Houston
Hardee HC (1982) Communications. Sandia Nat Lab Albuquerque, NM (April)
Sanyal SK, Meidav, HT (1976) Well logging in the geothermal industry. 17th Annu SPWLA Logging Symp Transactions
Schlumberger Services Catalog (1978) Schlumberger, with notes from H. Cloud. Schlumberger Well Services, Farmington, NM
US Geodynamics Committee (1979) Continental scientific drilling program. Nat Acad Sci, Wash DC, pp 161-171. Cont Sci Dr Res Los Alamos
Veneruso A (1982) Gearhart Industries. Fort Worth, Texas, Communication (March)

*Acknowledgment.* This work supported by the U.S. Department of Energy at Sandia National Laboratories under Contract DE-AC04-76DP00789.

# LOGGING TECHNOLOGY FOR HIGH-TEMPERATURE GEOTHERMAL BOREHOLES

Bert R. Dennis[1]

## Introduction

The Earth Science Instrumentation Group at the Los Alamos National Laboratory is developing instrumentation and new tools that can withstand downhole temperatures in excess of 300°C and pressures greater than 138 MPa (20,000 psi).

This group was formed in 1973 to provide geophysical measurement support for the Hot Dry Rock (HDR) Geothermal Project at Fenton Hill, New Mexico.

The HDR Project traces its beginning to Los Alamos geoscientists who believed there was a huge underground reservoir of hot, dry rock that could become another large source of energy if the technology to use it could be developed. Volcanic activity in the Fenton Hill area a million years ago had created the Valles Caldera when a volcano there collapsed into its empty magma chamber. Preliminary tests at the Fenton Hill site, located 55 miles from the Laboratory, confirmed the existence of a large volume of hot rock deep underground. It was an ideal place to test the HDR concept. Basically, the concept involved:

- drilling two holes deep into the hot rock,
- connecting them with fractures produced by hydraulic pressurization,
- pumping cold water down one hole and bringing the fluid up the other hole as super-heated steam, thus recovering heat (and energy) from the rock.

Funded by the U.S. Department of Energy, the first phase of the project was completed in 1977. One test hole was drilled to a depth of (2724 m) 8936 ft and a temperature of 200°C. The second hole reached a final depth of (3183 m) 9339 ft and a bottomhole temperature of 205°C. A subsequent fracture completed the loop, and an experimental heat-extraction system was established between the two holes in 1979. This research system demonstrated the technical feasibility and environmental acceptance of the HDR concept.

A second engineering system is under construction. It is larger, hotter, and deeper than the first research system. One borehole was drilled to a depth of 4633 m (15,200 ft), of which about 3962 m (13,000 ft) is in Precambrian granitic rock. Starting at a depth of 2926 m (9600 ft), the borehole was inclined up to 35° from vertical. Bottomhole temperature is 317°C. The other borehole was drilled to 4237 m (13,900 ft). Its inclined part is positioned directly over the first wellbore, with a vertical separation of 457 m (1500 ft)

[1] Los Alamos National Laboratory, Los Alamos, NM 87545, USA

Observation of the Continental Crust
through Drilling I (ed. by C.B. Raleigh)

between them. A series of hydraulic-induced fractures will connect the two wellbores and create an extremely large reservoir of hot rock. Upon completion, this engineering extraction system will produce enough heat to represent a useful energy source or demonstrate the costs of extracting thermal energy from the naturally heated rock beneath the Earth's crust.

The DOE funds the materials development program for the HDR Project that covers all aspects of geothermal energy extraction. Materials and components, high-temperature tools and instrumentation are needed to support research on drilling, hydraulic fracturing, and wellbore logging. Some of this equipment developed by the petroleum industry can be modified or upgraded for work in hot, dry rock.

When it is feasible, Los Alamos contracts with private industry and other commercial organizations to develop the geothermal equipment needed at Fenton Hill. The Earth Science Instrumentation Group is developing a wide spectrum of high-temperature equipment for geothermal research, as well as sophisticated techniques for geophysical measurements and data acquisition and analysis.

In addition to the Fenton Hill operations, work has been performed using the Los Alamos equipment on a number of other projects in cooperation with industrial contractors. Some of these projects include:

- Fluid Sample in the Oceanic Rift, East Pacific Rise
- Crosswell Acoustic Imaging in the Western Gas Sands, Rifle, CO
- Microseismic Mapping for Union Geothermal at the Baca Location in New Mexico
- Temperature/Fluid Velocity Survey at Ontario, OR for the ORE-IDA Food Company
- Fluid Samples/Temperature/Fluid Velocity Surveys for Geothermal Resources, Inc. at Thompson's Ridge, NM
- Temperature/VSP Surveys for Phillips Petroleum in Brawley, CA
- Microseismic Studies for Republic Geothermal in Beowawe, NV.

## Facilities

Offices, shops, and laboratories of the Earth Science Instrumentation Group are located at TA-33, a technical area that is outside the main Laboratory complex in the county of Los Alamos and borders the Bandelier National Park. Technical Area 57 is the technical area at Fenton Hill west of Los Alamos in the Jemez Mountains. Fifty-five miles separate the two sites. The group designs, builds, and tests the equipment at the TA-33 site before taking them to TA-57 for field assembly and operation. Well equipped facilities at both sites support the group's complete capability to develop unique, high-temperature geothermal tools and instrumentation.

At TA-33, engineers and designers routinely use sophisticated CAD (computer-aided design) systems for tool design. Machinists and technicians who are skilled in geothermal instrumentation and electronics work in two highly-specialized fabrication shops.

Technicians check out subassemblies and downhole equipment components in an Instrument Sonde Test Facility where conditions can be varied to simulate deep geothermal wellbore temperature and pressures. At the Cable Test Facility, high-temperature well logging cable is evaluated for physical and electrical characteristics in water over the temperature ranges up to 350°C and pressure ranges up to 17.2 MPa (2500 psi). Temperature and pressure in the 7.6 m (25-ft) long, horizontal autoclave are remotely controlled from an adjacent control room. Pressure-testing equipment in the Calibration Facility calibrates pressure transducers to 103 MPa (15,000 psi) and temperature-testing equipment calibrates thermocouples and other types of temperature sensors from up to 350°C. Strain-gage installation equipment is located there, in addition to measurement systems for materials testing and performance tests on accelerometers and geophones used in downhole acoustic instrumentation.

Data acquisition systems are located at both technical areas and in the group's Wellside Service Units, which are well-equipped logging trucks. Each logging truck carries 6096 m (20,000 ft) of TFE-Teflon insulated, armored, 7-conductor cable capable of operating at temperatures exceeding 300°C.

The data acquisition and control equipment for downhole measurements located at Fenton Hill uses a Hewlett Packard 9845B computer for the central processing unit. Up to three downhole instruments may be deployed simultaneously in three different boreholes. There are two trailer-mounted diesel-driven drawworks with 6096 m (20,000 ft) of high-temperature, 7-conductor, armored cable at the Fenton Hill Test Site in addition to the two Wellside Service Units. The data acquisition systems presently in use are designed for low-level signal measurements from the remote downhole sensors where the 7-conductor, armored, logging cable provides the signal transmission link. All signal-conditioning grounding and shielding uses guarded differential techniques to optimize maximum signal-to-noise and common-mode rejection. Data is digitized, normalized, and fitted to calibration information, then displayed and stored in engineering units.

The signal conditioning and data acquisition system for acoustic detection is unique. Signals detected by triaxial geophones and accelerometers are transmitted to several special purpose storage scopes with circulating memories. This unit triggers on the present amplitude of the incoming microseismic signal and provides a delay sweep that allows display of several milliseconds of signal obtained prior to the trigger. An interface bus designed at the Laboratory allows the computer equipment to talk to the scope units, control the entire acoustic detection system, and store the desired signals on disc units for online processing to determine source orientation and distance from the detector. Now fracture growth can be observed on-line in real time. The technology developed here to perform acoustic mapping in a geothermal environment is rapidly gaining interest in the industry.

A data acquisition and control system that duplicates the downhole instrumentation system installed at Fenton Hill is located at TA-33. This system is used to check out downhole tool operations in the laboratory and to write the computer programs required for each tool. With this system, members of the geothermal team design and develop special computer and peripheral equipment interfaces for such things as PCM, FM multiplex, and FFT analyzers. Programs are written and checked out here for the acoustic detection systems interfacing the data recording equipment.

Components

The most critical components required to produce an operational geothermal well-logging system are the cable, O-ring seals, and sensors that can withstand the high bottomhole temperatures and pressure. Their development paved the way for upgrading all the downhole instrument systems at Fenton Hill.

L'Garde, Inc. developed a high-temperature elastomer called EPDM-Y267 compound for O-rings and other components used in the Laboratory's high-temperature downhole tools. The O-rings were tested by temperature and pressure cycling to determine operating characteristics at 300°C and 65 MPa (10,000 psi). These O-rings are now commercially available from Parker Seal Company.

Small, high-temperature direct current motors were designed by American Electronics, Inc. The motors are used in various devices such as the fluid sampler, geophone tools, and tracer injector on an intermittent basis.

An armored, 7-conductor cable with TFE Teflon used for the electrical insulating material was manufactured for the Laboratory by the Rochester Corporation. TFE Teflon, although non-extrudable, has been shown by extensive testing to be the best wireline insulation material for use above 300°C at the present time.

Extensive frequency response tests were conducted on the new 7-conductor, TFE-Teflon insulated, armored cable. Tests conducted over 6100 m (20,000 ft) of the cable show a corner frequency response of greater than 10 KHz, which is a factor of 10 better than either the standard Tefzel or the PFA-Teflon insulated logging cables. The data also shows that the response is down only 20 dB at 100 KHz. Frequency response enhancement may be achieved by careful selection of particular conductors for signal lines coupled with proper cable termination impedance.

One of the most critical components of the downhole instrument package is the cablehead. The cablehead provides the downhole termination of the armored, 7-conductor, logging cable and completes the transistion to the instrument sonde. The cablehead ensures water tight integrity of the electrical wire terminations so very important in the high-temperature environment. It also allows a disconnect from the armored cable should the sonde become stuck in the borehole. The cablehead, designed by the Laboratory, has been subjected to numerous logging runs in geothermal environments in fluid temperatures exceeding 300°C.

A significant improvement in electrical connectors for downhole instrument packages was made by Reynolds Corporation under contract to the Laboratory. The new connectors are configured like the standard Amphenol but uses a ceramic insert. The ceramic insert is unique in that it is made from a mold and is machineable during the curing process. This insert will accept the crimped pins that are readily inserted and removed from the ceramic feedthrus. This feature allows the connector to be reused many times. The connectors have been tested to 350°C with no observable degradation.

Acoustic tools developed at Los Alamos were designed to "map" fractures and characterize geothermal reservoirs. The series of downhole instruments include geophone acoustic detectors, acoustic transceivers, accelerometers and explosive detonators to simulate seismic sources for calibration purposes.

The high-temperature geophone technology developed at Los Alamos has been transferred to industry. High-temperature geophones may be purchased from Mark Products in Houston, TX. Low-frequency geophones (corner frequency below 30 Hz) are affected by tilting the axis more than 14° from the vertical. However, high-frequency geophones may be used in inclined positions greater than 80° from the vertical.

An accelerometer acoustic detector may be used to complement the triaxial geophone array. Results of extensive testing has shown that the high-temperature Endevco piezoelectric accelerometer is best suited for the tools developed at Los Alamos. This accelerometer has a greater frequency range, the required electrical isolation, and proper size for mounting in the limited space provided in the downhole sonde.

The explosive detonator system provides acoustic signals from a known fixed location in one of the wellbores. These signals are used to calibrate the high-temperature triaxial geophone array deployed in adjacent wells. The downhole sonde is modular in design and is used to fire up to four detonators in sequence. Each of the detonators contain pellets of hexanitrostilbene (HNS) explosive. The detonator is a modified version of the Reynolds Industries RP-84 slapper initiator used in other high-temperature explosive devices. The detonator is initiated by a downhole capacitive discharge firing unit, which is controlled through the armored instrumentation cable from the surface control unit. The firing circuit and detonator has operated successfully in wellbores where temperatures exceed 300°C.

Other explosive devices use a thermally stable explosive shaped to perform specific functions. One such tool leaves a bell-shaped cavity in hard rock for sidetracking the wellbore. A second tool uses a shaped charge to cut off drill pipe leaving a clean circumferential edge for overshot tools and a fracture initiation tool is used to detonate a small explosive charge in a preselected zone in the openhole section of the wellbore to initiate the fracture system by reducing the wellbore impedance. The thermally stable explosive has been tested to temperatures exceeding 340°C.

## Downhole Instruments

A variety of high-temperature downhole instruments have been developed for use in the HDR Fenton Hill boreholes. These downhole systems have been designed to operate at temperatures above 300°C and pressures up to 20,000 psi. A brief summary of the downhole instruments that have been used for numerous wellbore surveys is given below.

### Temperature Probes

Temperature probes are for borehole temperature surveys to provide diagnostic measurements and geothermal reservoir characterization. The measurements determine thermal gradient along the borehole, locate borehole/fracture intersections, and calculate thermal drawdown and recovery rates of the circulating systems. Temperature probes give diagnostic data during drilling, cementing, pressurization, and fracturing operations.

The high-temperature probe, fabricated in various sizes, is one of the simpler wellbore survey tools that is readily fielded to allow on-line analysis of changing conditions. It uses a thin-walled, stainless steel thermistor probe as the sensor. The thermistor is essentially a semiconductor that behaves as a temperature-sensitive electrical resistance, providing a high degree of resolution that is well suited for continuous measurement in temperature environments exceeding 300°C.

## Borehole Fluid Sampler

One of the first high-temperature tools developed for the HDR Project. Using three conductors of the high-temperature armored instrument logging cable, the tool takes two separate samples of 270 cc or one large sample of 780 cc. Chemical analyses of the fluid samples, which are taken at various depths in a borehole, show geothermal reservoir properties.

## Downhole Injector and Gamma-Ray Detector

Radioactive tracers injected at specific locations in the borehole show fluid circulation in the fractured geothermal reservoir. Radioactive $Br^{82}$ or any other tracer is injected into the injection well with the injector system developed at Los Alamos. A gamma-ray detector is mounted in the same sonde as the injector. When the sonde is positioned in the borehole at a predetermined location, a motor-propelled rod smashes the quartz ampoule that contains the tracer. The integral gamma-ray detector tracks the radioactive tracer as it leaves the injection well into the fracture system. A second gamma-ray detector monitors the arrival of the tracer in the production well. This tool also shows fluid flow velocities and residence time in the fracture system.

## Fluid-Velocity (Spinner Probe)

The fluid velocity (spinner) probe yields accurate data about complete fluid-flow patterns in a borehole. A typical borehole will have fractures in several zones. The fluid velocity, as measured by the sensor, can help characterize the man-made reservoir by determining the nature and locations of the fractures, where and how much fluid leaves the borehole reservoir, and the relative contribution of each fracture to the total reservoir.

## Caliper Tool

An independent multiarm caliper tool designed to define the profile of geothermal boreholes such as those at Fenton Hill. Development began with the design and fabrication of a prototype 3-arm sonde, which can readily be modified to have six independent arms. The arms are evenly spaced circumferentially and are capable of measuring radii from 2-1/2 to 7 in. at a common depth. Arm length can be varied to measure hole diameters up to 30 in. or can be designed for maximum sensitivity at given diameters. The arms are activated by a motor and can be extended or retracted on command from the surface.

## Acoustic Transceiver System

The acoustic transceiver system consists of a transmitter and associated electronics consisting of two magnetostrictive transponding heads. One head transmits a frequency of 8.5 KHz, and the second head transmits at 17 KHz. The receiver electronics include an automatic gain control system operating over a range of attenuation from 1 to 100 in eight steps (6 decibels/step). The assembled sonde incorporates a suitable pressure housing and thermal protection to ensure operation in the geothermal borehole environment of up to 20,000 psi external pressure and 275°C external temperature.

## Geophone Acoustic Detector

The geophone acoustic detector incorporates a triaxial geophone array to map source locations of microseismic signals generated during pressurization and hydraulic fracturing. The sonde also includes a downhole switching system to allow multiplexing of additional pertinent data to monitor internal dewar temperature, sonde orientation, and downhole power supply voltages.

## Downhole Acoustic Detonator

The downhole acoustic detonator is used to calibrate geophone acoustic detectors deployed in adjacent boreholes. A geophone sonde, used in conjunction with the detonator package, registers the acoustic signals generated by the explosive. The two tools are lowered into separate boreholes. The detonator can set off up to 4 charges sequentially at any desired location. The firing system consists of a downhole firing module, surface control unit, detonator rack, and high-temperature detonators.

## Explosive Sidetracking Tool

The explosive sidetracking tool rubblizes rock and leaves a bell-shaped cavity to facilitate sidetracking in hard rock at a specific location. The entire system is rated for operation at boreholes temperatures about 260°C. The downhole firing unit uses a capacitive discharge system (charged from the surface after insertion) to fire the high-temperature detonators. A dual firing unit gives redundancy. The firing unit is housed in a Laboratory-designed dewar flask of the type incorporated in many high-temperature packages to protect sensitive electronics.

## Explosive Cut-Off Tool

The explosive cut-off tool severs the drill pipe string. The device uses a high-temperature/high-explosive shaped charge, a high-temperature detonator, and a dewared firing unit. The tool's shaped charge is designed to leave a clean circumferential edge where it cuts off pipe.

## Explosive Fracture-Initiation Tool

Deploys and detonates a small explosive charge at a preselected position in the open borehole section of a geothermal wellbore. Detonation of a shaped-charge explosive can reduce the wellbore impendance of existing fractures or initiate fractures in the selected zone, after which hydraulic pressurization extends the fractures into the formation. The device is rated for wellbore temperatures of 300°C, a pressure limit of 15,000 psi, and incorporates a downhole firing module, high-temperature detonator, and shaped high-explosive charge. The tool's diameter is 3-1/4 in. and it is 18.5-ft long.

## Conclusion

State-of-the-art for geophysical measurements in high-temperature geothermal environments is presently limited to wellbore temperatures of 300°C. Development of well-logging systems in the temperature ranges of 400°C to 500°C will require research in materials for transducers, components, electronics and data transmission links. Several developments on the horizon include metal sheathed logging cable that could advance data transmission capability. A program now in the advance development stages at Los Alamos using integrated thermionic circuits that have proven successful for electronic systems that will operate continuously at 500°C.

## Bibliography


Archuleta J, Todd B (1978) Hot Dry Rock Geothermal Energy Development Project, Cablehead Assembly, Equipment Development Report. Los Alamos Sci Lab report LA-7325-MS, August

Archuleta J, Fink C, Kurtenbach J (1978a) Equipment Development Report, Downhole Fluid Injector. Los Alamos Sci Lab report LA-7151-MS, February

Archuleta J, Fink C, Kurtenbach J (1978b) Equipment Development Report, Borehole Fluid Sampling Tool. Los Alamos Sci Lab report LA-7152-MS, February

Bennett GA, Sherman GR (1983) Analysis and Thermal Design Improvements of Downhole Tools for Use in Hot Dry Wells. Los Alamos Nat Lab report LA-9671-HDR, February

Lawton R, Kolar J, Todd B (1982) Fluid Velocity Sensor, Development Report. Los Alamos Nat Lab report LALP-82-16, September

Smith MC (1983) Major Accomplishments of the Hot Dry Rock Program, 1970-1982. Los Alamos Nat Lab report LALP-82-37, August

Smith MC, Ponder GM (1982) Hot Dry Rock Geothermal Energy Development Program, Annual Report, Fiscal Year 1981. Los Alamos Nat Lab report LA-9287-HDR, April

Smith MC, Nunz GJ, Ponder GM (1983) Hot Dry Rock Geothermal Energy Development Program, Annual Report, Fiscal Year 1982. Los Alamos Nat Lab report LA-9780-HDR, September

# USE OF GEOPHYSICAL LOGS FOR QUANTITATIVE DETERMINATION OF FRACTURING, ALTERATION, AND LITHOLOGY IN THE UPPER OCEANIC CRUST

Roger N. Anderson,[1] Honor O'Malley,[2] and Robin L. Newmark[1]

## Introduction

The upper layers of the oceanic crust are characterized by large changes in geophysical properties both sub-areally and with submarine depth which belie their volcanic modes of origin. Feeder dikes connecting surface-flow basalts with the magma chamber below, produce new sea-floor with similar volcanic structure worldwide within the spreading systems active on all mid-ocean ridges. All ocean crust consists of pillow basalts overlying sheeted dikes which in turn rest upon gabbroes which are the frozen remains of old magma chambers. Even the chemistry of the oceanic crust is essentially the same from ocean to ocean. It is the rare ocean basalt which differs from MORB, or mid-ocean ridge basalt composition. Yet geophysical properties of Mid-Atlantic Ridge flank crust can be very different from East Pacific Rise crust. For example, the seismic velocity in the very uppermost crust is often slower in the Atlantic than in the Pacific (Houtz and Ewing 1976). The most highly altered metamorphic rocks are invariably found in the Atlantic.

Since the volcanic origin of all sea floor assures similar initial composition and general structure, studies of the causes of geophysical differences quickly lead to interesting conclusions concerning (1) the amount of faulting and fracturing the crust undergoes after emplacement, (2) the degree of hydrothermal circulation and metamorphic alteration the crust experiences during aging, and (3) the permeability structure of the crust which ties (1) to (2). Put another way, physical properties such as the propagation rate of elastic waves, or the variation in permeability and porosity with depth, are controlled by the degree and interconnectedness of fracturing in the crust, and by the degree and extent of alteration that metamorphoses the crust and fills the fractures.

Direct determination of sonic velocity, electrical resistivity, and porosity have been made in situ in the oceanic crust for several years now by geophysical logging in Deep Sea Drilling Project (DSDP) boreholes. Below we report results of several new analyses and measurement techniques that extend geophysical logging through the use of multichannel sonic tools which record seismic waveforms over short intervals, ultrasonic borehole imagers which produce an accurate picture of the wellbore, and active nuclear source-counter arrays which separate hydroxyl mineral content from free-water porosity in the for-

[1]Borehole Research Group, Lamont-Doherty Geological Observatory, Columbia University, Palisades, NY 10964,USA

[2]Teachers College, Columbia University, New York, NY 10027, USA

Observation of the Continental Crust through Drilling I (ed. by C.B. Raleigh)

mation. These three techniques are then combined to locate fractures and quantify the extent to which they are filled by alteration products.

We use data from the deepest hole yet drilled into the oceanic crust at site 504B, in 6 Ma sea-floor on the south flank of the Costa Rica Rift, eastern equatorial Pacific, as well as data from hole 556 in 17 Ma sea-floor on the west flank of the Mid-Atlantic Ridge, to demonstrate the applicability of the new techniques for identifying differences in lithology, alteration and fracturing found in the two oceans. Our ultimate aim is to characterize the permeability, degree of alteration, and structure of these sites in hopes of understanding the physical and chemical processes which produce very different geology at these disparate sites.

## The Experiment at Hole 504B

In November-January, 1981-1982, Leg 83 of the Deep Sea Drilling Project reoccupied Hole 504B on the south flank of the Costa Rica Rift, and extended drilling begun previously during Legs 69 and 70, to over one kilometer into the upper oceanic crust (Fig. 1). Three layers were

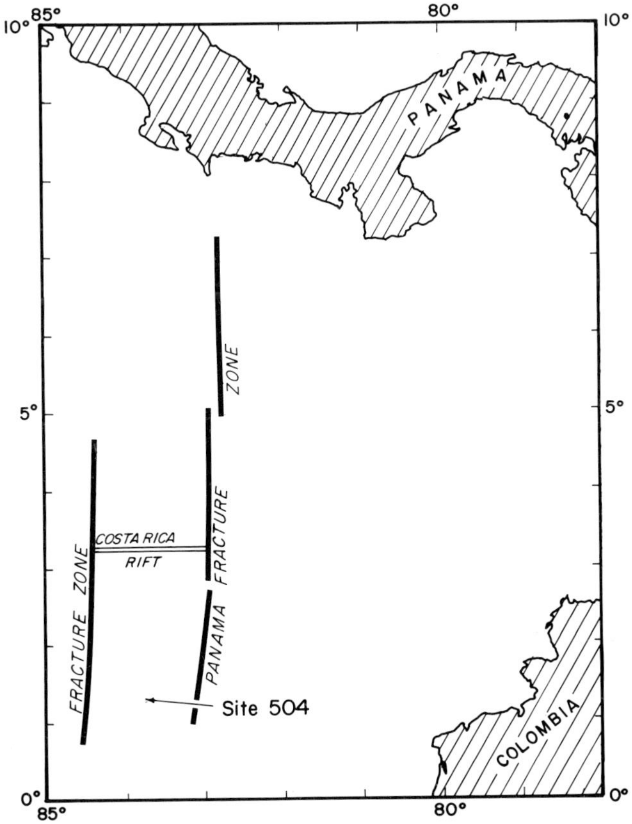


Fig. 1. Location of DSDP hole 504B in the eastern equatorial Pacific

drilled; an uppermost rubbly zone made up of intermixed pillow basalts, basaltic breccia, and flow basalts in the form of sills (called layer 2A), an intermediate zone of more thoroughly altered and cemented pillow basalts (layer 2B), and a lower unit of sheeted basaltic dikes (layer 2C). The following Schlumberger Well Services logs were run over the entire basement section of 504B: active-source neutron-porosity and gamma-ray density logs, spherically focused electrical laterlog, caliper, and four-channel, long-spaced sonic logs (Fig. 2). An additional run of the sonic tool was made which windowed-out the usual first-arrival compressional wave and recorded instead the shear wave velocity (Newmark et al. 1984). Two additional specialty logs were run in the hole: a large-scale electrical resistivity log (Becker 1982), and a United States Geological Survey (USGS) borehole televiewer log (Newmark et al. 1984).

Virtually every log registered a change in gradient (see Newmark et al. 1984) from the layer 2A pillow basalts in the upper 150 m, to the layer 2B pillows and breccias of the next 400 m, and finally into the sheeted dikes of layer 2C over the lower 500 m of hole 504B (Fig. 2). For example, note the change in slope at 4300 m in the compressional velocity,

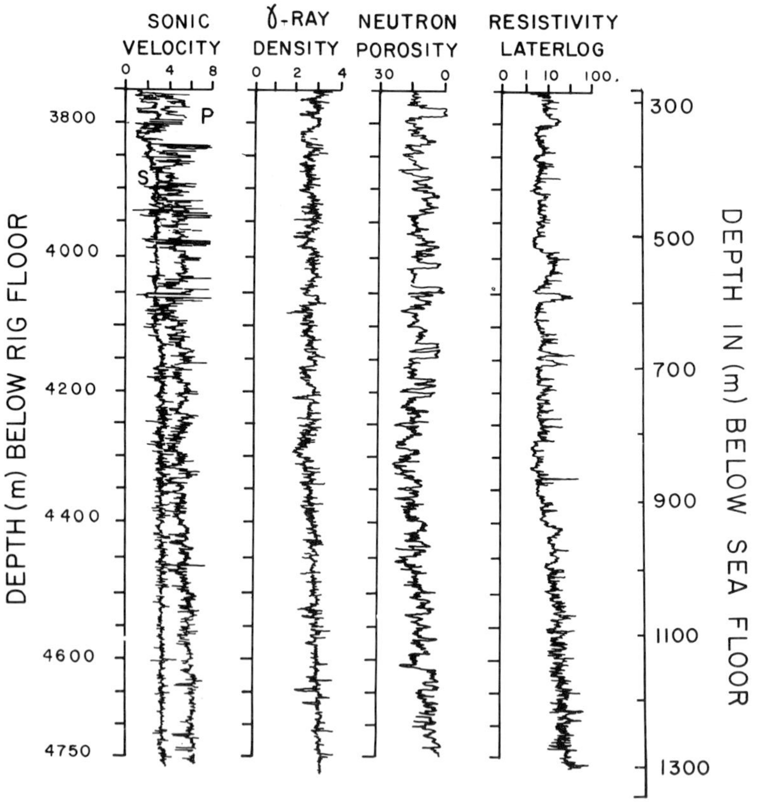


Fig. 2. Geophysical logs from Hole 504B. Sonic, nuclear and resistivity logs were from Schlumberger Well Services. Gradient changes in all the logs at 3900 m (400 m below sea floor) and at 4300 m (900 m) are layer 2A-2B and 2B-2C boundaries respectively

porosity, and resistivity of the crust. This boundary marks the deepest occurrence of pillow basalts and the first recovery of dikes within the cores at hole 504B, (Anderson et al. 1982).

## Estimate of Hydroxyl Mineral Content

The core recovery within oceanic basement on the Deep Sea Drilling Project averages 30% worldwide, and in the sheeted dikes at the bottom of hole 504B, it falls off to less than 15%. Even when the recovery is excellent, it is not easy to quantify either the amount of alteration or its detailed variation with depth in the crust. Bulk chemical analyses accurately measure hydroxyl mineral content, but can only be run on a few selected samples from the cores. The active-source nuclear logs offer a method of calculating the quantity of hydroxyl present over the entire wellbore length; a method soundly based in the nuclear physics of neutron and gamma-ray propagation.

With a method commonly called Density Logging, a Cesium-137 chemical source is lowered into the borehole and pressed against the wallrock with a hydraulic pad. Medium-energy gamma rays emitted by the source at several hundred keV interact with the formation by Compton scattering with electrons. Two gamma-ray counters spaced one and two feet, respectively, away from the source on the pad measure the gamma-ray intensity at those points. Two counters allow for the correction for borehole geometry effects by differencing the far-field counter and the closest counter, which will be more severely affected by near-field, borehole effects. Compton scattering is not dependent upon specific elements, but upon electron density of the formation.

The electron density of the formation has three components, the bulk density of the rock, the water present as porosity in pore spaces, and the water bound into alteration products as hydroxyls. The density log does not distinguish between the density of the alteration products versus that of unaltered rock. Instead, a bulk density of the formation plus porosity is determined by measuring the formation electron density. Electron density can then be converted to bulk-formation density as follows:

$$C \sim I\ e^{-d\ \mathrm{RhoB}}, \tag{1}$$

where C = count rate

I = a constant related to the strength and geometry of the source
d = a scaling constant
RhoB = formation bulk density.

Since cores recovered on the <u>Glomar Challenger</u> are routinely put through a gamma-ray density measurement in the laboratory aboard ship immediately upon sampling, and since the cored material is representative of the density of the country rock between fractures in the oceanic crust, this laboratory density minus RhoB from the gamma-ray log equals the porosity of the formation. Porosity is calculated as the ratio of Rho measured on cores minus RhoB to Rho measured on cores minus Rho sea water, since sea water is both the drilling fluid and the approximate composition of the formation fluid in pore spaces. The density log measures the porosity of the formation accurately, but does not determine hydroxyl mineral content, which would contribute to lowered density measurements of the laboratory samples.

Separately, an americium-beryllium neutron source is mounted with two helium 3 neutron detectors on another skid-pad within the nuclear tool configuration and run downhole as well. This source emits fast neutrons with average energies of about 4.5 meV. These neutrons are scattered and lose energy during collision with formation nucleii. Energy loss is inversely proportional to the mass of the struck nucleus. Stated simply, when a neutron hits something its own size, maximum energy transfer occurs. By analogy, if a marble hits a bowling ball, little energy is transferred. However, if a marble hits another marble its own size, maximum energy transfer occurs. Hydrogens have mass approximately equal to neutrons, and their mass is much less than any other element in the formation; so if we measure the energy state of near-field and far-field neutron counters (for borehole size compensation), we have a quantitative measure of the total number of hydrogens in the formation. Hydrogen present as hydroxyl in alteration minerals is measured just as accurately as hydrogen in free water in pore spaces. The neutron log then gives a fair representation of water present as porosity Plus that bound chemically as hydroxyls within alteration minerals.

Commercial logging services offer methods of calculating "Hydroxyl mineral content" from various cross-correlation techniques involving the entire suite of geophysical logs. Even though we recover only 15-30% of the rock cut by the bit, chemical analyses of these cores place a lower limit on the alteration present in the crust (it is assumed that the hardest, least altered rock is preferentially cored while the highly altered rock crumbles before the drillbit). The alteration content from the neutron and active gamma-ray logs using the commercial "dual-water model" was first calculated for hole 504B (MSI, or minimum shale index in Fig. 3). This curve is calculated from an AVERAGE of neutron and density-log porosities (NPHI and DPHI) at each depth point. The lesser of two values, this average (called PHIA) or neutron porosity alone (NPHI), is selected at each point to construct the MSI curve.

This curve did not resemble the degree of alteration observed in the borehole cores. Specifically, a mineralized stockwork was drilled at 4225 to 4300 m below the rig floor. This stockwork was composed of up to 60% hydroxyl minerals. However, the MSI curve shows this zone to be unremarkable. Also, the cores reveal that alteration steadily increases from layer 2A through layer 2B to peak at the stockwork. The MSI curve does not show such an alteration increase, (Fig. 3).

The averaging and minimizing scheme which results in the MSI curve does not make physically sound use of the differences between porosities detected by the neutron versus gamma-ray counters. We have devised a differencing scheme to identify hydroxyl mineral content of the formation, (Table 1). We difference NPHI and DPHI and scale to the hydroxyl mineral % present in the rock using the lab-log density difference. Our alteration curve faithfully reproduces both the extent of alteration within the stockwork and the increase in hydroxyl-mineral content toward the bottom of layer 2B, (Figs. 2 and 3). A slight increase in alteration at the base of the dike sequence is also seen both in our curve and in the cores, (Alt and Laverne 1984).

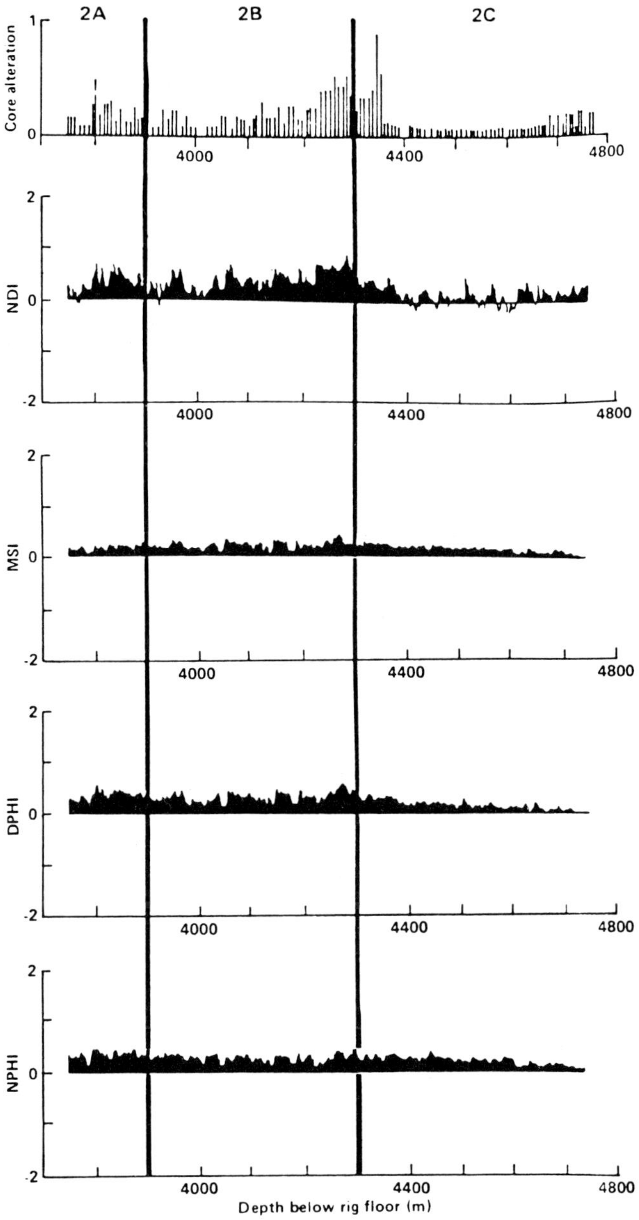


Fig. 3. NPHI is the porosity calculated from the neutron log, DPHI is the density calculated from the gamma-ray density log, MSI is the minimum shale index from a dual-water CYBERLOOK program, and NDI is the hydroxyl mineral content log calculated from the differencing scheme from Table 1

## Fracture Log from the Borehole Televiewer

Fractures intersecting the borehole are clearly seen by the borehole televiewer, as is the ultrasonic acoustic reflectivity of the wellbore which is useful for identification of void spaces and pillow boundaries (Zoback and Anderson 1982). Fractures were counted in 1-m overlapping intervals down the hole. A "wiggle" display of these and the apparent reflectivity of the formation as measured by a photo-

Table 1. Calculation of NDI, the neutron-density log index of hydroxyl mineral content

| | | |
|---|---|---|
| X | = | NPHI - DPHI |
| RGHAX | = | (1 + 1.71[ 1-2PHIA+NPHI] ) / 1 - X |
| SHLNX | = | RGHAX - GDCL / GDSH - GDCL |
| NDI | = | SHLNX x ( 2-SHLNX ) |
| where | | |
| NPHI | = | neutron porosity |
| DPHI | = | gamma ray density porosity |
| PHIA | = | NPHI + DPHI / 2 |
| GDCL | = | density of clean, unaltered rock |
| GDSH | = | density of alteration material |

graphic light meter are shown in Fig. 4, along with the other logs run in the hole. A discussion of the borehole televiewer fracture and reflectivity log as well as the wellbore images is given by Newmark et al. (1984).

Zoback and Anderson (1982) found that over the uppermost 500 m of the hole, the intensity of observable fracturing imaged by the televiewer actually increased, in contrast to hypotheses that burial pressures would close fractures as lithostatic pressure increased. They further found that an increase in sonic velocity with depth correlated with an increase in clay infilling of these fractures instead of with a decrease in the degree of fracturing.

We have extended the Zoback and Anderson (1982) analysis another 500 m into the sheeted dikes of layer 2C, and find that in the dikes the degree of fracturing does begin to decrease as does the amount of alteration (Fig. 4). A corresponding increase in the sonic-velocity gradient accompanies these decreases in fracturing and alteration level, but we doubt these changes are related to the closing of fractures by increase of overburden pressure. Rather, a lower degree of primary fracturing upon cooling is likely.

It should be emphasized, however, that the borehole televiewer sees only large fractures with apertures at least 1 mm wide. The extent of microfracturing often seen in cores returned to the surface from the oceanic crust never shows on borehole-televiewer logs. To the first order, the correspondence between fracture and alteration changes seen in the logs and seismic and sonic-velocity gradients observed in the crust suggest that macro- rather than micro-fractures affect these velocity determinations. The similarity of velocities determined from the very different wavelengths of seismic and sonic energy transmission led Moos and Zoback (1983) to a similar conclusion in wells drilled through crystalline basement on land.

## Geophysical Log Cross-Correlation Using the Phase Coherence Technique

Now that we have developed the methodology for the determination of quantitative measures of the change in extent of fracturing and degree of alteration change in the borehole, we can then turn to the correlation of these Geological processes to the Geophysical observations made in the well. For example, qualitative correlation can be seen between the extent of fracturing and degree of alteration in the hole, and variations in all the other logs. The gradient change at the layer 2B, 2C boundary is seen both in the borehole televiewer logs and in the alteration content log, (Fig. 4).

Another qualitative correlation which is obvious from a visual inspection of the log results is that the amount of attenuation apparent in the sonic waveforms is related in some manner to the degree of alteration seen in the borehole by the nuclear-tool differencing scheme (Fig. 5). Consider the sonic waveform change at 4050 m, or approximately 550 m subseafloor. Here, a massive, solid flow basalt has little hydroxyl-mineral alteration, and the waveforms are much higher amplitude with a longer wavetrain than above or below. But the amount of "clay" also correlates somewhat with the fracture-count log from the televiewer (e.g., at 4300 m or about 850 m below sea floor, where a drop in degree of fracturing corresponds to a drop in clay content as well). So, are the waveform changes that we see responding to alteration changes downhole, or to fractures, or to fractures filled with clay? If we could determine which, we would have a powerful new diagnostic tool for use in oceanic crust.

## Phase Coherence

A method is needed to cross-correlate two logs which may have variable degrees of correlation at different wavelengths down the well. We know that the various logs all show a gradient change at the layer 2B, 2C boundary, so they would all correlate strongly at wavelengths approaching the depth of the well (500-1000 m wavelengths). The question of interest here is how well do the various logs correlate at the shorter wavelengths where extent of fracturing or alteration changes happen in the formation. We can place a lower limit on our scale of detection, because heave of the ship produces inaccurate depth determinations at a few meter wavelengths and smaller. Cross correlation at these small wavelengths is hampered by inaccurate depth location.

A technique which produces a quantitative correlation coefficient at each wavelength between two wavetrains is Phase Coherence. This method is only useful for portions of logs where it is suspected that one process is affecting two logs to be cross correlated. Fast Fourier Transforms of the two wiggle plots versus depth are taken. Then the *Phase-At-Each-Wavelength* between two curves is compared. The percent of correlation between phase is then displayed at each wavelength. Wavelength here is a harmonic of the length of well under consideration. At each wavelength with 50% correlation, the two curves are in phase half the time and out of phase half the time. This corresponds to a random correlation coefficient of 0.0. If the two curves are exactly in phase all the time, the phase coherence at all wavelengths is 100%, or the correlation coefficient is 1.0. If however, the two curves are exactly out of phase everywhere, then the phase coherence is 0%, which corresponds to a correlation coefficient of -1.0.

The phase coherence method first breaks each log into its harmonic components, then correlates each wavelength sequentially, displaying the correlation coefficient at each wavelength as a %. The correlation coefficients are then connected by a line to present the degree of correlation over the complete suite of wavelengths contained within the two logs. Thus, if over all wavelengths the correlation is 50%, then the two curves in question are randomly related.

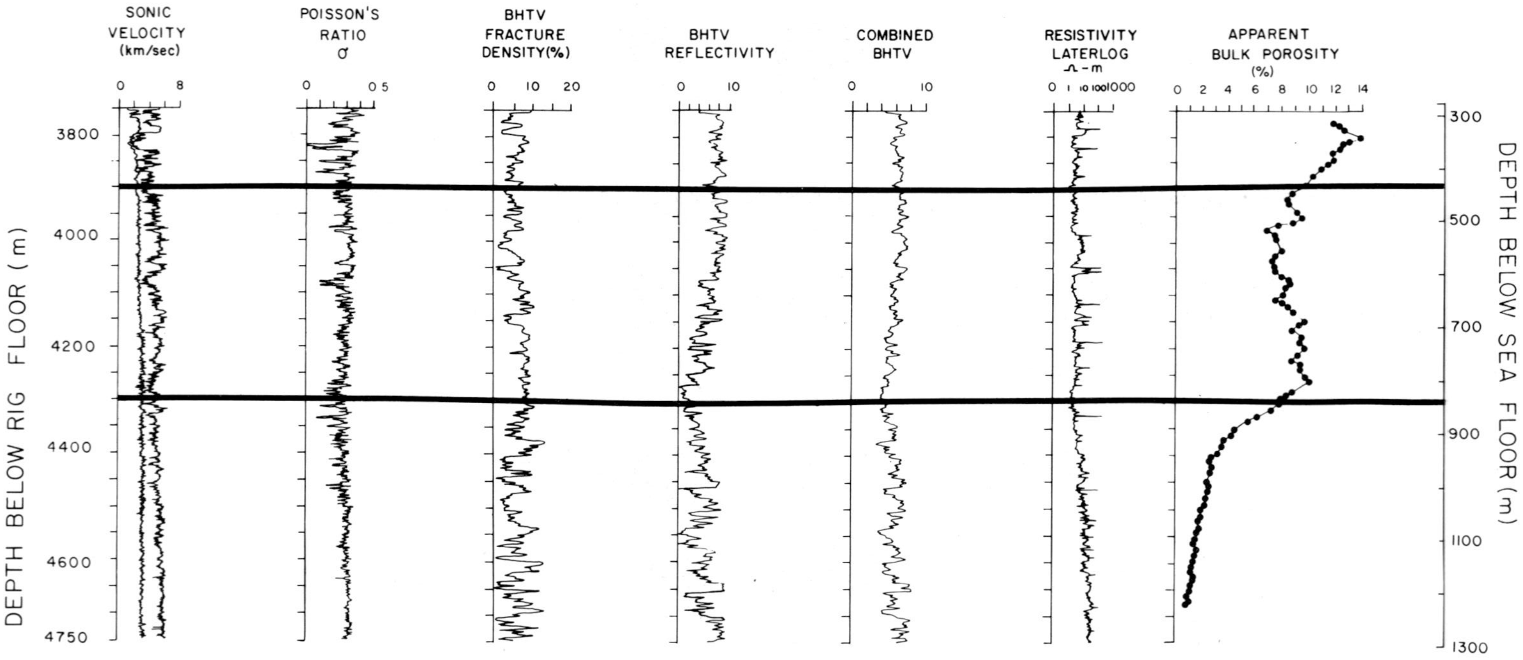


Fig. 4. Specialty logs from hole 504B. Poisson's Ratio calculated from compressional and shear velocity logs. BHTV logs were from a United States Geological Survey borehole televiewer. (Newmark et al. 1984)

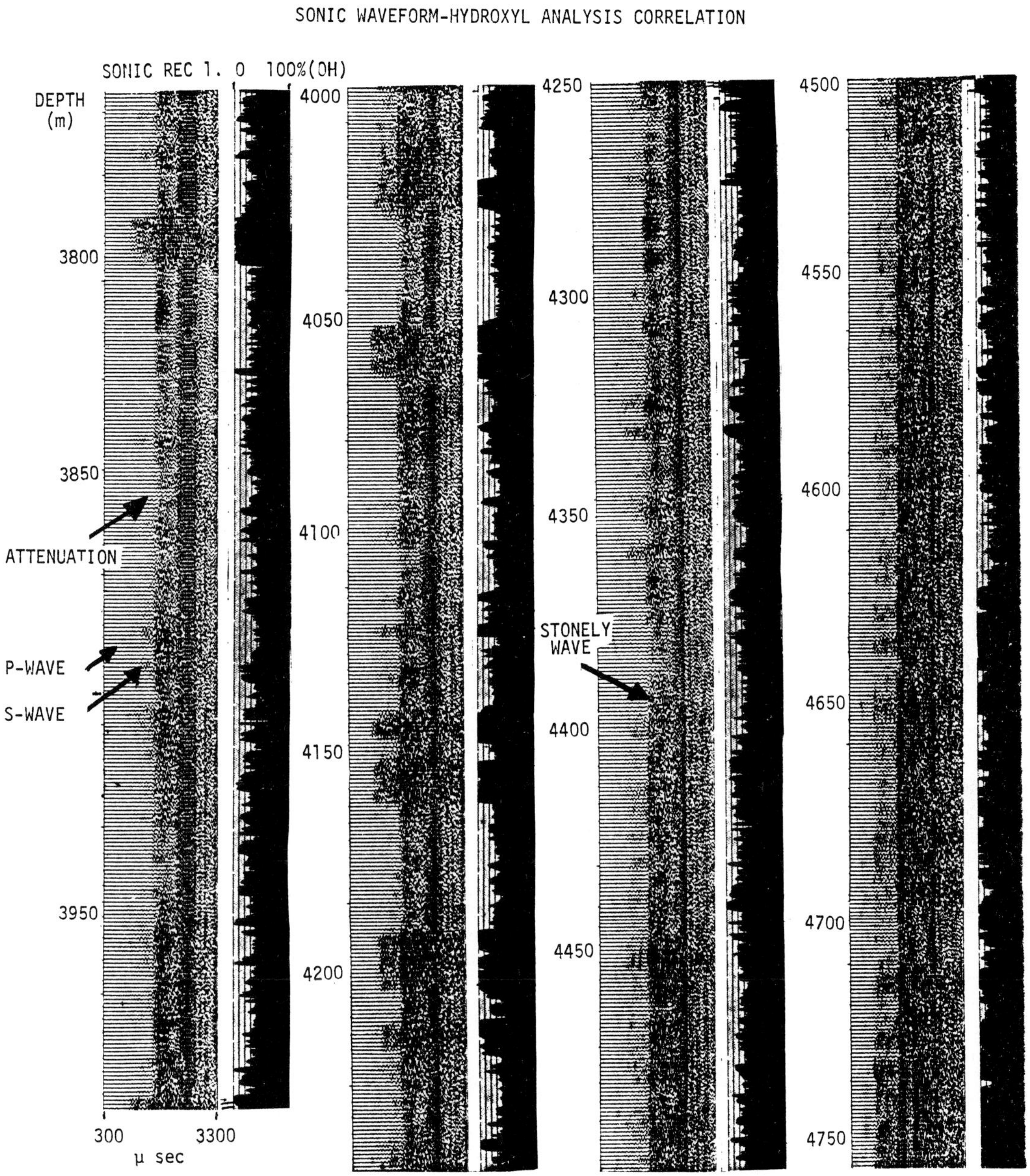


Fig. 5. The raw sonic waveforms from the first receiver located 8 feet above the source are displayed with the hydroxyl-content log NDI from Fig. 3. Blackness is the percentage of matrix material, horizontal bars are every 10% of alteration products present (e.g., at 4100 m depth, 30% altered clays and zeolites and 70% unaltered basalt are present in the rock)

Consider now the pillow basalts in layers 2A and 2B of hole 504B. We can question whether there is correlation between alteration content and velocity in layers 2A and 2B, as might be indicated by Fig. 5. First, the phase coherence between alteration in these layers and the fracture pattern shows only random correlation at all wavelengths, (Fig. 6). Such a truly random-number-generated set of curves for the 550 m interval under question here shows that we should ignore wavelengths of greater than 100 m. At harmonics with 100 to 550 m wavelengths, there are not enough cycles to give accurate phase coherence information. Ship's heave and other depth-determination inaccuracies between different logging runs makes wavelengths less than about 5 m inaccurate as well.

Shear wave velocity also is randomly correlated with alteration content, but compressional velocity shows weak positive correlation at 10-15 m wavelengths and weak negative correlation at 50-100 m wavelengths, (Fig. 6). We expect the effect of alteration to be to lower the velocity of rock, but Zoback and Anderson (1982), showed that velocity increased from layer 2A into layer 2B at hole 504B because the upper crust was heavily fractured with open porosity in layer 2A, but lowered porosity in layer 2B from increased filling of fractures by alteration products.

## Bandpass Filtering

We can isolate the alteration effect on velocity by removing the effects of the large porosity drop in layer 2A by removing the trend of both the velocity and alteration content curves, then bandpass filtering the detrended data sets to pass only the long wavelength harmonics. From the phase coherence plot (Fig. 7), we place the bandpass filter designed to bring out possible negative correlation between alteration and velocity at 50-150 m wavelengths only, (Fig. 7). Use of detrended data at long wavelengths only eliminates the positive correlation caused by decrease in porosity over layer 2A and reveals the extent to which low velocity zones correlate with zones of high alteration mineral content. As can be seen, this correlation is predominant in layer 2B, and much poorer in layer 2A, as expected. It appears that 50-150 m wavelength variations in compressional velocity in layer 2B are caused by variations in the degree of alteration, as suggested by Zoback and Anderson (1982).

## Power Spectra, Attenuation and Fracturing in the Dikes of Layer 2C

The sonic waveforms contain much more information about the geology of the rock through which they pass than just velocity. The sonic-logging tool has a source which emits a broadband acoustic click with significant energy from 5 to 40 kHz. The energy travels from the tool through the borehole drilling fluid (sea water in all DSDP holes) as a compressional wave. Upon contact with the wellbore, complex waveform conversions occur. Major compressional and shear headwaves propagate into the formation, but the largest energy propagates along the wellbore as a tube wave called the Stoneley wave (analogous to the Rayleigh wave except that the Stoneley wave travels along a rock-liquid surface). Fluid waves called normal modes travel down the borehole as well.

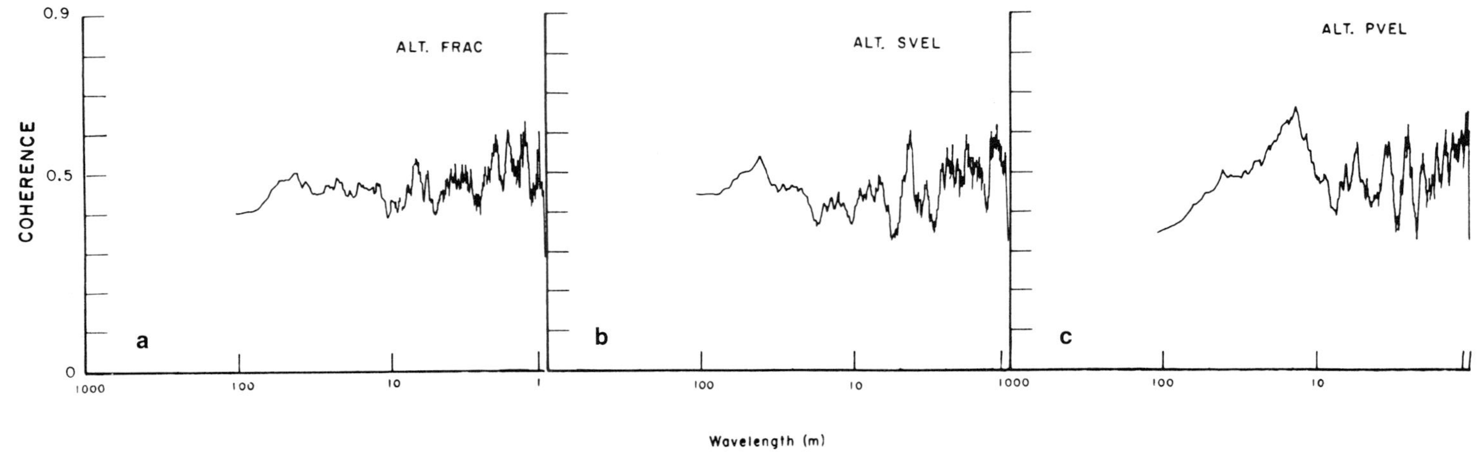


▲ Fig. 6. Phase coherence analysis (see text) of the cross correlation of a the alteration content and borehole telewiever combined BHTV logs; b alteration content versus shear wave velocity, and c alteration content versus compressional wave velocity. Phase coherence method fits FFT harmonics to depth curves in Fig. 4, then determines the degree to which two curves are in phase at each wavelength. Wavelength here is depth in the hole in meters. 50% correlation is random

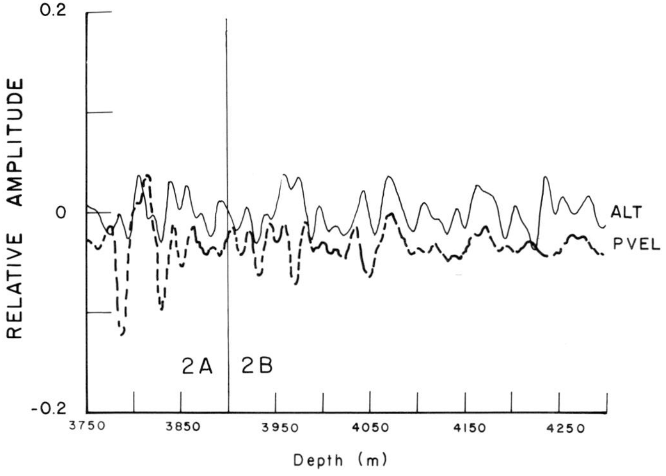


◀ Fig. 7. The compressional velocity (Fig. 4) and alteration content (Fig. 3, NDI) are normalized, detrended, and bandpass-filtered to display only wavelengths between 50 and 150 m. The alteration curve has been inverted to show correlation in layer 2B between high velocities and zones of high alteration mineral content

## The Fracture Spectrum of the Oceanic Crust at Hole 504B

In order to extract more quantitative information about what exactly the sonic waveforms are responding to, we took a Fast-Fourier Transform of compressional and shear-waveform energy packets called "CODA" (Fig. 8). The Schlumberger full-waveform sonic-logging tool used on DSDP Leg 83 returns waveforms recorded at 8-feet, 10-feet, 10-feet, and 12-feet spacing from source to receiver. The duplication is required because two different sources are used, although they are acoustically paired as closely as possible. The "Shear coda" contains not only a shear wave but other surface waves as well, such as a "Rayleigh wave" which is about 5% slower than the shear wave, but faster than the Stoneley wave, (Richards, 1984, pers. commun.).

By placing time windows across the waveforms so that first the compressional and then the shear coda con be isolated, power spectra of each coda from four waveforms at each depth interval down the well can be determined. It is then necessary to determine particular frequencies among the waveforms which might be noisy, and eliminate spurious energy from the analysis. This "noise" is of unknown origin; known causes are roughness of the wellbore, mechanical banging of the tool centralizers against the rock, conversion interference, coupling changes, or geological variations. The spacing between receivers is two feet, so we arbitrarily discriminate against differences in the spectra which occur over intervals of less than two feet, even if caused by real geology. We calculate the cross-product of the Fourier transform of four pairs of waveforms at each depth. The technique calculates the power spectrum of each waveform of the pair, emphasizing frequencies having large amplitudes common to both waveforms. That is, the cross correlation passes common energy in the time domain.

We then display the wideband *integrated energy* in decibels (dB) between 10 and 20 kilohertz of each waveform spectrum, and energy of peak cross-spectral frequencies as wiggle plots versus depth. We also display the spectra and cross spectra on grayness plots (the greater the energy in decibels at each frequency, the blacker the image) (Fig. 9, A, B, and C). We show both compressional and shear slownesses (microsec/ft) as well. *Center frequency* is automatically picked and marked by a white bar on the grayness plot. Again for information, borehole televiewer images are shown in Fig. 9 for selected portions of the hole.

This cross-spectral technique shows dramatic variations up and down the hole. Specifically, the entire pillow basalt interval has too low cross-spectral energy on all receiver pairs to yield consistent velocity estimates among all waveform pairs. We require an integrated energy of at least 20 decibels above background noise level from the cross spectra or we consider the individual spectra to be too different to yield useful cross-spectral information. A partial explanation for this observation is that fracturing and zones of alteration vary too greatly on a scale smaller than the two-foot smapling interval, so that the sonic receivers record very different waveforms. Consistent first motion detection required by the slowness estimater was also difficult to achieve. A sonic tool using many more than four receivers spaced at closer than two-foot intervals will be required to quantify sonic spectral changes in layers 2A and 2B of the oceanic crust at Hole 504B.

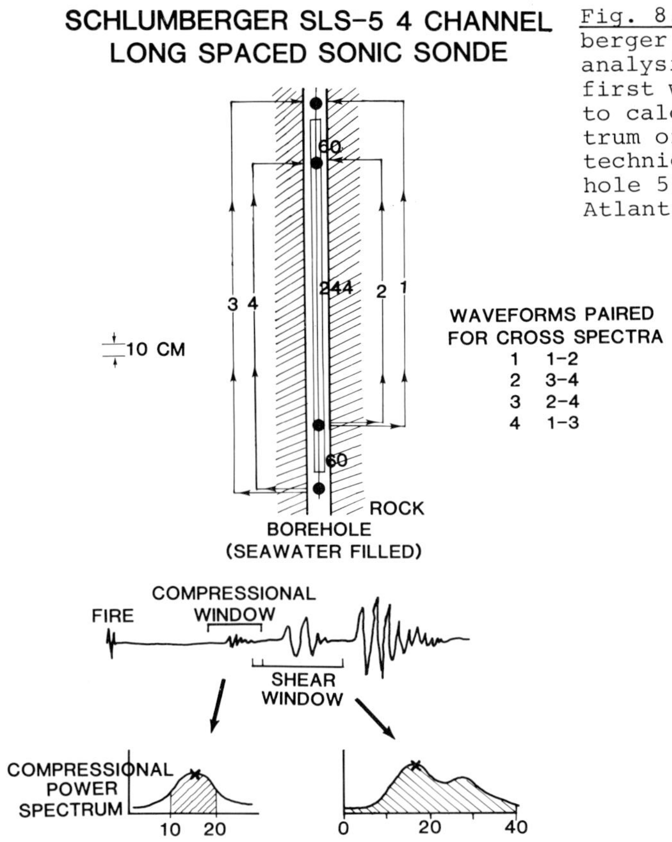


Fig. 8. Schematic of Schlumberger sonic logging tool and analysis technique used to first window waveforms then to calculate the cross-spectrum of waveform pairs. This technique was used at both hole 504B and at 556 in the Atlantic. x Center frequency

However, in the dikes at the bottom of 504B, there are significant energy and frequency shifts which must be explained. Consider, for example, zones 1-4 in Fig. 9C and D; zone 1 has low P (compressional) and S (shear) energy levels at all frequencies, zone 2 has high P and S energy, but at low frequencies; whereas zone 3 has low P energy at high frequencies but high S energy also at high frequencies. Zone 4 has high P but low S energy, both at low frequencies.

What causes these energy shifts? If we examine the qualitative correlation with borehole-televiewer imagery, we add somewhat to a possible solution. For example, zones 2 and 3 both appear as solid, highly reflective rock on the televiewer image (Fig. 9C). Yet zone 2 appears to have several large fractures crosscutting the image (black horizontal lines), whereas zone 3 rock has few fractures. Again zones 1 and 4 look alike in that both are mostly low-reflectance, highly fractured, (altered) rock.

The problem becomes more interesting when we consider the frequency shifts which accompany the energy changes. Consider zones 2 and 3 again. Zone 2 with its large fractures and high energy levels has very low P and S center-frequency, particularly when compared to zone 3 (Fig. 9D). The scattering effect of large fractures in crystalline media should be to act as a low-pass filter (e.g., Menke 1983), and this high energy but low-frequency behavior within zone 2 is exactly

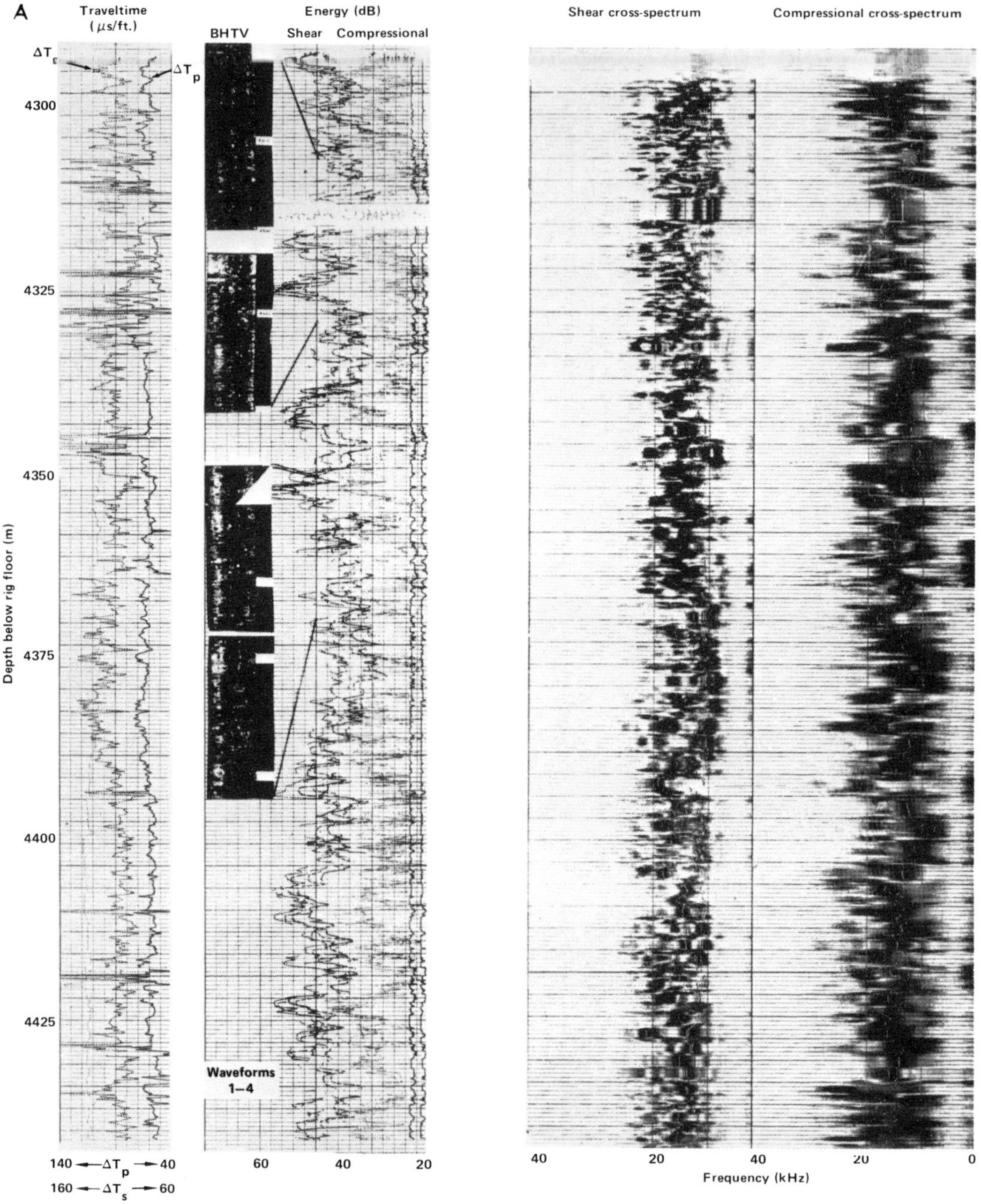


Fig. 9A-D. Integrated energy of windowed p-wave and s-wave coda displayed for all four cross-spectral pairs from DTCS program, a mark of Schlumberger Ltd., as in Fig. 8 in the dikes of hole 504B. On the right is the actual spectrum at each depth with grayness keyed to power in dB. The white tick at peak power is the center frequency. On the left, representative borehole televiewer images (Newmark et al. 1984) are displayed. In C, four zones are identified in which major changes in energy and center frequency can be related to televiewer indicated zones of fractures. In D, stylized schematic power spectra of compressional and shear energy in zones 1-4 are shown

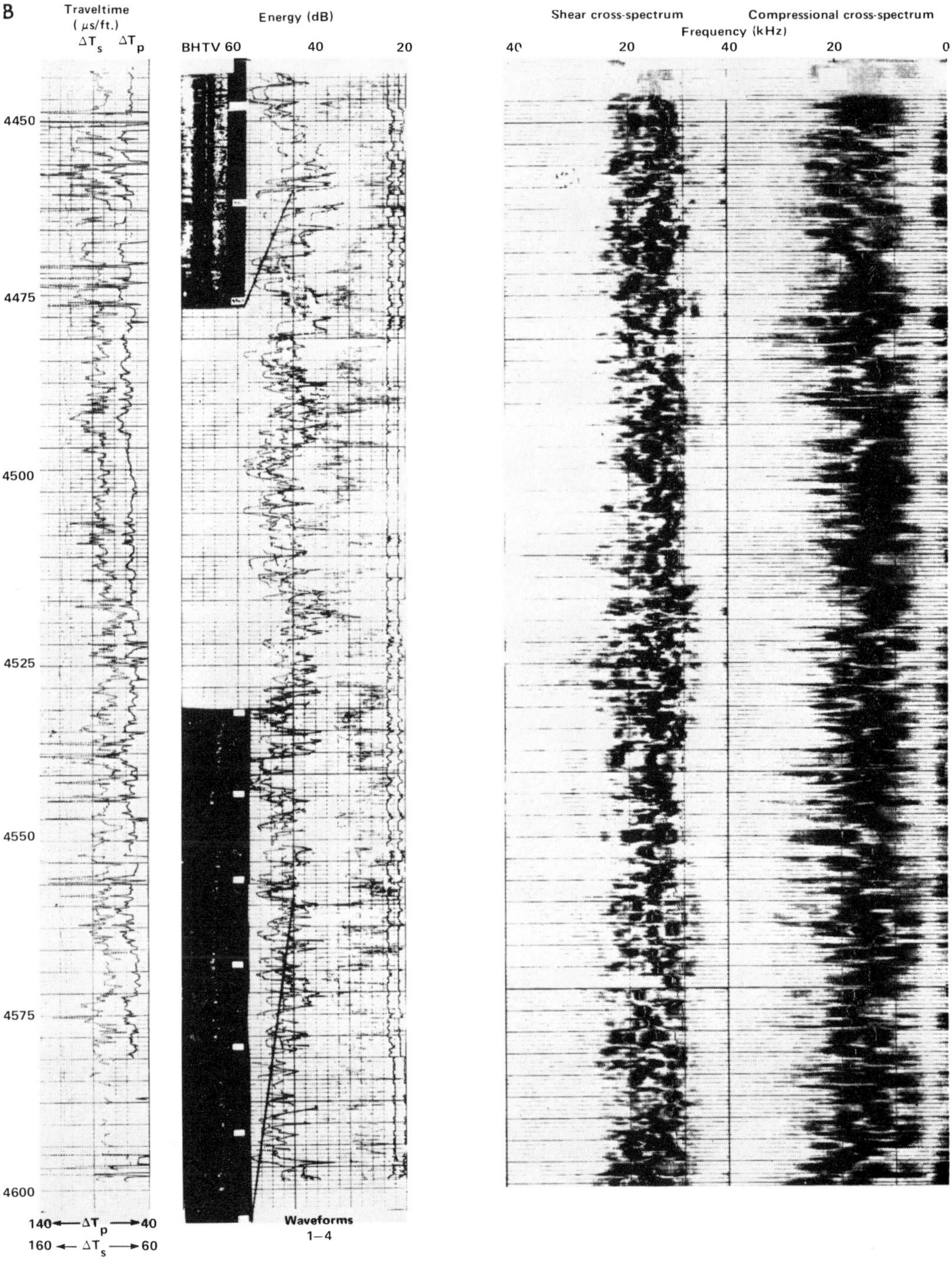


Fig. 9B. (Figure 9C,D see page 198 and 199)

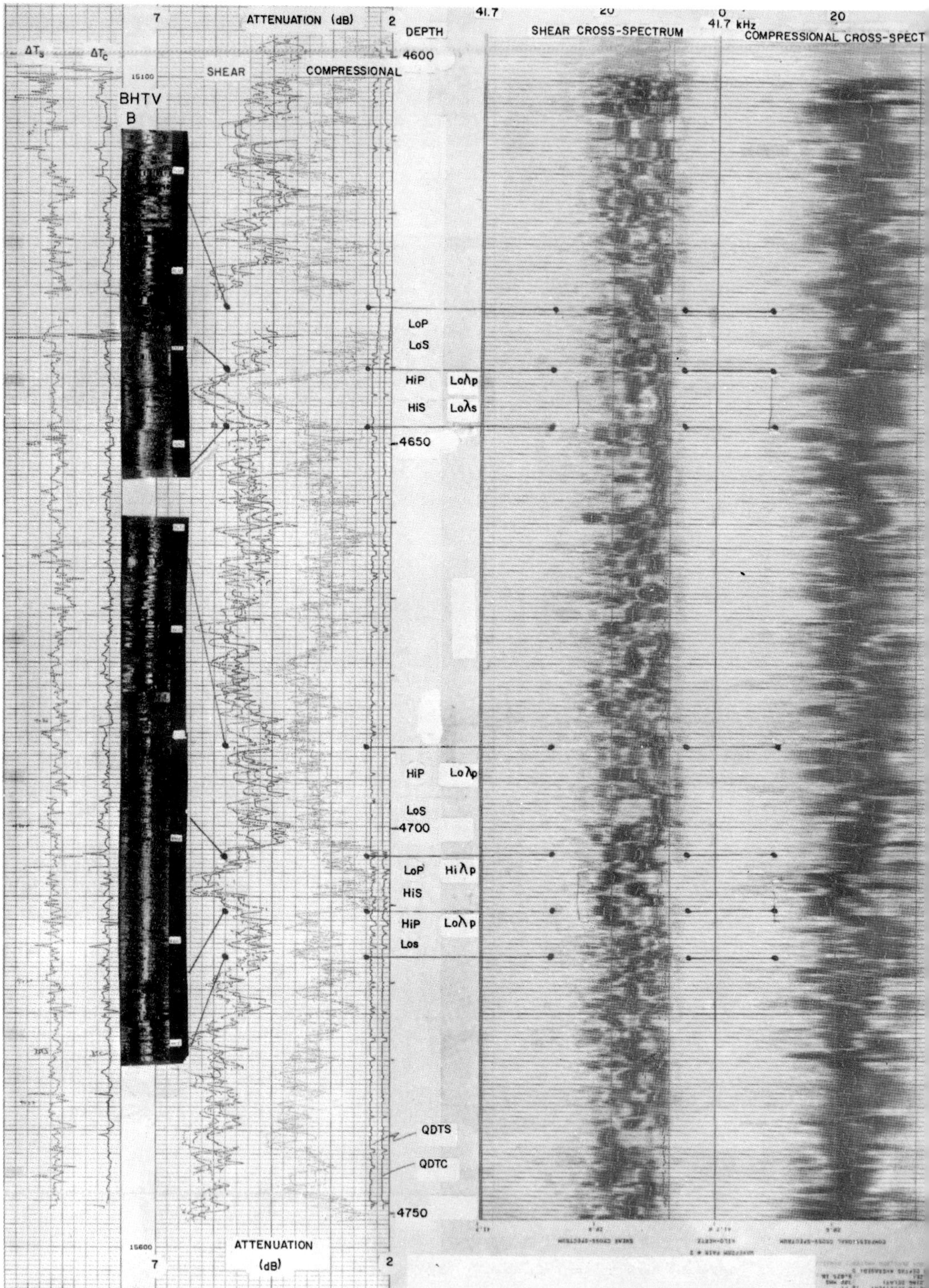
ATTENUATION (dB)
7
2
DEPTH
41.7
20
SHEAR CROSS-SPECTRUM
0
41.7 kHz
20
COMPRESSIONAL CROSS-SPECT
ΔTs
ΔTc
4600
SHEAR
COMPRESSIONAL
BHTV
B
LoP
LoS
HiP
Loλp
HiS
Loλs
4650
HiP
Loλp
LoS
4700
LoP
Hiλp
HiS
HiP
Loλp
Los
QDTS
QDTC
4750
ATTENUATION
(dB)
7
2

Fig. 9C.

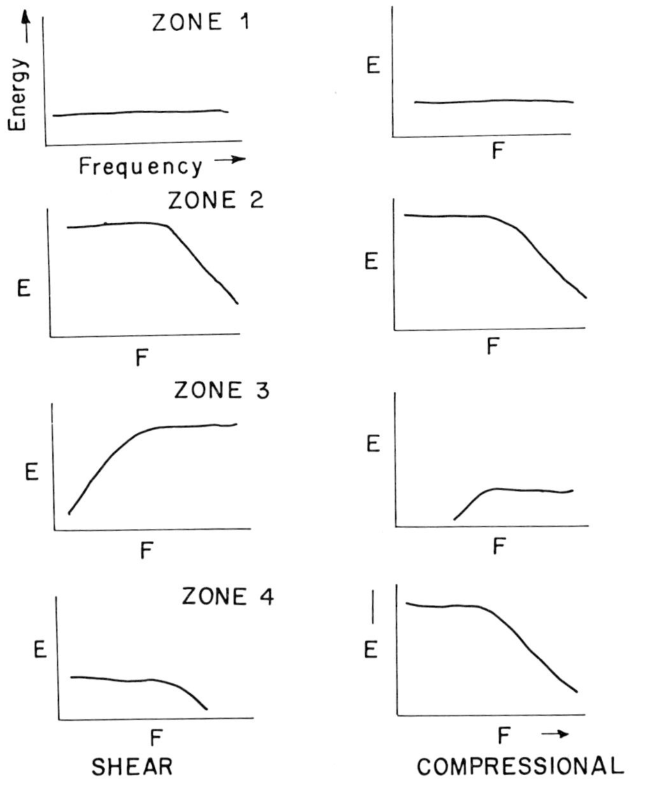


Fig. 9D.

as if from such a *low-pass filter*. Zone 3 is dominantly high-frequency P, but has low P-energy levels as well; and shear energy is high at high frequencies. Zone 3 is being *high-pass* filtered. No fractures are evident in the televiewer log. By contrast, zone 4 has low S-energy, and also low S-frequency, but has high P energy at low frequencies. This is as if from a low-pass filter (Fig. 9D). What physical property could be causing the impedance changes which result in some zones being high-pass filtered but others being low-pass filtered?

Another interesting observation from the cross-spectra has to do with the resonance character of the borehole/rock system. While the compressional cross-spectra show resonance about a clearly defined center frequency, the shear spectra are very different. Multiple resonances are seen throughout the dike interval, particularly in highly attenuated zones (e.g., in Fig. 9C, single-peaked resonance is found in zone 2, but multiple resonances are found in zones 1,3, and 4).

It becomes quite apparent that a means of systematically examining the relation of fractures to the cross-spectra must be devised if we are to make much sense of the changes we see in the waveform spectra. For this purpose, we constructed a "fracture-spectrum" log (Fig. 10). Here, we display percent fractures as a wiggle plot versus depth as in Fig. 2, but not only with P and S velocity and caliper, but also with P and S integrated energy and center frequency plotted as well. We take the center frequency within zones of multi-resonance as the frequency at which the highest energy occurs, regardless of the energy lobe within which it occurs (Fig. 8).

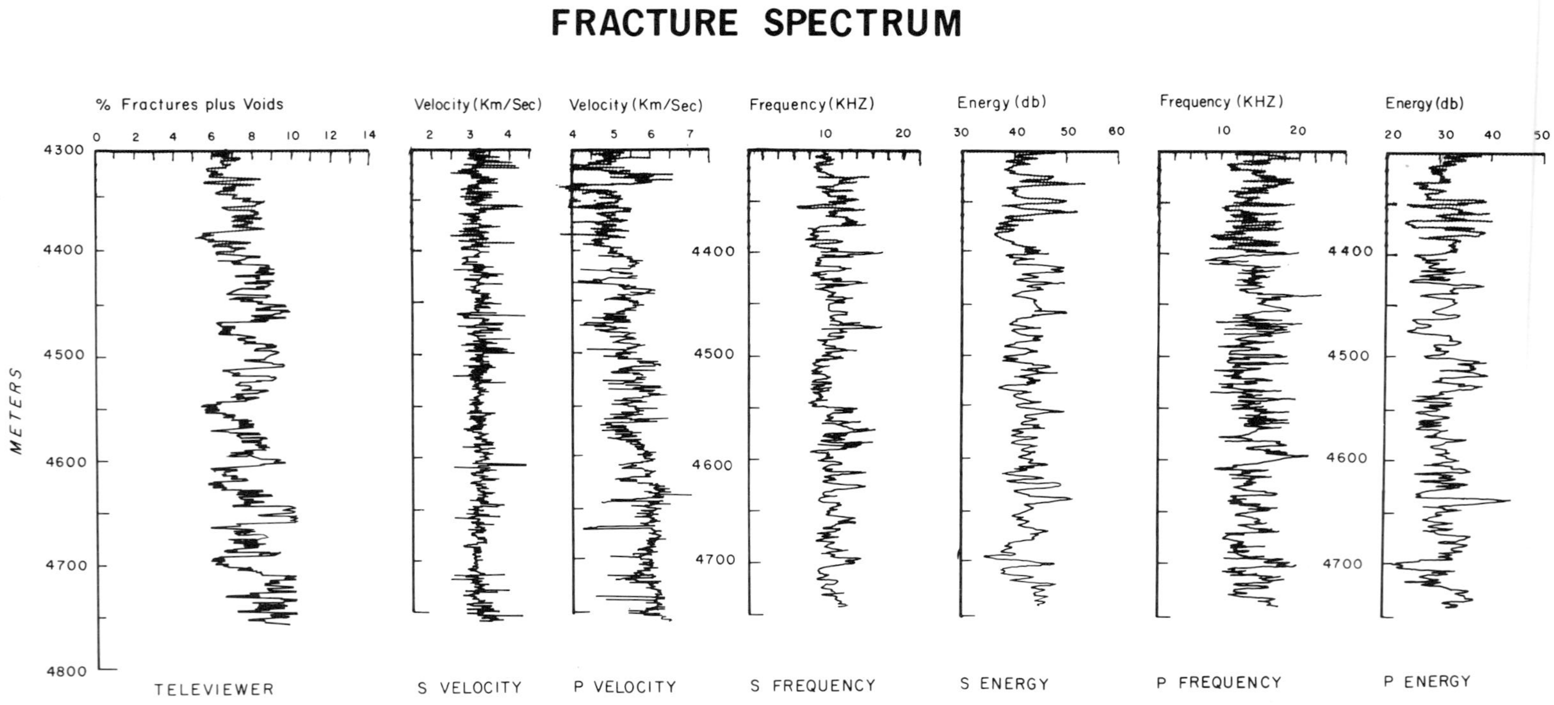


Fig. 10. Fracture spectrum logs: p-wave and s-wave velocities, caliper, and borehole-televiewer fracture and void count from Fig. 4 for hole 504B, and energy and frequency of p-wave and s-wave coda from Fig. 9 are shown in the dikes and transitional basalts at the top of the dikes (shaded depths). Only within the shaded zone is the hole caliper showing deviation from gauge

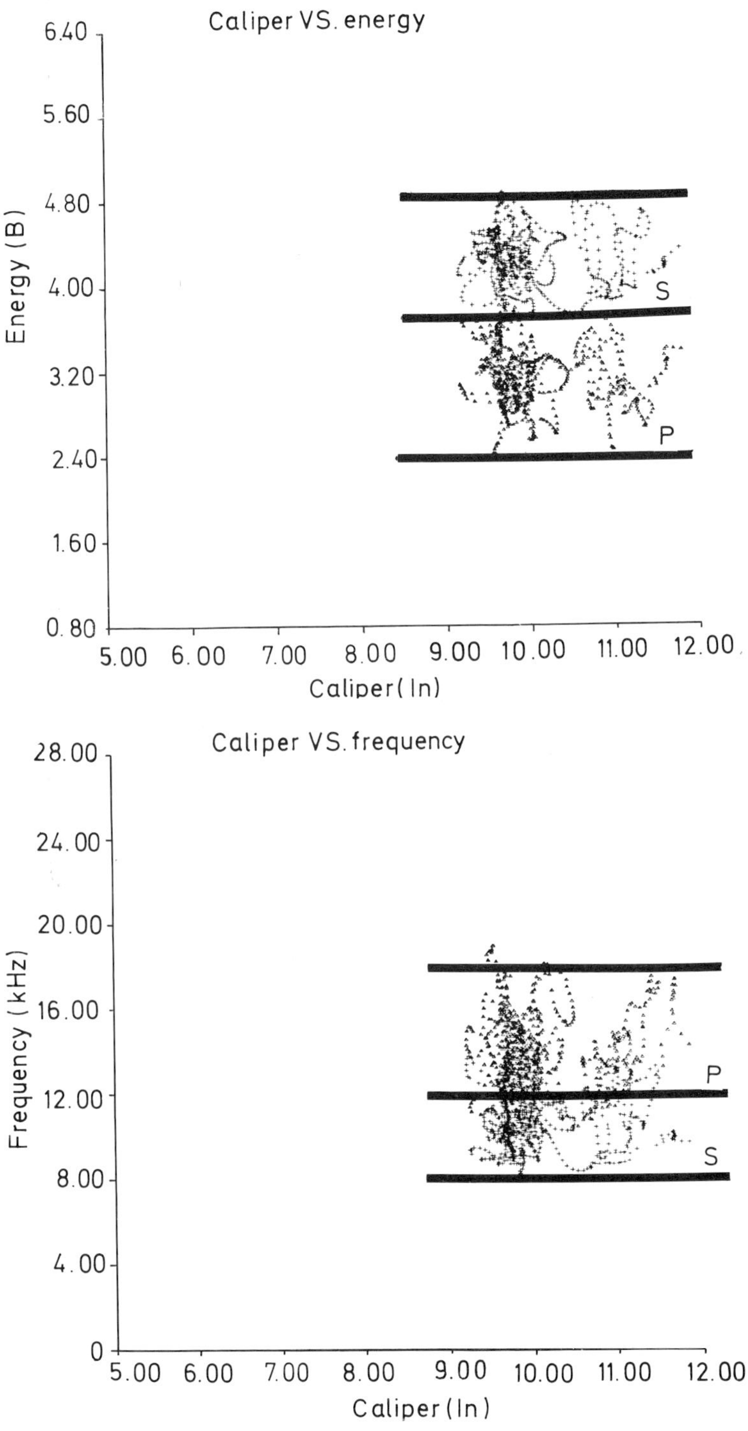


Fig. 10A. Caliper in inches cross-plotted against compressional and shear coda energies and frequencies to demonstrate no correlation in the dikes between borehole size changes and spectral changes

First of all, we find no correlation between compressional and shear coda center frequency or energy and caliper log variations (Fig. 10A). We do find correlation between coda-energy attenuation and borehole fractures from the televiewer log, however. We can verify that this correlation exists even though the data are variable by again using the phase-coherence technique (Fig. 11). In these phase-coherence plots, positive borehole televiewer correlation with high P- and S-energy is found in zones in layer 2C; this translates to a correlation between depths with high P- and S-energy levels and zones with a small degree of fracturing from the borehole televiewer log (that is, the fracture log has been inverted so that zones of fracturing are negative numbers and zones of solid rock are positive numbers. Thus positive correlation implies high energy at zones of low fracturing).

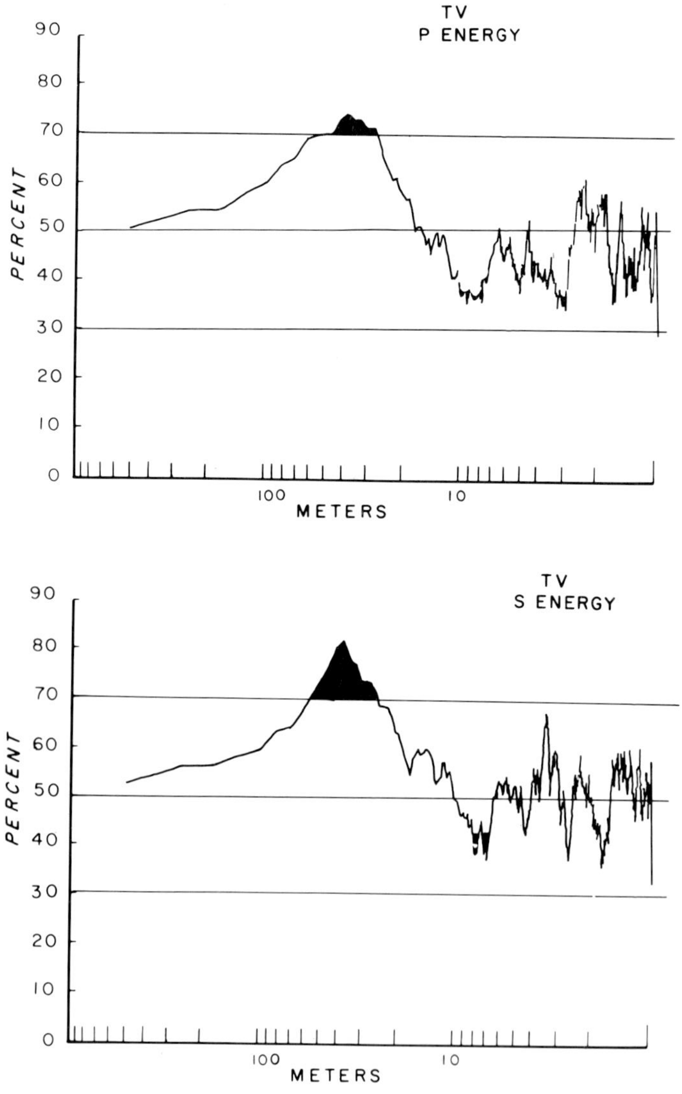


Fig. 11. Phase coherence of tv (borehole-televiewer fractures plus voids) and P and S energy at hole 504B from Fig. 10. Positive correlation is between high energy and zones of low fracture level

We find no correlation between degree of alteration and compressional and shear coda energy in the dikes, (Fig. 11B). The stronger correlation between energy and fractures than energy versus alteration-extent implies that fracturing is the dominant process within which the spectra are changed in the dike section of the hole. Such might not be the case in the heavily altered layer 2B pillow basalts, but we were not successful in examining the spectra within this layer, as detailed above.

Returning to the "fracture-spectrum" data presentation, we can make traditional log analysis cross-plots as well (Fig. 12). We find rather surprising results. P energy is low when P frequency is high, as would be expected from low-pass filtering caused by scattering due to fracture interfaces (Menke 1983), but shear coda energy is low when S frequency is low, the opposite to that expected from scattering.

Shown another way, we find positive coherence at the 60 to 70% level between Shear frequency and energy, and 30 to 40% coherence between compressional energy and frequency at about 7 to 70 m wavelengths (Fig. 13). The latter means that 60 to 70% of the time P energy is high when P frequency is low. Correspondingly, 60 to 70% of the time shear energy is high when shear frequency is high. These phase coherences, while only in the 60 to 70% correlation range, corroborate what our eye sees in the cross-plots (Fig. 13 versus Fig. 12).

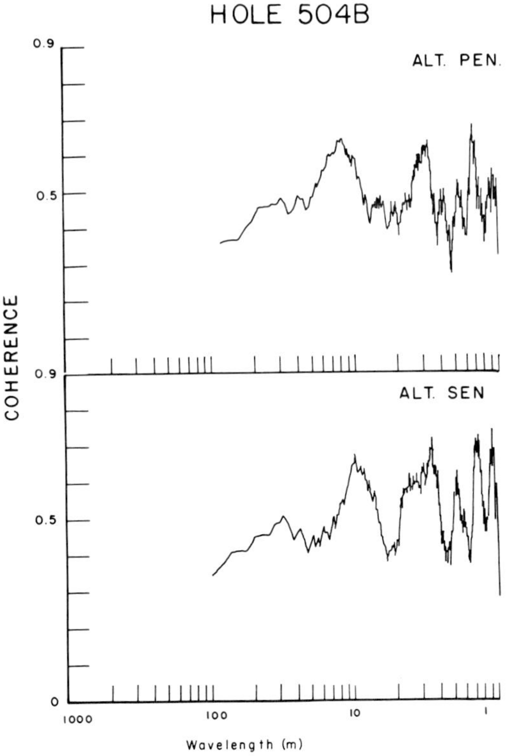


Fig. 11A. Phase coherence between alteration content and compressional and shear coda energies over the dikes from 4300 to 4750 m into hole 504B

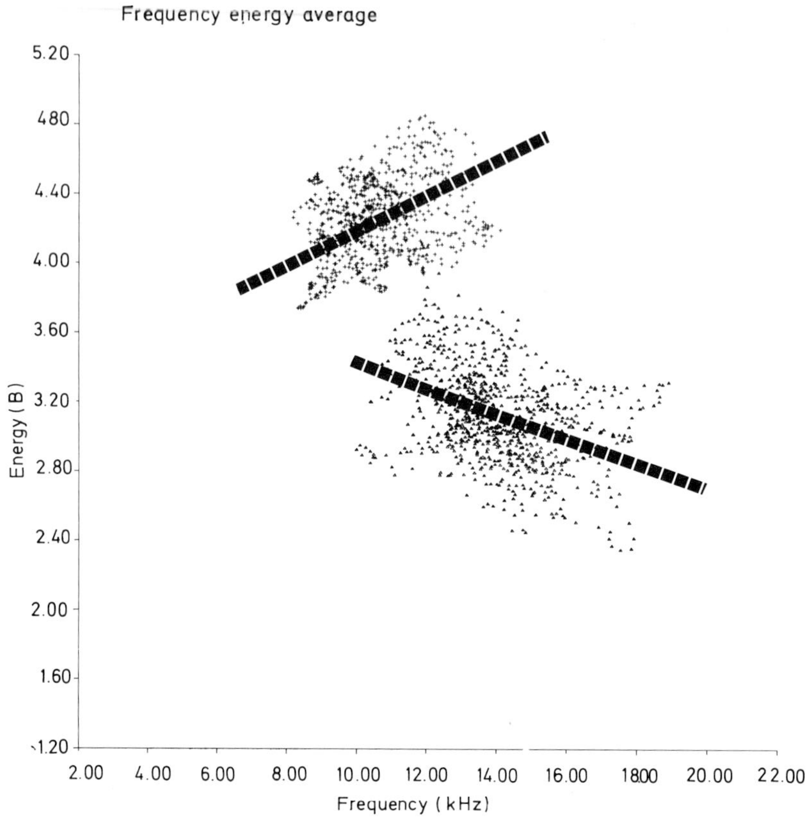


Fig. 12. Both compressional (Δ) and shear (+) energy cross-plotted against frequency from Fig. 10 for hole 504B. While the data scatter considerably, trends indicated by the heavy dashed lines are evident

This difference between the spectral responses of compressional and shear wave propagation becomes easier to explain if we transpose our energy-frequency spectra to energy-wavelength space. This wavelength is that of the sonic energy itself, and as such is completely different from the wavelength of phase coherence up and down the hole discussed previously. Here, wavelength is calculated by dividing velocity by frequency at each measurement point. If we then display energy versus this sonic wavelength, we see a clear deliniation of the spectral response we have observed in the dikes at the bottom of Hole 504B, (Fig. 14). An anti-resonance or energy loss at characteristic wavelengths within the dikes is observed. Energy propagating at 30- to 40-cm wavelengths is being attenuated by the rock. Since we also observe a correlation of low-energy levels with zones of fracturing, (Fig. 11), we can say that energy with these wavelengths is being lost into the fractures intersecting the wellbore within these zones of intense fracturing. We can then discuss the length of fractures which would be required to produce such a natural anti-resonance, *if and only if we are certain that the anti-resonance is caused by geology and not by tool interference with the wellbore or other geometric effects*. Dimensions of fracturing carries information about permeability, both hydraulic and acoustic, since intersection of the wellbore by fractures of a certain characteristic length might be required to produce a naturally occurring anti-resonance. How then can we eliminate the possibility that tool geometry is causing the observed loss of 30-40 cm wavelength energy

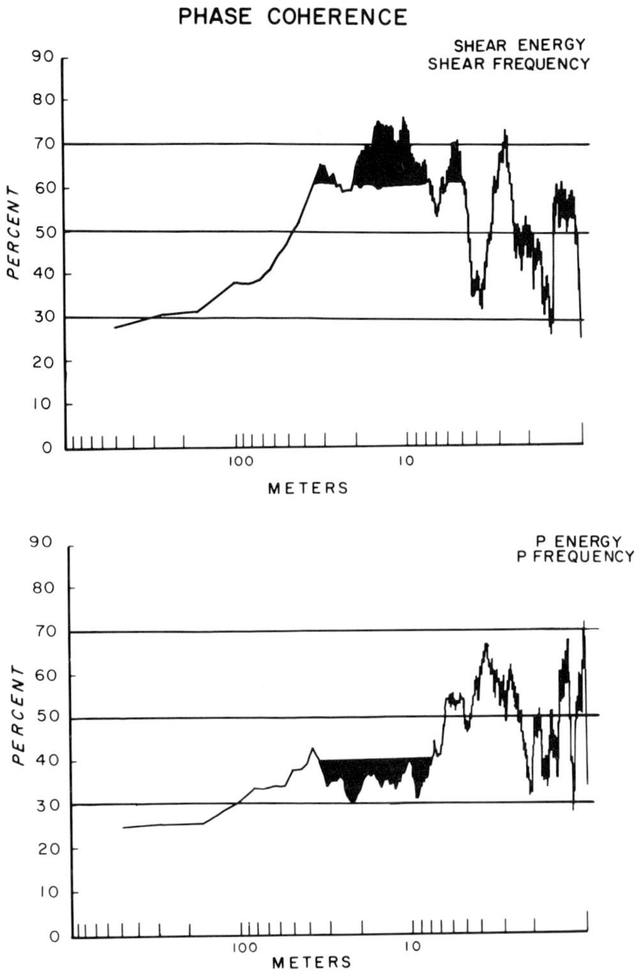


Fig. 13. Phase coherence analysis of the cross-plot shown in Fig. 12 for hole 504B, verifying a 60 to 70% correlation rate between high shear energy and high frequency and a similar correlation rate between high p energy and low p frequency. Wavelengths greater than 70 m are distorted by the ends of the log – that is, the log is not long enough for accurate phase coherence calculations at these wavelengths

One way is to turn to another hole drilled into a different geology, but one logged by the same Schlumberger tools. If the same anti-resonance is found in that hole as in 504B, then we are almost certainly looking at a tool-generated phenomenon; if however, the filter characteristics at such a well are different from those at 504B, then geology becomes the suspected cause of the anti-resonance.

## Geophysical Logs from 17-Million-Year-Old Mid-Atlantic Ridge Crustal Hole 556

As surprising as it may seem, Hole 556 from the Mid-Atlantic ridge flank (Fig. 15), is the only other borehole with both significant penetration into the oceanic crust, and a set of high quality, Schlumberger geophysical logs comparable to those run in Hole 504B in the over 620 holes drilled by the Deep Sea Drilling Project. (We also

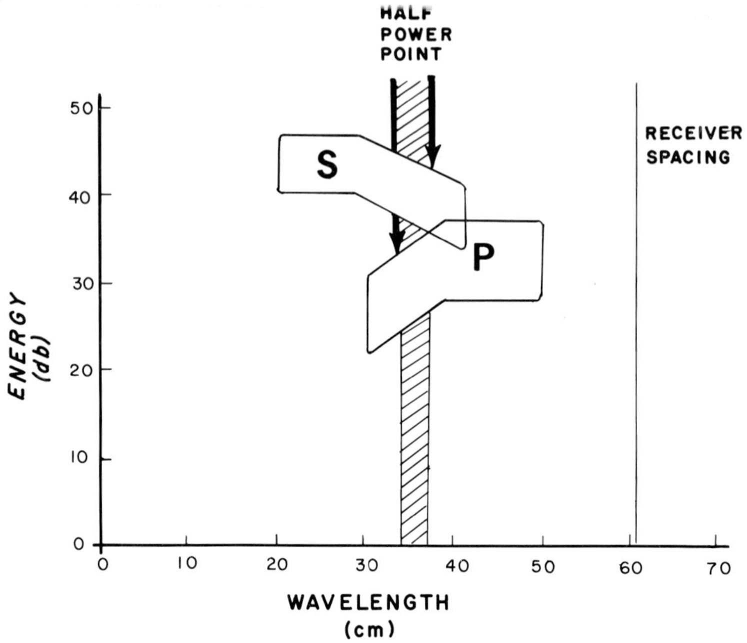


Fig. 14. The energy-frequency crossplot of Fig. 12 can be transposed to an energy-wavelength crossplot by calculating velocity divided by frequency at each data point. The data for hole 504B then fall within the envelopes shown above. The half-power point is the 3-dB-down point of each data cluster. We interpret the energy drop at 30 to 40 cm wavelengths to be from anti-resonance

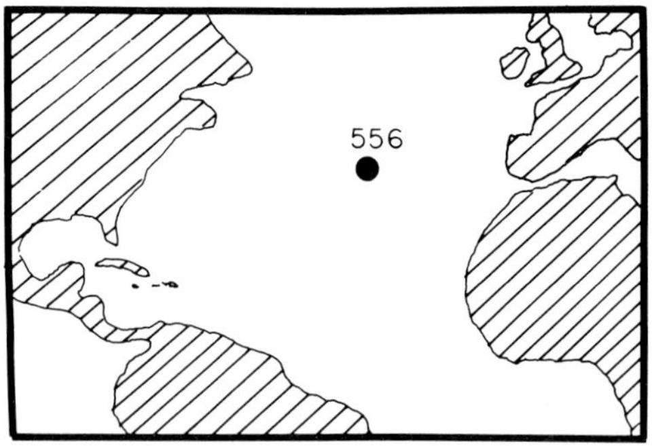


Fig. 15. Location map for DSDP hole 556 drilled at 38 56.38 North latitude and 34 41.12 West longitude in 3672 m of water depth. Total penetration was 639 m, 461.5 m into calcareous sediments and 177.5 m into basalt and gabbro

examined holes 558 and 564 drilled on the same leg, but found the crustal section in each to be too short to generalize – 65 m and 50 m respectively). Yet the geological contrast provided by this well could not have been chosen better. First of all, exactly the same logging tools were used to log hole 556 on leg 82, as were used at hole 504B on leg 83 2 months later. Second, the same model and gauge of drillbit was used. Third, an entirely different basement lithology sequence was penetrated in the 177.5 m of oceanic crust drilled at hole 556.

Beneath 461.5 m of mostly carbonate sediments, a complex lithology sequence was encountered. First, 10 m of basalt breccia and intermixed carbonate sediment was encountered, then 35 m of pillow basalts overlay a massive basalt unit and 40 m of basaltic breccia. Then for the first time, a significant thickness of gabbro and serpentinized gabbro were penetrated. Interlayered gabbro, gabbroic breccia and serpentinized gabbro extended from 90 m subbasement to the total depth 177.5 m into the crust. Obviously, the gabbros are more characteristic of layer 3 rocks and should not have been encountered so close to the surface. Their occurrence allows us to contrast tool responses to differing metamorphic lithologies within hole 556 as well as with hole 504B.

## Geophysical Logging in Variable Metamorphic Lithologies

The variable lithologies identified above were from identification of core, recovery of which dropped from close to 50% in the basalts to less than 20% in the gabbros, (Bogault and Cande 1982). They were located stratigraphically from major changes in geophysical log characteristics at the approximate depths within which the cores were taken (Hill 1983). It is these log-changes which we wish to examine in detail. We must understand tool responses to the different lithologies if we are to compare logging results with those at hole 504B.

Porosities calculated from the neutron, gamma-ray, resistivity, and sonic logs in the standard way are displayed with raw resistivity, density, natural gamma-ray, compressional and shear velocity, Poisson's ratio, and caliper logs in Fig. 16. Major lithology boundaries are evident from the density log, the resistivity log, and the differences between the various porosity logs at each depth (Fig. 16). For example, the very bottom of the hole is a gabbroic breccia with relatively low density and only a small difference between neutron and gamma-ray porosities, whereas the serpentinized gabbro just above has very high density, but also large differences between neutron and gamma-ray porosity.

The existence of variable alteration-lithologies at hole 556 allows us to test our Hydroxyl mineral content differencing scheme with more sophisticated commercial cross-correlation shale index techniques. Specifically, five "clay content" curves were generated for hole 556 using Corban answers mark of Schlumberger, Ltd. (analyses by J. E. McDonald, Schlumberger Offshore, New Orleans). Their averaging scheme again selects the minimum of these five curves at each depth point downhole, (Fig. 17). The Corban interpretation improperly identifies much of the bound water as free water porosity because of very low clay volumes calculated from the resistivity and natural gamma ray logs. This points to the difficulty in using standard oil field techniques for the very different scientific logging requirements of the oceanic crust.

Our neutron-gamma ray differencing scheme gives hydroxyl contents very similar to the neutron-density cross-plot values calculated by Corban, (Fig. 17, far right for ours versus far left). The two serpentinized gabbro zones made up of over 60% hydrated minerals show up nicely on either our differencing scheme or the Corban N-D cross-plot curves. Both predict a marked increase in hydroxyl-mineral content from the gabbroic breccia to the serpentinized gabbro units, as was verified by the chemical bulk analyses of the cores, (Bogault and Cande 1982).

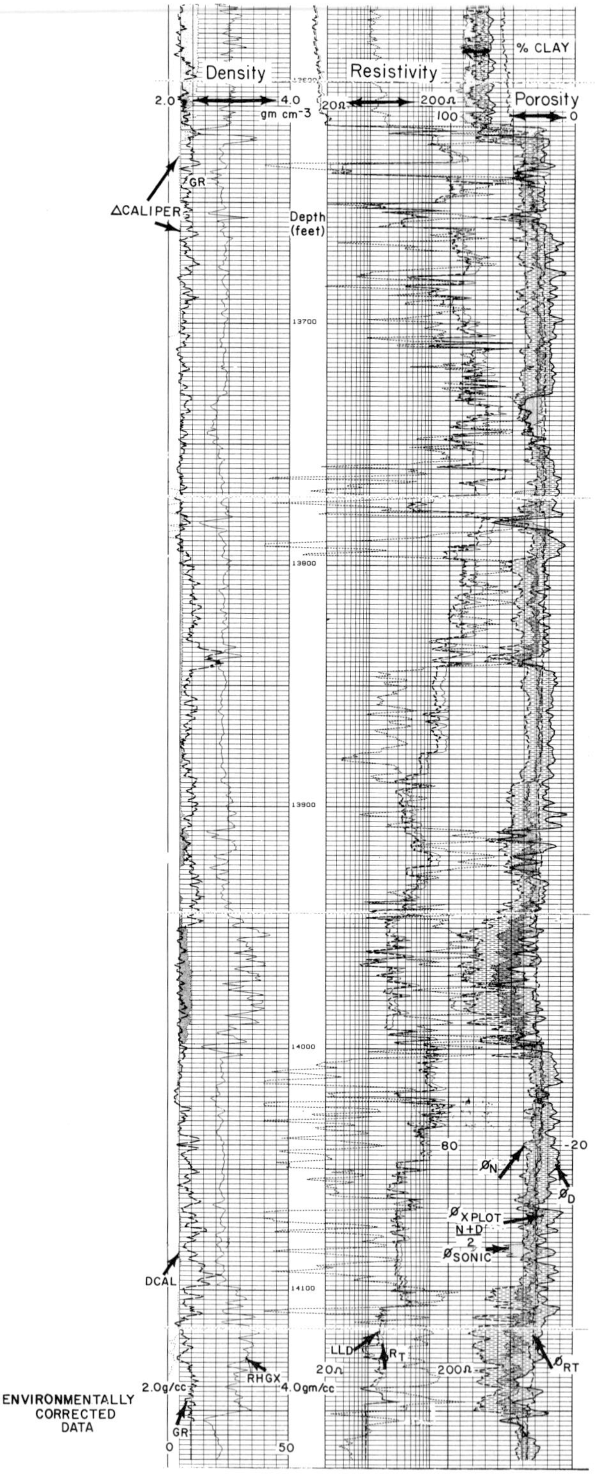


SITE 556
17ma in Mid-Atlantic
Ridge flank (west)

Fig. 16.

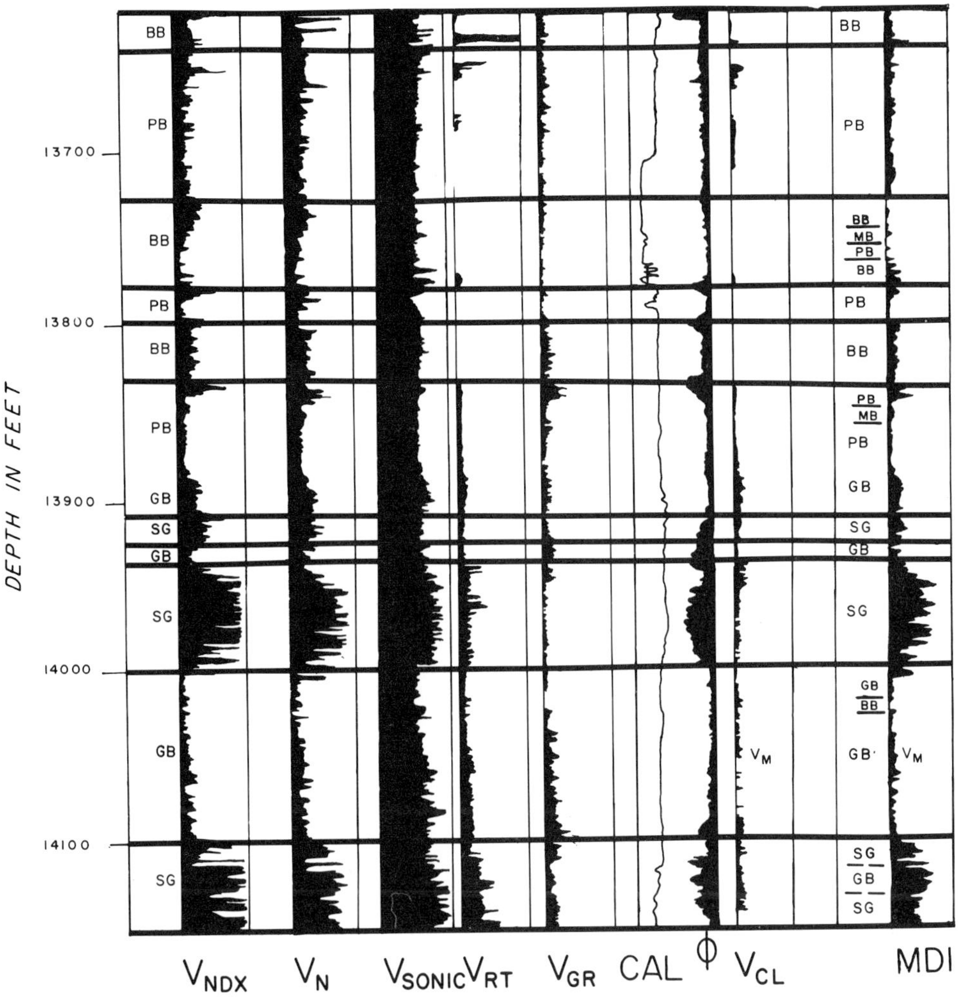


Fig. 17. Clay-content log for hole 556 as in Fig. 3. Free water porosities for this hole are probably low, but the ratio is correct. Alteration has bound more water in hydroxyl-minerals than is free in pore spaces in this hole. Fully 60% of the minerals in the serpentinite zones are hydrated. Clay-content log shown at right against Corban answers program, mark of Schlumberger Ltd., volumes of clay on left. Lithology boundaries as in Fig. 16

Fig. 16. Composite logs run at hole 556 by Schlumberger Well Services. On the left are DCAL (caliper hole size minus bit size), and gamma-ray density. In the center track is laterlog electrical resistivity, and on the left are several porosity determinations. N neutron; D density; RT resistivity; xplot is N+D/2. Lithology boundaries were generally placed by the core recovery which decreased from about 50% in the basalts to less than 20% in the gabbroes. Specific lithology boundaries were then determined from the changes in the porosity-sensitive logs. Several stringers are inferred where no core was recovered. Environmental corrections were for hole size, temperature, and pressure. Porosities are not corrected for bound water. At far left, compressional and shear wave velocity on right, Poisson's Ratio in bold line on left

We find that free water residing as porosity in the basement drilled at hole 556 is small compared to the amount of water bound within hydroxyl-minerals in the heavily altered gabbro and basaltic breccia sections of the hole.

An excellent lithology identification scheme seems to be an oilfield technique called M-N crossplotting. M and N are calculated as follows:

$$M = [\ (\Delta tf - \Delta t) \ / \ (\rho b - \rho f)\ ] \times 0.01$$
$$N = (\emptyset nf - \emptyset n) \ / \ (\rho b - \rho f), \qquad (2)$$

where

$\Delta t$ = travel time
pb = gamma ray density
$\emptyset n$ = neutron porosity
f = fluid.

Such a crossplot identifies characteristic regions for each lithology. Serpentinite, for example, has M values ranging from 0.7 to 0.8, and N values of about 0.3; whereas gabbro has M of 0.8 and N of 0.5. Pillow basalts are closer to an M of 0.7 and an N of 0.45 (Fig. 18). We know these values in this well because cores were recovered, but lithologies of future wells could be determined solely based upon the M-N crossplot once we have confidence in the lithology locations on the crossplot. For example, some oil and gas is found in igneous reservoirs in Argentina, and consequently Schlumberger scientists have previously located the basalt and diabase points on the M-N crossplot (Khatchikian 1983). The oceanic crustal points are a little higher than these land points for both M and N (Fig. 18). An ultimate end to such a log calibration would be lithology identification without core being available.

## The Spectral Response of Hole 556 Oceanic Crust

Once we have identified lithologic units within the geophysical logs from hole 556, we can return to our original problem: does an anti-resonance exist in the sonic waveform spectra from hole 556?

We are lucky that exactly the same multichannel sonic-logging tool was used at 556 as at 504B. We windowed the compressional and shear wave coda, did cross-spectra and displayed the energy and frequency in exactly the same manner as at 504B (Fig. 19, A and B). We then display the same frequency-energy crossplot of compressional (Fig. 20A), and shear spectra (Fig. 20B). The attenuation of both the compressional wave and the shear wave coda suggests passage of the energy through a medium which is acting as a frequency-dependent attenuator. In the case of compressional energy, this observation is opposite to that from hole 504B in that frequency is low when energy is low (Figs. 20A versus 12). No anti-resonance characteristic is evident in the waveform spectra from hole 556.

In fact, alteration and lithology-change can account for part of the "filter characteristics" observed in the spectra. If we observe spectral response to lithology changes, we find that the gabbros are high-energy, high-frequency formations which act as a high-pass filter (Fig. 19C). However, the serpentinized gabbros are low-energy and low-frequency rocks (Fig. 19C). For example, consider the bottom of the hole again. We can even identify a gabbroic breccia stringer within

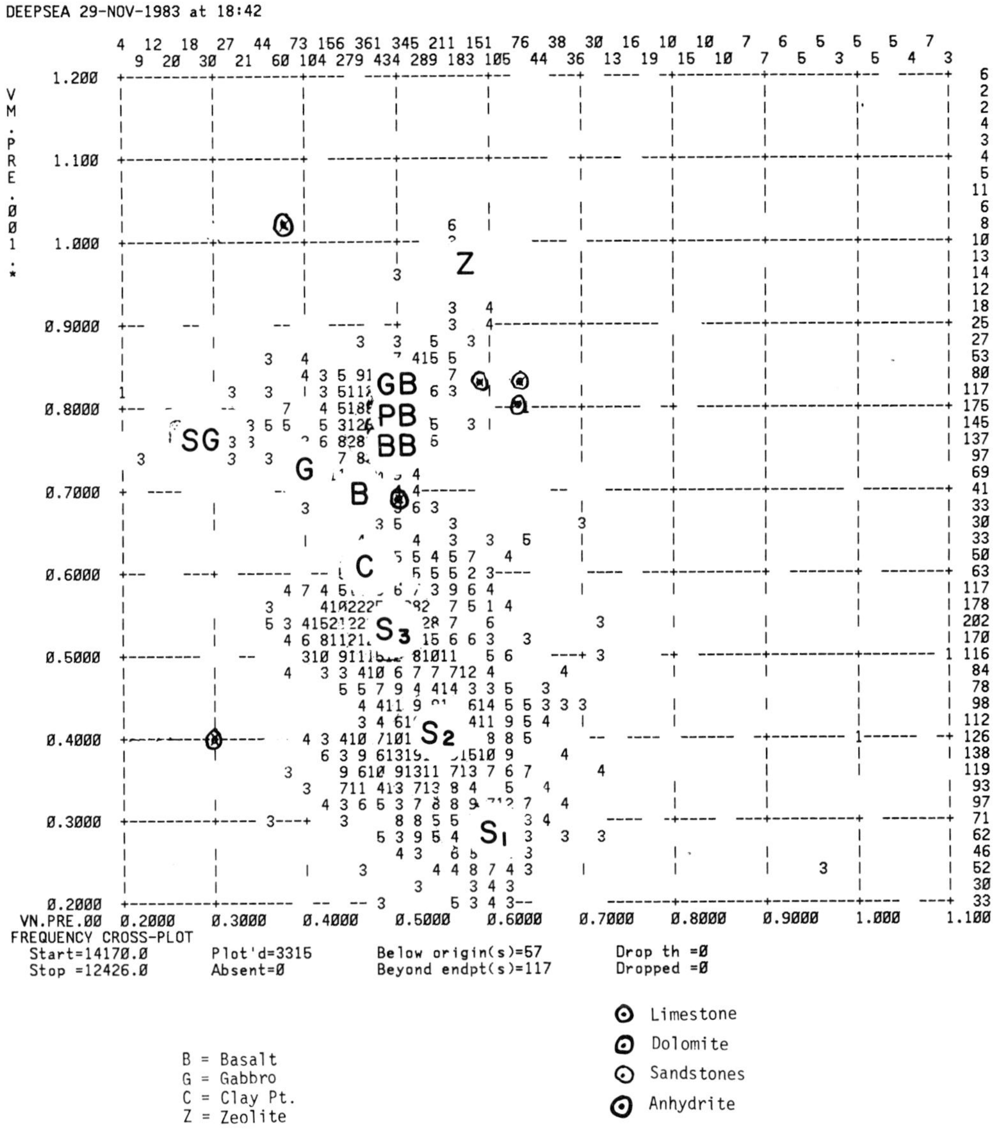


Fig. 18. Lithologies are identifiable on a M-N crossplot at hole 556. S1, S2, and S3 are different sedimentary units. The igneous lithology locations other than those from 556 are from oil-bearing formations in Argentina (Khatchikian 1983), and the evaporite, clay zeolite and pyrite points are from McDonald (1983, pers. commun.)

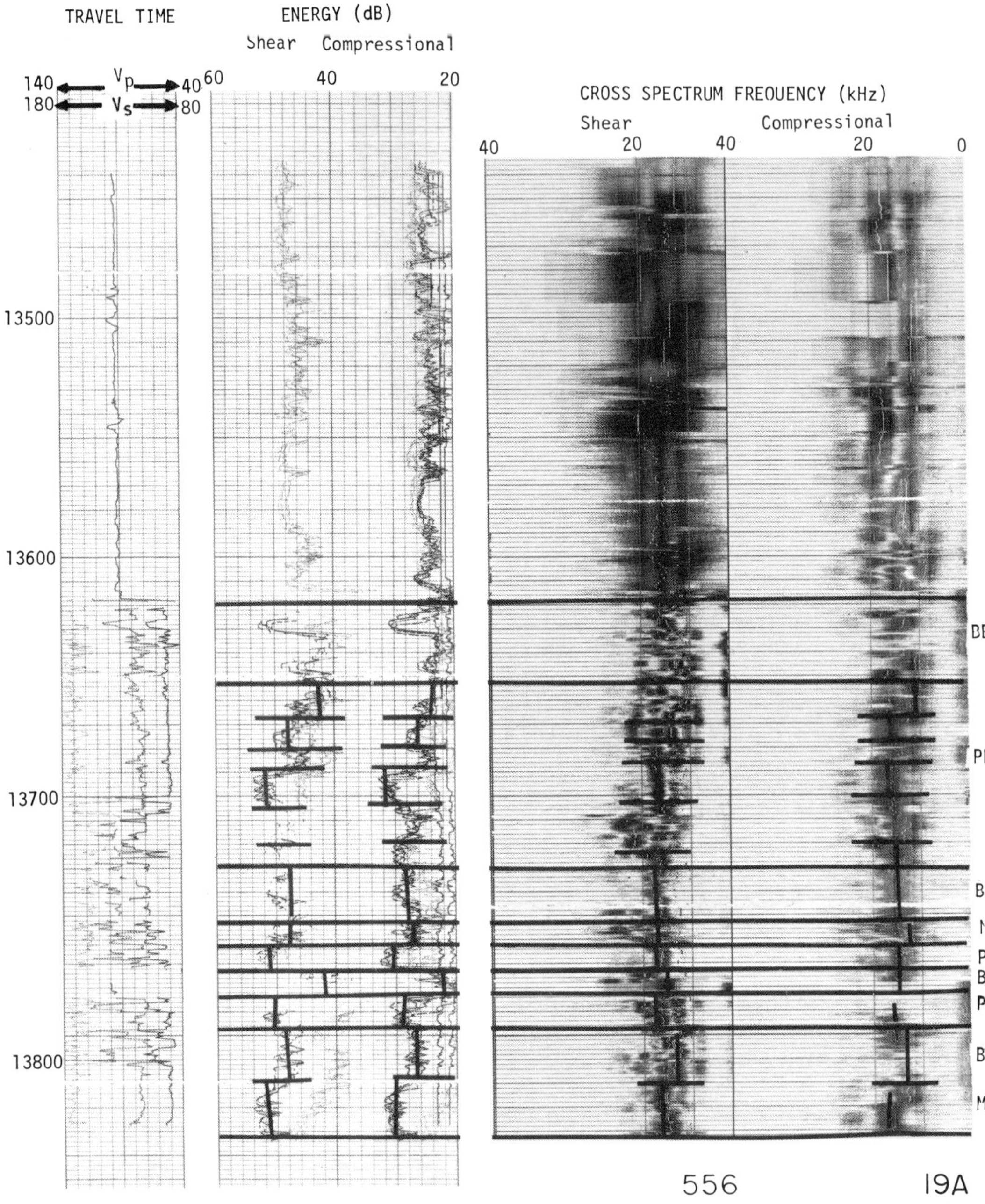


Fig. 19A-C. Energy-frequency cross-spectra from DTCS program, a mark of Schlumberger, Ltd., as in Figs. 8 and 9 for hole 556. Lithology units from Figs. 16-18. Notice the correlation between high energy and high center frequency and low energy with low center frequency. This is clearly lithology-related. C is a stylized schematic of the power spectra for compressional and shear energy in the four main lithology units of hole 556 basement

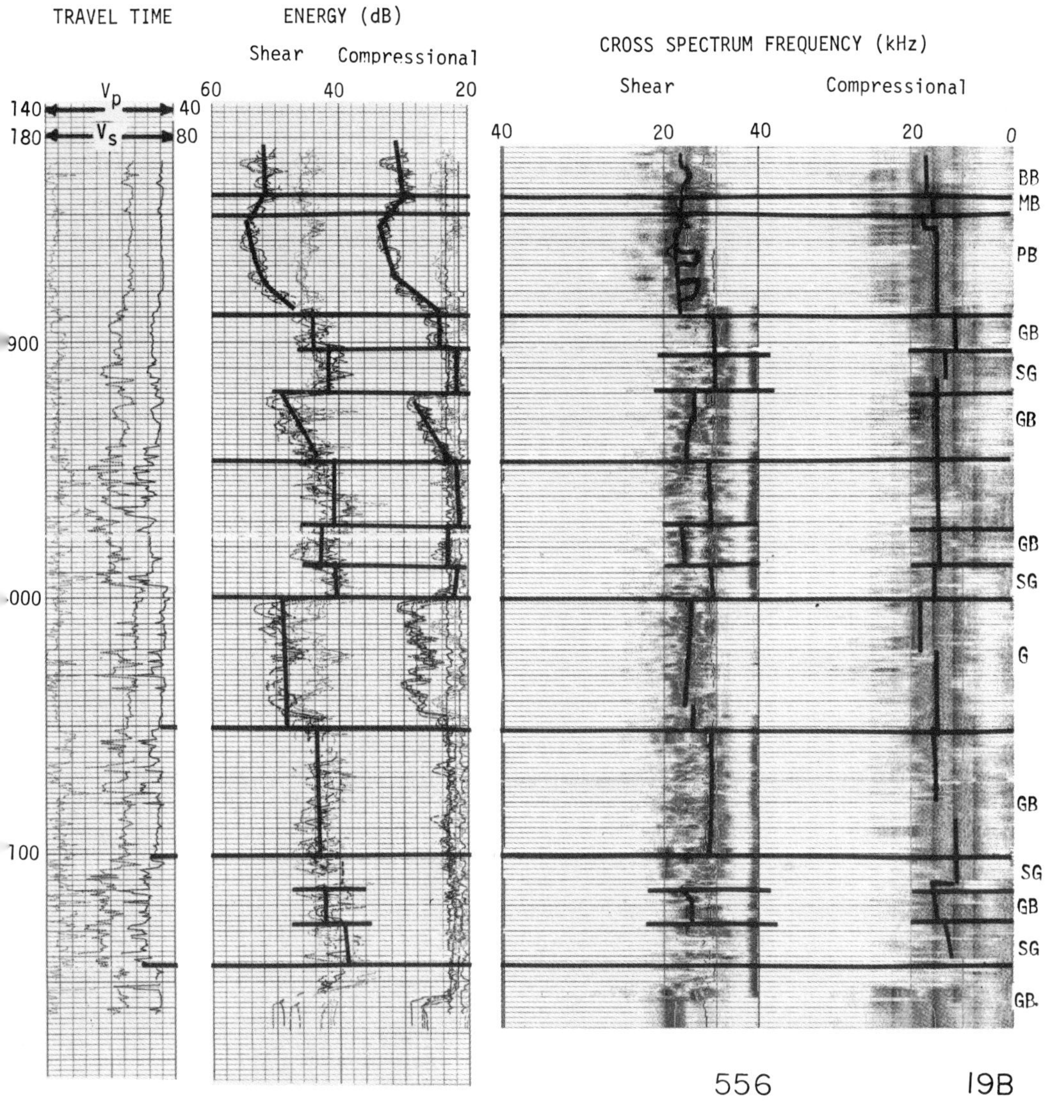


Fig. 19B. (Figure 19C see page 214)

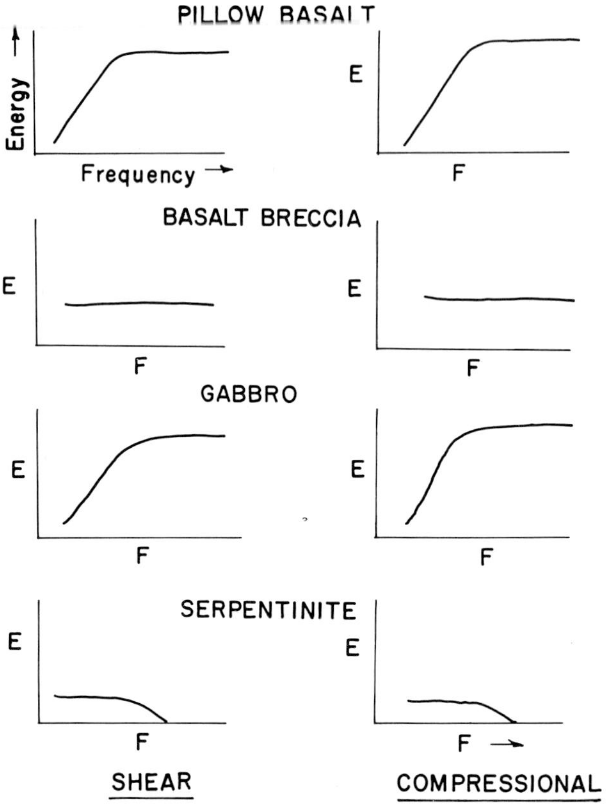


Fig. 19C.

the serpentinized gabbro layer from a small energy and corresponding frequency increase at 10 to 15 m above total depth (Fig.19B).

We can quantify the observation from Fig. 19C of frequency-dependent attenuation by producing a log of compressional and shear coda center frequency and integrated energy as at hole 504B, (Fig. 21). Phase coherence analyses as before reveal strong positive coherence between compressional versus shear energy levels, and positive but weak coherences at wavelengths of 20 to 100 feet between both compressional and shear frequency versus energy, (Fig. 22). That is, there are zones of high and low energy level in the well which also appear filtered, in that high energy zones are also high frequency zones, and vice versa.

In this well, we have no information on the distribution of fracturing within the different lithologies because no borehole televiewer runs were made. Fracturing might be expected to control the shorter wavelength coherences because the crust appears to be so highly fractured.

We can examine, however, the relation between alteration and these long wavelength spectral changes in the well. Phase coherence between the hydroxyl-mineral content which we calculated earlier and the spectra shows weak negative coherence in the compressional and shear energies at long wavelengths (Fig. 22).

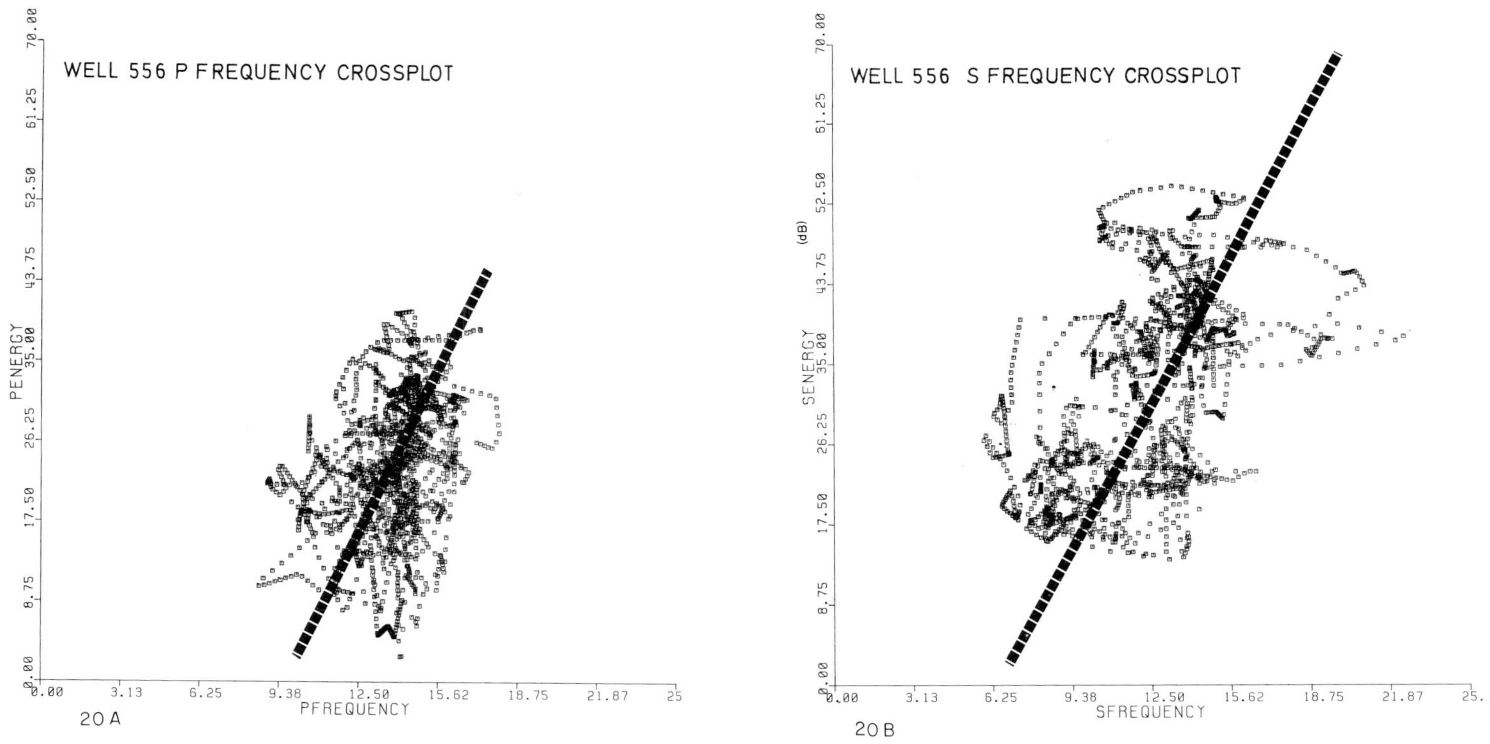


Fig. 20A,B. Energy-frequency crossplots for shear and compressional coda from Fig. 19 for hole 556. While the data are scattered, the bold dashed lines indicate the trends apparent in the data

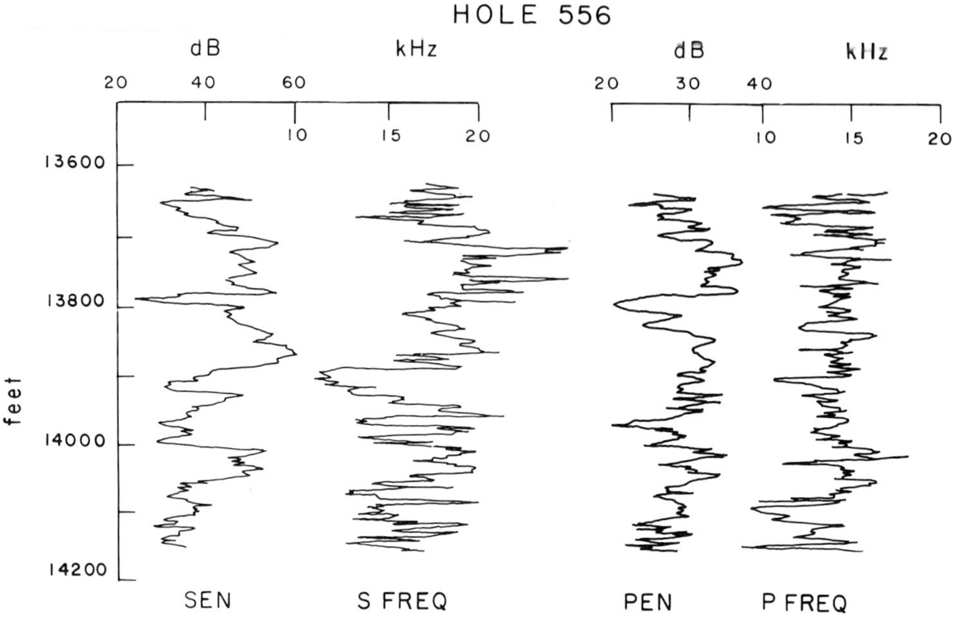


Fig. 21. Spectrum log of compressional and shear coda energies and center frequencies for hole 556 (as in Fig. 10)

To demonstrate this correlation further, we de-trend and bandpass filter the spectral and alteration content logs passing only energy with 20-100 feet wavelengths (Fig. 23). We invert the bandpassed alteration log so that high hydroxyl-content now correlates with low energy or frequency in the plots of Fig. 23. We see that the compressional energy variations down the well correlate well with the shear energy changes, and to a lesser degree with the alteration log. Correlation is most striking in the gabbros of the lower half of the well.

The extent to which the long wavelength components of the spectra correlate with degree of alteration can be further quantified by calculating a correlation coefficient for the bandpassed data. There is a -0.42 correlation between the bandpassed shear energy log and alteration content (Fig. 24). This is equivalent to a phase coherence of 29% in Fig. 22. Compressional energy does not correlate as well, with a coefficient of -0.32, or 34% phase coherence. However, most of the alteration change in hole 556 occurs in the gabbroes at the bottom of the well. If we correlate only the alteration and spectral changes in this portion of the hole, we calculate a correlation coefficient of -0.55, equivalent to a phase coherence of 22% for shear energy versus alteration and -0.45, or 28% for compressional energy / alteration, (Fig. 24).

These weak correlations are to be expected since impedance changes which control the overall spectral response of the formation are caused by more than just alteration changes. We can examine some of these other effects qualitatively, if we consider the formulation of impedance in the crust.

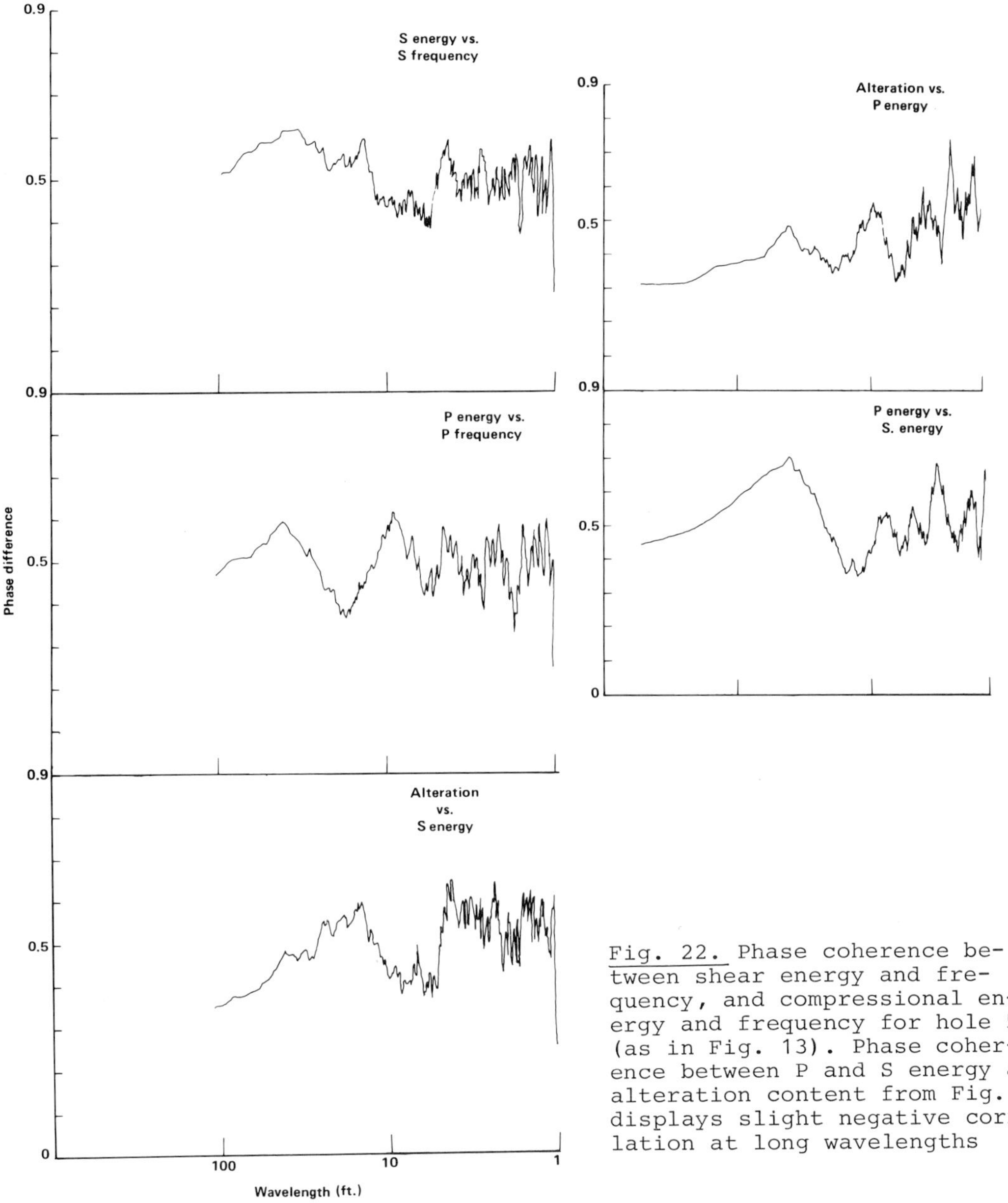


Fig. 22. Phase coherence between shear energy and frequency, and compressional energy and frequency for hole 556 (as in Fig. 13). Phase coherence between P and S energy and alteration content from Fig. 17, displays slight negative correlation at long wavelengths

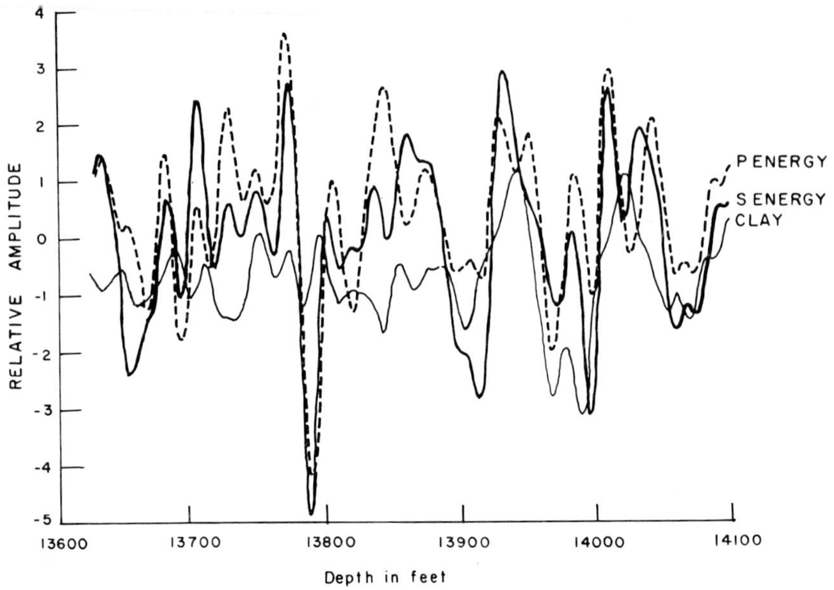


Fig. 23. The P energy, S energy, and alteration logs from hole 556 were then de-trended, normalized, and bandpass filtered to pass only wavelengths of 20-100 feet. Note particularly strong correlation between P and S energy

## The Impedance of the Oceanic Crust

Impedance is the opposition to the flow of energy occurring in a sound-conducting medium. The spectral changes which are caused by variable alteration-lithology might give important insight into impedance changes occurring in the oceanic crust. The sonic spectra respond in a very complex way to the impedance changes which occur downhole in the crust. We have isolated different responses from two holes in very different geology. In the Atlantic hole 556, alteration and lithology changes resulted in frequency-dependent attenuation. In the Pacific, the lithology within the sheeted dikes of layer 2C is absolutely constant. Yet there, the degree of fracturing varies down the hole. Anti-resonance must be a suspected result of some process occurring within this fracture system.

In fact, the impedance of the oceanic crust is too complex a function of too many variables for us to make much further progress in deciphering quantitative impedance cause and effect relationships. But we can isolate two factors which are clearly of importance for future logging and spectral studies of the oceanic crust.

First, we can define the impedance (Z) of the borehole-rock system as:

$$Z = \{ (pc)^2 + 2\ ftp - k/2\ ft\ ]^{1/2} \qquad (3)$$

where

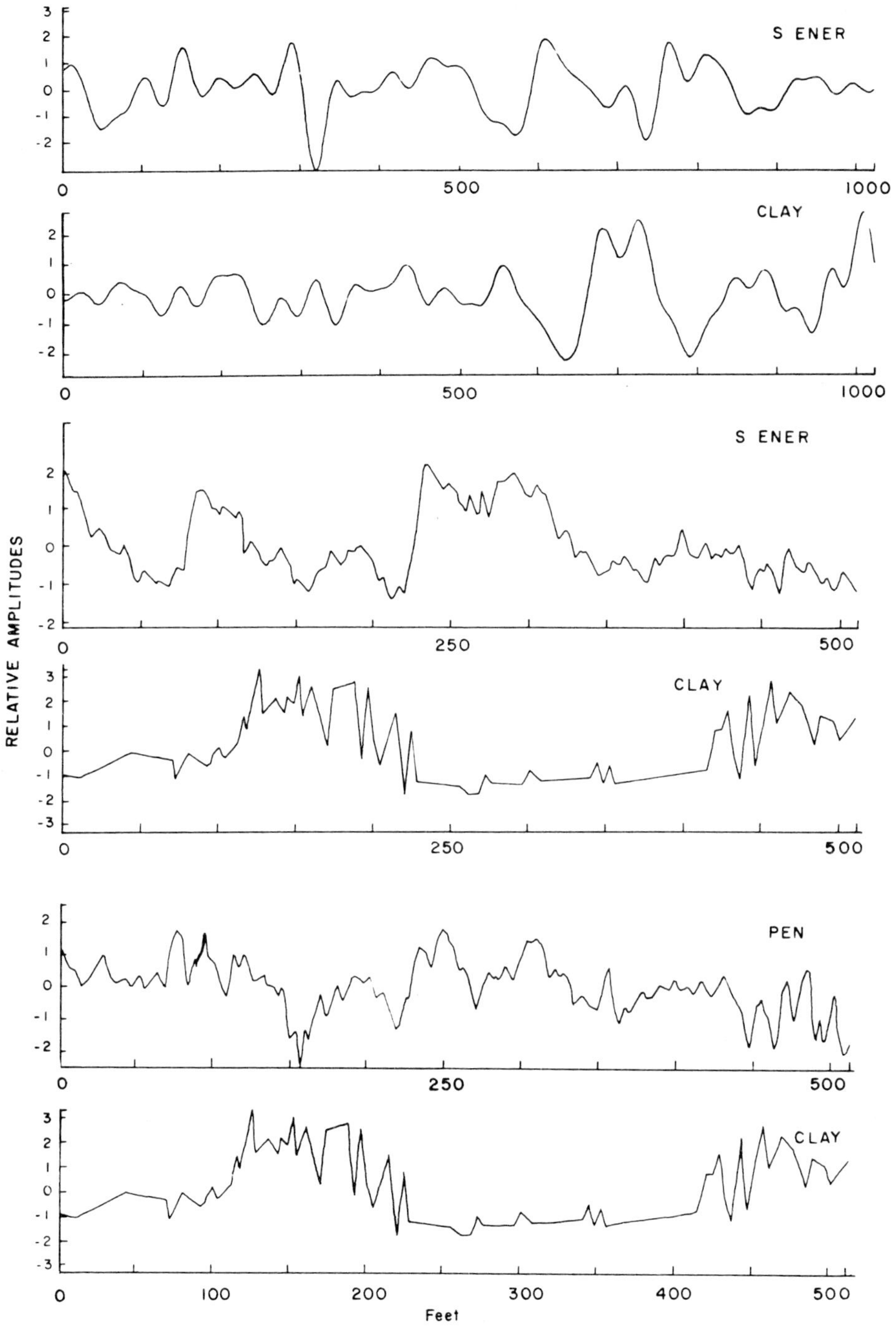


Fig. 24. Correlation coefficient of 0.42 between S energy (bandpassed as in Fig. 23) and alteration log in Hole 556 (also bandpassed). However, most of the correlation occurs in the gabbroes at the bottom of the well, so correlation coefficients of 0.5 between S energy and alteration content and P energy and alteration were calculated by isolating only the gabbroic bottom of the hole

p = density in our case; mass in general, which equals the density measured by the logging tool because the volume sensed by the tool is constant

c = velocity
f = frequency
t = time
k = spring constant

It is evident from this equation that direct observations of sonic velocity and rock density from geophysical logging do not provide enough information to constrain the impedance variations of the crust. Specifically, there are real (resistance) and imaginary (reactance) components to the impedance. Only the imaginary reactance terms are frequency dependent; the resistive real term generally has no frequency dependence. At hole 556, we have observed variations in the power spectra of compressional and shear coda which appear to be dependent on alteration and lithology. Although frequency-dependent impedance changes are suggested, we do not yet have enough information to determine cause and effect. For example, there are two forms of imaginary reactance, mass reactance which is equivalent to electrical inductance, and compliant reactance which is analogous to capacitance. As frequency increases, mass reactance which is density-dependent, increases; and, as frequency increases, compliant reactance which is dependent on the spring constant of the medium, decreases. We cannot yet determine which physical parameters in the crust of hole 556 are affecting which reactance term. Other alternatives such as frequency dependent coupling, scattering or geometrical effects are also possible. We look forward to the day when we might be able to describe the transfer function of the oceanic crust.

We can speculate that part of the alteration effect on impedance within the crust at hole 556 must be from the complex mass reactance term in Eq. (3). For example, serpentinized gabbro must have higher mass reactance than gabbro because of its higher density (Fig. 16). Yet its high degree of alteration has also lowered its Poisson ratio, which has some relation to the spring constant of the formation affecting the compliant reactance term. Somehow within Eq. (3), serpentinized gabbro must offer less opposition to the flow of low-frequency energy, but more opposition to the flow of high-frequency energy, in order to account for the observed changes in the spectra.

However, the absolute level of energy passing through the system is also affected. Recall that the spectra for the serpentinized gabbro zones exhibit low energy as well as low frequency (Fig. 19C). This overall drop in energy level may be a result of the real resistance increase caused by serpentinized gabbro's higher density, although the velocity of both is less than that in the basalt zones, (Fig. 16).

In addition to lithology, we can identify other major observational components to impedance changes in the oceanic crust (Fig. 25). Alteration changes within any of the crustal lithologies will certainly affect impedance. We have not been successful in isolating spectral changes due solely to alteration, though the use of a multichannel-sonic-logging tool in the highly fractured layers 2A and 2B at site 504B offers some promise. There, the lithology is constant, and the degree of fracturing is severe in both layers, whereas alteration definitely increases from layer 2A into layer 2B.

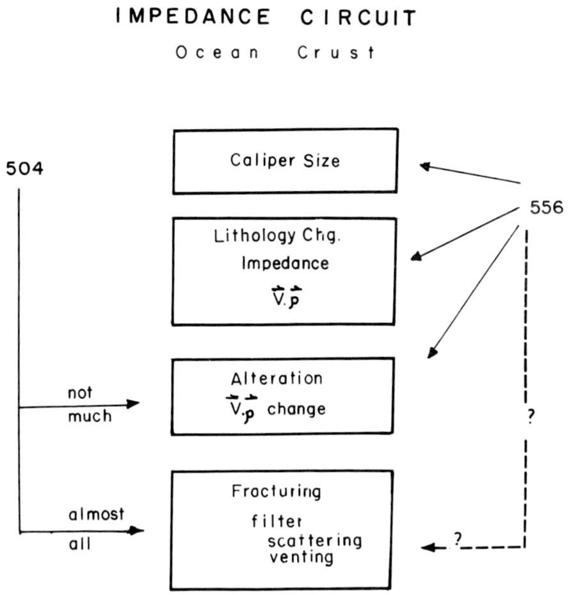


Fig. 25. Schematic box diagram of the physical properties which affect formation impedance, and how the amount of information we have about Hole 504B compares to that at Hole 556

Another variable which clearly affects impedance is mechanical; caliper size of the hole (Palliet 1981). Zones of poor hole have very different energy conversion characteristics than those at the boundary of bit-sized hole. Our best policy with the study of the oceanic crust is to begin only in zones of good hole. Fortunately, with the exception of one serpentinite zone in hole 556, both the dikes of hole 504B and the rest of hole 556 are gauge-sized holes (Fig. 16).

## Anti-Resonance and Venting in the Sheeted Dikes at Hole 504B

Whereas the spectra from hole 556 appear to be affected by each of the major impedance changes in Fig. 25, the waveforms from the dikes at the bottom of hole 504B offer us a chance to isolate a single parameter. To be specific, caliper changes, lithology boundaries, alteration zones, and an unknown fracture pattern complicate the study at hole 556, (no borehole televiewer information exists at this site), but in hole 504B, sheeted dikes have a gauge hole in a constant lithology rock with little alteration present at the bottom of the hole (Figs. 2, 3 and 4). The effects of fracturing on the acoustic impedance can be deciphered, at least to some degree. We are then interested in the implication of impedance changes for other physical properties of the crust such as permeability.

We can build a quantitative, though simplistic acoustic model for the sheeted dike section of hole 504B, since fractures appear to dominate the impedance structure over this portion of the hole. Since we know the anti-resonance frequency within the system, we can calculate rather simply the length of fractures responsible for the anti-resonant behavior of compressional and shear waves in the dikes of layer 2C at site 504B. Though this is an admitted gross oversimplification of the complex acoustic system operating in the crust, we can approximate the energy propagation in the dikes as occurring in interconnected tubes (fractures intersecting the borehole). The dikes then become a half-wavelength resonator.

Experimental and modeling results in air-filled tubes show that the effect of intersecting tubes is to vent energy from one tube out to the other (Egolf et al. 1978). The effect on the energy spectrum is to create a resonance within the vent tube. Low frequency energy is preferentially attenuated within such a vent system (Fig. 26). The vent tube-length then controls the anti-resonance frequency within the half-wavelength resonator, as well. That is, maximum energy loss within fractures intersecting the wellbore will occur at the wavelength at which anti-resonance is set-up within the intersecting fractures, which then vent energy away from the main energy propagation path. The frequency at which the first anti-resonance occurs is equal to the following:

$$f_a = (s + 1 - 1/2)\ (c/2L) \qquad (4)$$

where

$f_a$ = frequency of the anti-resonance
$s$ = harmonic number
$c$ = velocity of sound
$L$ = length of the vent.

The half-wavelength resonator has resonances at whole numbered harmonics and anti-resonances in between. For shear waves in the dikes, $f_a$ is 9 kHz, s equals 1 (the first harmonic), and the velocity is 3 km $s^{-1}$; the vent fracture length is then 25 cm. We then suspect why the energy-wavelength crossplot shows dramatic energy loss at 30- to 40-cm wavelengths (Fig. 14). Fractures intersecting the borehole appear to have a characteristic lengths of about 25 cm, so that energy with wavelengths of 30 to 40 cm corresponds to the first anti-resonance within these fractures.

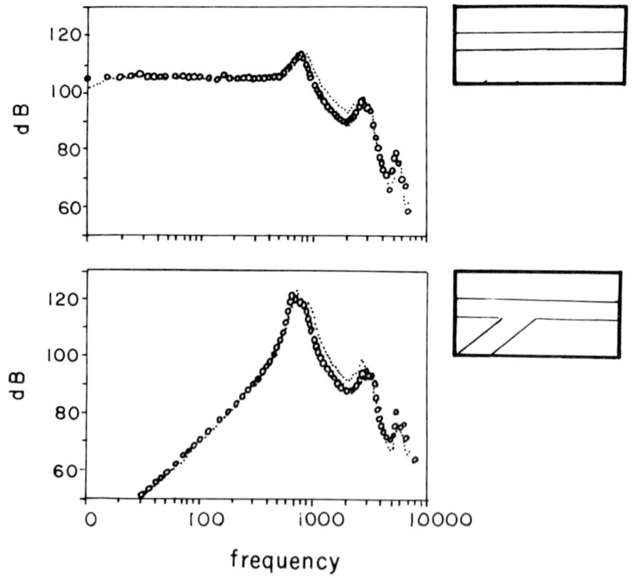


Fig. 26. The change in the power spectrum for transmission of sound in an air-filled tube (top), versus the same tube with a vent intersecting it (Egolf et al. 1978). The energy peak is caused by resonance set up in the vent. Open circles are observations, light dashing is prediction from a venting model

## Filter Characteristics of the Crust and the Venting Model

Such a venting model makes specific predictions about the filter characteristics particularly applicable to the normal mode spectra. Specifically, the fluid waves should display high energy at high frequencies and low energy at low frequencies, (Fig. 26).

The simplest quantitative analysis of the tube and fluid waveforms we can do is to Fast Fourier Transform (FFT) the entire wavetrain and display the power spectrum (Fig. 27). Most of the compressional and shear coda energy is concentrated in a bandwidth of 10 to 25 kHz, the Stoneley energy is narrow band and centered at about 8 kHz (the highest amplitude, blue band at 8 kHz in Fig. 27). A source model with the Schlumberger sonic tool in a uniform cylindrical borehole predicts spectra for the normal modes which show a low frequency cut-off at 6 kHz for the first mode, 12 kHz for the second, and 24 kHz for the third mode, (Fig. 28). Therefore, the predominant energy variation observed at 30-40 kHz (Fig. 27) is from these late arriving normal modes.

These energy-level changes in the high frequency normal modes can be qualitatively correlated with the presence or absence of fractures seen by the borehole televiewer. For example, consider the zone of attenuation at 4715 m, which correlates with a fracture zone seen on the borehole televiewer records (Fig. 27).

We can more quantitatively examine the possible filter affects of fractures on the Stoneley and normal modal spectra by placing a time window across these arrivals, (Fig. 29). The window begins at a fixed time after source discharge set for a Stoneley velocity of 190 microseconds per foot. The window has a duration of 450 microseconds; also the frequency spectrum was low pass filtered from 0 to 10 kHz with a linear ramp to 15 kHz to isolate the predominantly 8 kHz Stoneley energy.

Again concentrating on the dikes at the bottom of the hole, energy variations in the Stoneley energy correlates well with the variation in fracturing observed by the borehole televiewer log, (Fig. 30). Phase coherences are the highest for any data pairs of this study, with a correlation of over 80%. These spectral changes are not caused by caliper changes, as can be seen in Fig. 30. Nor are these changes in the spectra are clearly not related to alteration changes, since no phase coherence between alteration and normal mode energy exists at the bottom of the hole, (Fig. 30).

But the Stoneley energy is not dispersive, (Fig. 28), and shows little if any change in frequency across fracture zones even though the energy level clearly drops. We have therefore plotted the peak energy and frequency of the normal modal coda within the same time window by excluding all energy below 10 kHz. These fluid wave arrivals are highly dispersive, (Fig. 28), and should show frequency-decrease corresponding to energy-attenuation in the zones of heavy fracturing if the venting model correctly predicts the filter characteristics of the dikes.

The peak frequency of the normal modes varies as predicted by the venting model, (Fig. 31). When the energy level is high, the peak frequency is high and when the energy level is low, the peak frequency is also low. The normal modal energy changes also correlate strongly with the Stoneley wave energy levels as would be expected, (Fig. 31).

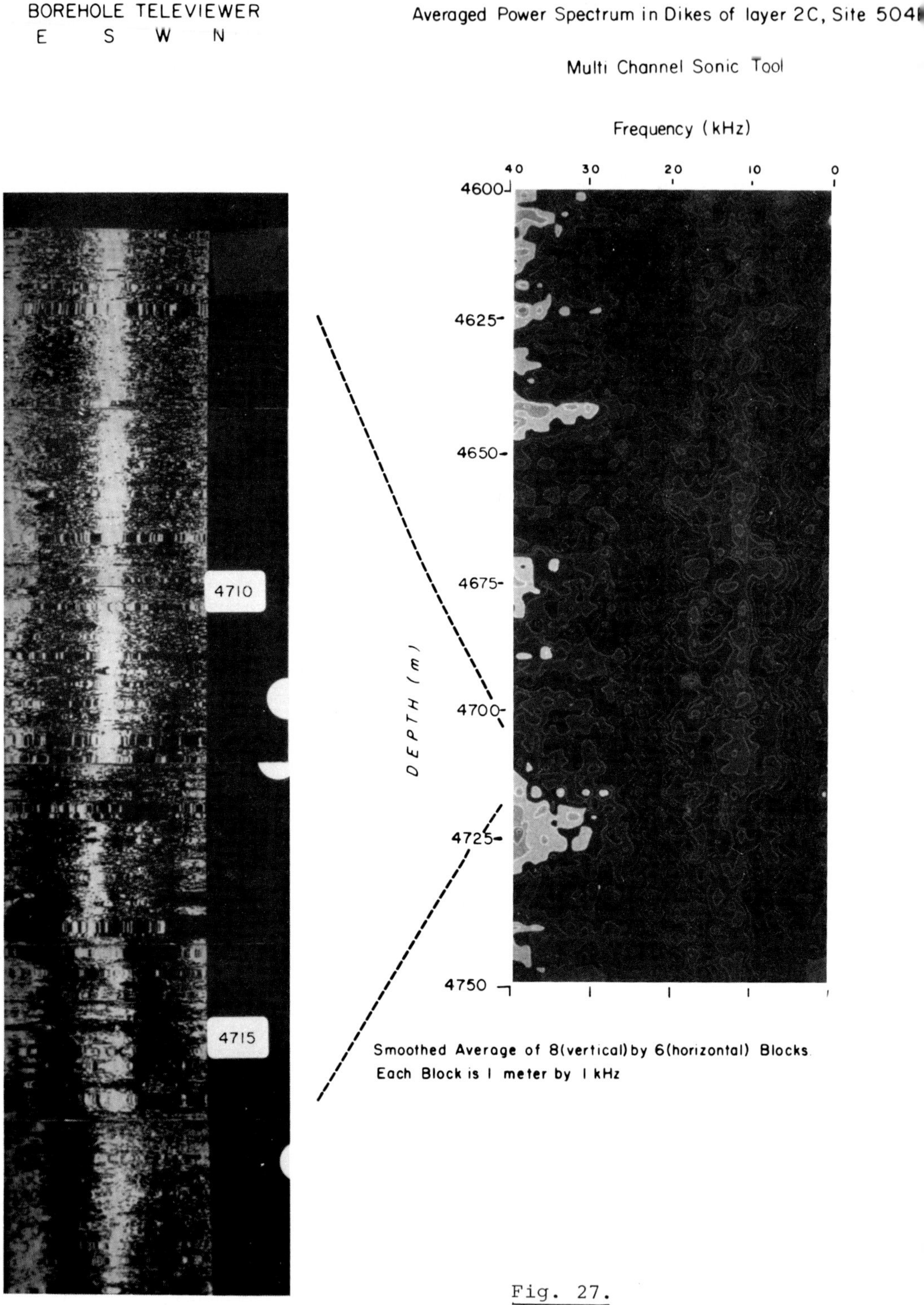
BOREHOLE TELEVIEWER
E S W N
4710
4715
Averaged Power Spectrum in Dikes of layer 2C, Site 504
Multi Channel Sonic Tool
Frequency (kHz)
40
30
20
10
0
4600
4625
4650
4675
4700
4725
4750
DEPTH (m)
Smoothed Average of 8(vertical) by 6(horizontal) Blocks
Each Block is 1 meter by 1 kHz

Fig. 27.

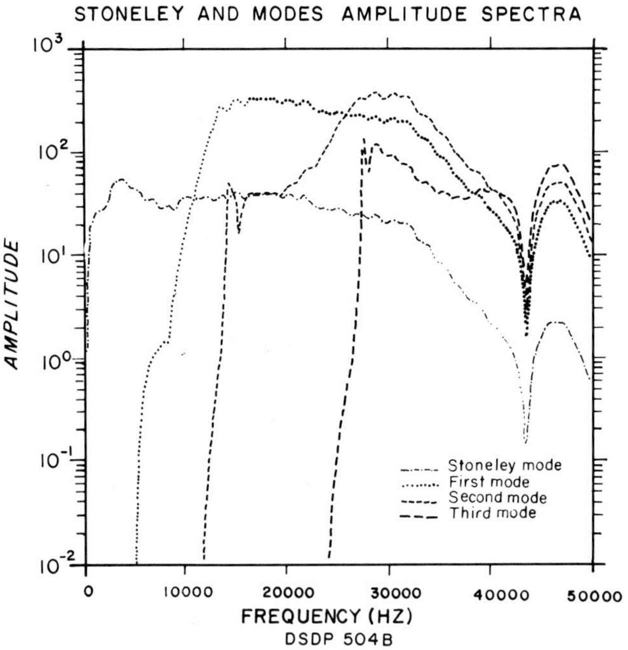


| LAYER | DTC | DTS | DEN | RADIUS | Q(COMP) | Q(SHR) |
|---|---|---|---|---|---|---|
| Tool | | | | 1.81 | | |
| Fluid | 186.0 | | 1.00 | 4.67 | 10000 | |
| Solid | 50.0 | 100.0 | 2.85 | 4.67 | 10000 | 10000 |

Rigid Tool Model, SLSSC UPPER 20 June '83

Fig. 28. Tool response model for upper transmitter of the Schlumberger SLS-5 sonic logging sonde in a uniform cylinder. Note that Stoneley energy is non-dispersive, whereas normal modal energy is highly dispersive

Fig. 27. The power spectrum of the bottom 75 m of sonic waveforms at Hole 504B smoothed and displayed against a borehole televiewer reflectivity image of the borehole from 4705 m to 4715 m. Power (in dB) is relative to the highest value within the interval. The televiewer image is oriented to N,E,S,W,N from left to right from a fluxgate magnetometer in the tool. Whiteness is a measure of reflectivity of the wellbore from a 1.2 mHz source. Vertical bars are from ship's heave. The horizontal fractures at 4715 m cause the energy drop in 30 to 40 kHz normal-mode energy observed in the spectrum (yellow colors)

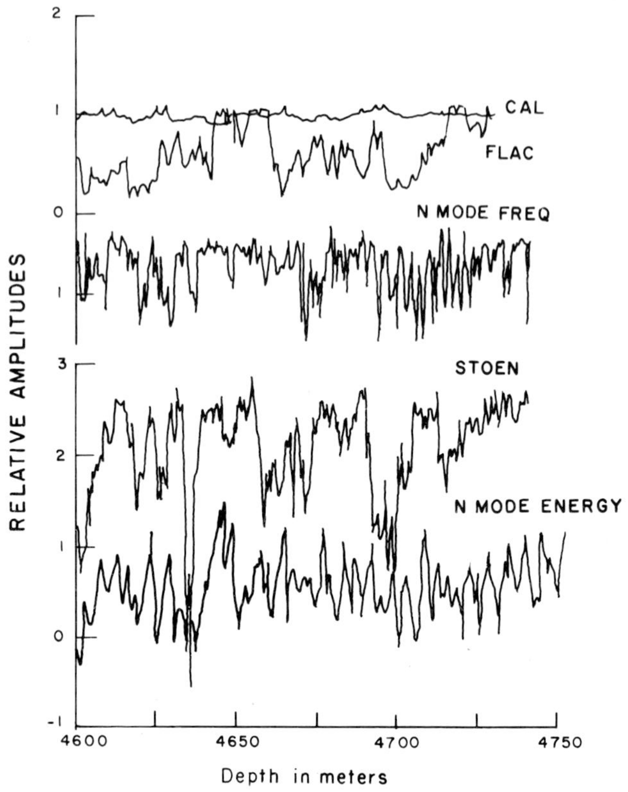


Fig. 29. The fracture log over the lower 150 m of Hole 504B is displayed along with the amount of Stoneley wave and Normal mode energy in dB and peak frequency of the normal modal energy in kHz calculated from windowed sonic waveforms

Best of all, the normal modal energy and frequency correlate strongly with the fracture pattern seen in the bottom of the hole. When the hole is only lightly fractured, the normal modal energy and frequency are high, but when the well is heavily fractured, the modal energy and frequency are low, (Fig. 31).

These correlations can be seen by bandpass filtering the Stoneley energy and normal mode spectral and fracture logs to pass only energy with strong phase coherence, (Fig. 32). It is interesting that this filter passes energy of 10 to 100 m wavelengths. Earlier we thought that fracturing would produce a shorter wavelength filtering effect. The fracture pattern is clearly consistent at longer wavelengths within the dikes than previously thought. This fact, that fracturing is concentrated in zones which have a long wavelength component, was also concluded from a different logging data analysis by Newmark et al. (1984). Zones of intense fracturing of 15 to 25 m thickness are separated by zones of relatively solid basalt of similar thicknesses in the dikes at the bottom of hole 504B.

It is important to note that the Stoneley wave itself does not contribute much to these spectral changes. Although it is the predominant energy wave within the window, its frequency remains relatively constant in the test interval. It is the normal modes which provide the major energy and frequency shifts shown in Figs. 29-32.

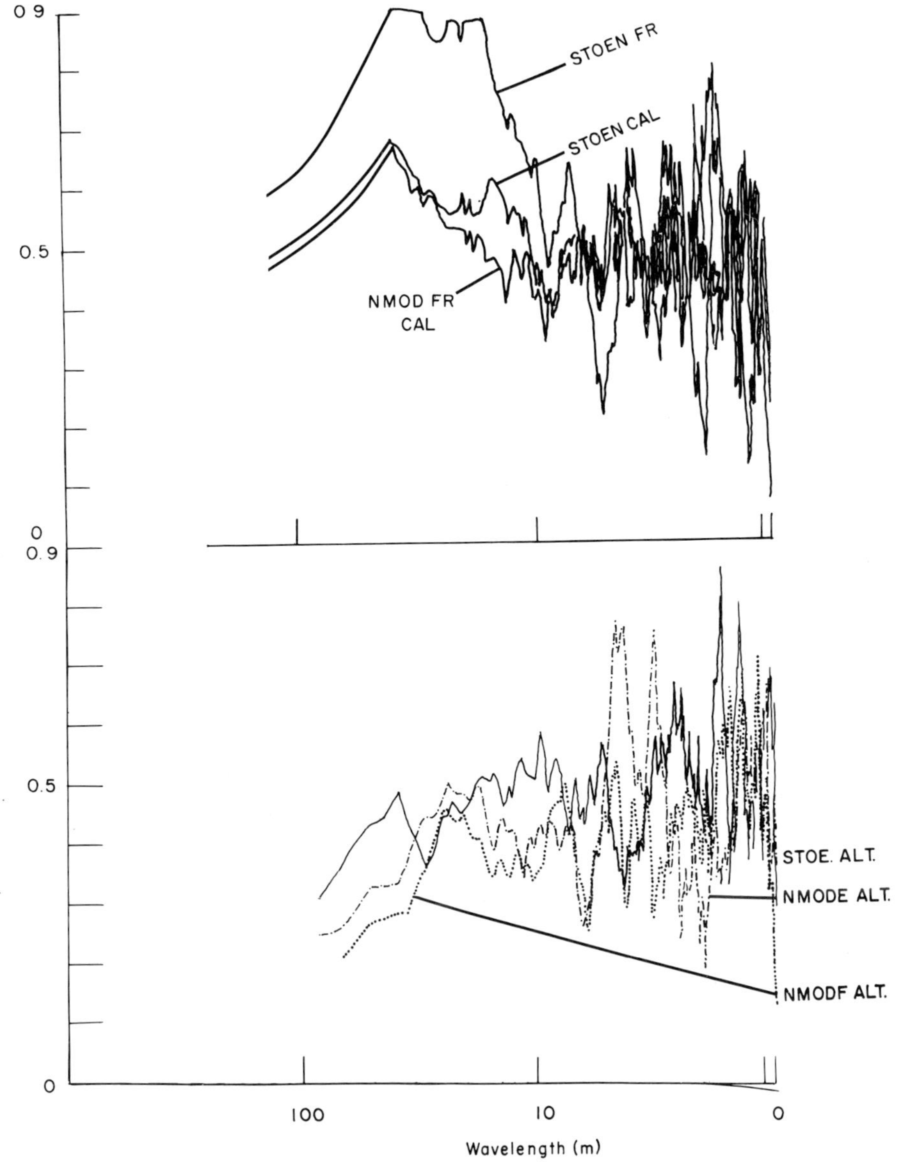


Fig. 30. Phase coherence of Stoneley energy against fracture log and caliper in Hole 504B showing much stronger correlation with fractures than with hole size. Normal mode frequency shows an even weaker correlation with caliper size. Stoneley energy, normal modal energy and frequency are correlated with alteration content below. No correlation is seen

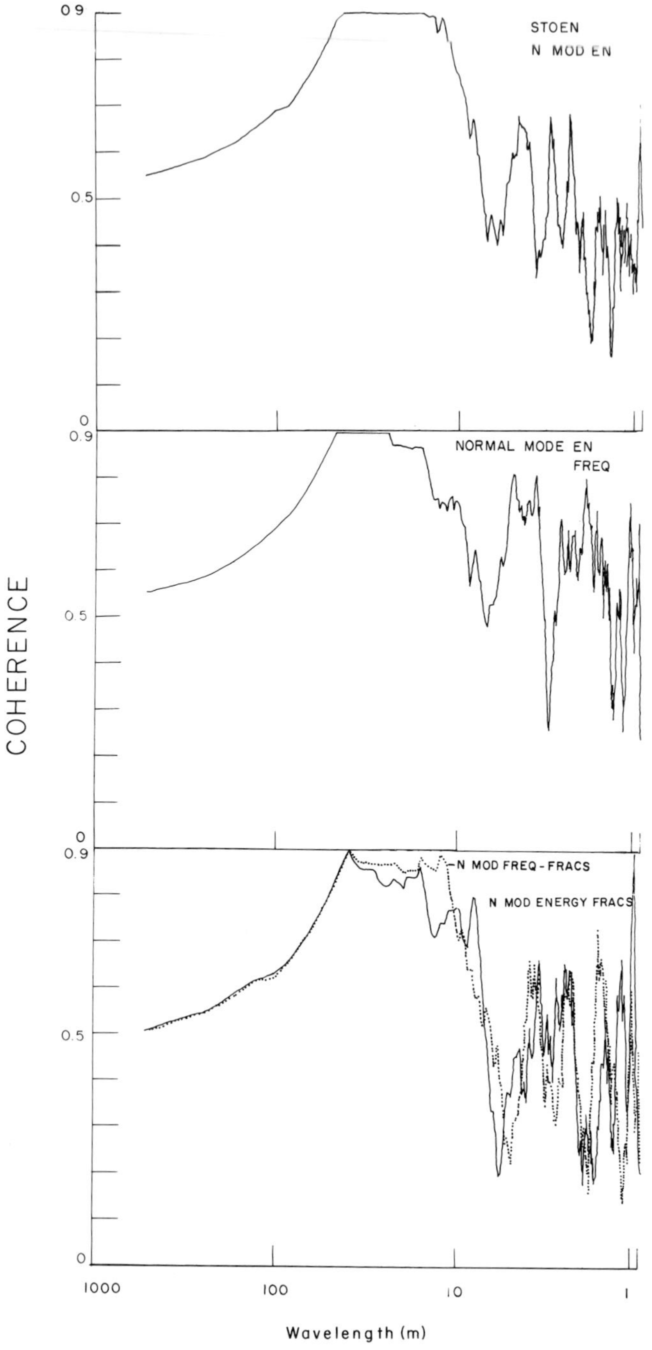


Fig. 31. Phase coherence between normal modal energy and Stoneley energy, normal modal energy and frequency, and both of the latter versus fracture pattern in the bottom 150 m of the dikes in hole 504B

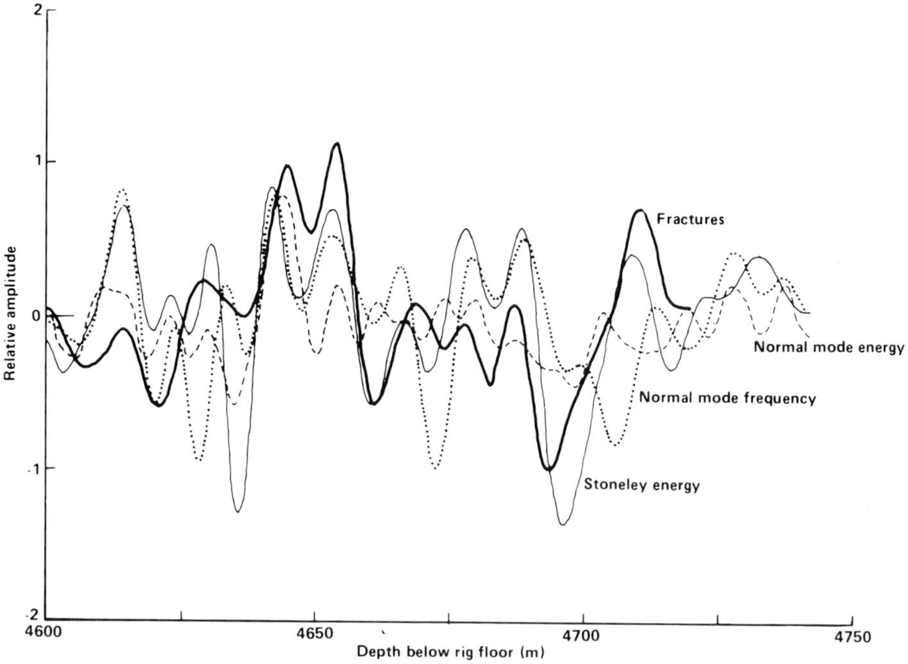


Fig. 32. De-trended, normalized, and bandpass filtered normal mode energy and frequency logs correlate with the fracture log and show pattern to fracturing in dikes of layer 2C. 15-25-m thick zones of intense and highly attenuating fractures are separated by 15-25-m thick zones of relatively solid basalt

We can now ask what aspect ratio of fractures within the zones of fracturing would be predicted by the venting model to produce the observed filtering effect on the normal modes. In order to better understand the effects of varying the size of vent fractures, we have conducted a series of model experiments in air-filled main and vented tubes of the same diameter in a lightly damped medium. The vent-opening away from the intersection with the main tube was changed to present variable diameter vent constrictions (Fig. 33). The purpose of this experiment is to map the effects of fractures with very small vent diameters relative to the main transmission path versus those with relatively large vent diameters. We hope this model represents adequately the case of variable constriction diameters for fractures which intersect the borehole in the oceanic crust.

The experiment consists of a constant voltage, pure tone source which is stepped upward in 48 frequency steps from 100 Hertz to 4 kHz. The energy output at the end of a main tube is measured at each step for each of five vent-constriction-to-main-tube ratios. The five different vent-constriction-to-main-tube ratios ranged from 0.17 : 1 to 1.06 : 1.

We discover that the effect of increasing the width of the vent constriction relative to the main fracture diameter is to shift the resonant frequency to higher frequencies and to increase the overall resonance and Anti-resonance of the system (Fig. 33). Peak frequency drops as the overall energy level drops in the manner observed. The borehole-

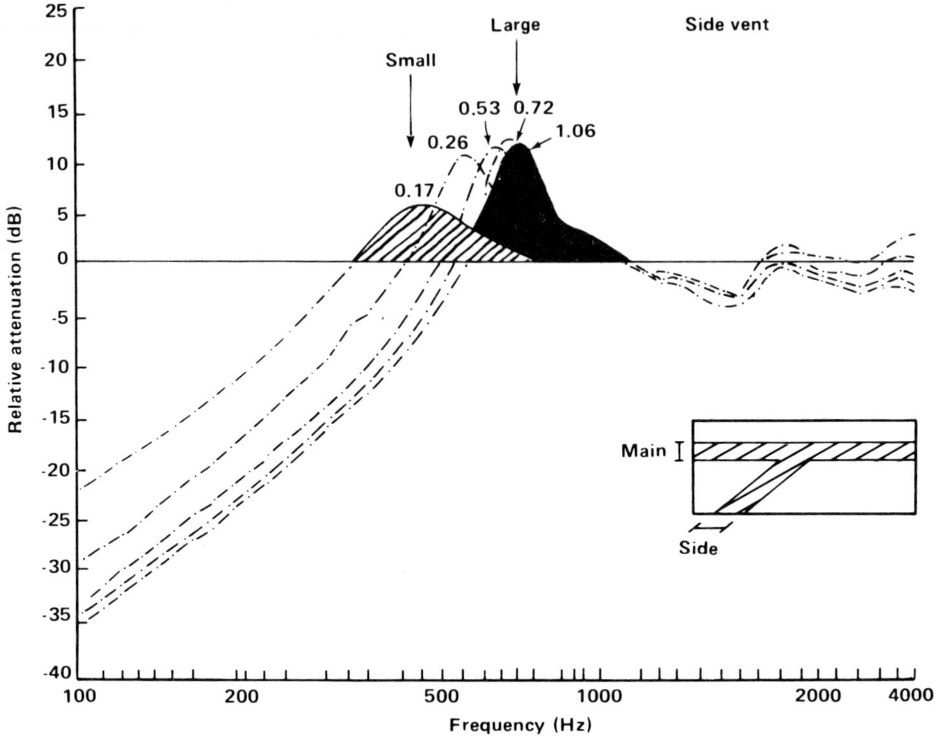


Fig. 33. Experimental results from an air-filled, vented tube for variable vent-constriction-to-main-tube diameter (ratios are indicated on each curve) in lightly damped matrix. Center or peak frequency changes and overall energy levels depend upon the vent-diameter ratio. In the oceanic crust, variable vent-diameter ratios would be expected. Perhaps this phenomenon accounts for some of the scatter observed in Figs. 12 and 20. If zones of intense fracturing seen in Figs. 31-32 consisted of smaller diameter vents than the solid, relatively unfractured zones, then this vent model would explain the spectral shifts observed: when fracturing is intense with many small diameter fractures, normal mode energy is low and frequency is also low. Zones of few fractures, each with larger diameters, would produce sections of the well with high energy and frequency normal modes

vent model analogy would require, however, that zones of intense fracturing be composed of many, many more small diameter fractures than zones of little fracturing, which must have a few relatively large diameter fractures. Then the spectral shifts seen in Figs. 31-32 would match those observed in Fig. 33.

In the oceanic crust, we expect to encounter a variably fractured medium so that constriction ratios of intersecting fractures would be highly variable. It is not surprising then that large scatter in Fig. 12 might result from the proposed "venting" mechanism if it is applicable to the oceanic crust. Maximum Anti-resonance occurs when the fractures are all about the same diameter, however.

We have not provided any direct proof that venting occurs in the oceanic crust. We have instead shown that a reasonable occurrence in a highly fractured rock of intersecting "vents" would provide the proper boundary conditions for the anti-resonance effects observed in the sonic waveform spectra within the dikes of layer 2C at hole 504B. Also, the Stoneley and normal mode filter-characteristics of the dikes might be explained by the venting hypothesis. This venting effect is well known in acoustics, and must be occurring to some degree in the highly fractured oceanic crust.

Assembling the anti-resonance, venting, phase coherence, and bandpass filtering analyses into a final conclusion from this study, we find that zones of intense fracturing are periodic down thw well. The dikes appear to be jointed into 15-25 m thick zones of intense fracturing. Such zones are characterized by the occurrence of subhorizontal, but relatively small diameter fractures with average lengths of about 25 cm, separated every 15-25 m by zones of more solid basalt with fewer, but larger diameter fractures. We reach these conclusions from the cross-correlation of the geophysical logs with the spectral changes seen in the sonic waveform logs. We predict that similar structures should be observable in the sheeted dikes of ophiolite complexes on land.

## Summary: The Effects of Permeability Changes on the Impedance of the Oceanic Crust

We have shown that impedance changes are complex and difficult to quantify in the oceanic crust. These changes most certainly come about as a result of physical property changes in the oceanic crust, yet specifying cause and effect will require more boreholes with more sophisticated logging than presently available.

As an example of the importance of fully describing the impedance transfer function of the crust, consider permeability. Fracturing, and the degree to which alteration products fill those fractures control crustal permeability. Physical property changes caused by the plugging of fractures with clay mineralization will produce impedance changes which, though complex, are beginning to be understood. It appears likely that in the future we will be able to work backwards from quantification of major impedance changes using geophysical logging to reconstruct the permeability history of the oceanic crust.

Hole 504B is a good locale to begin because permeability has been measured over the entire section of crust by flow-test methods, (Anderson and Zoback 1983; Anderson et al. 1984). Because of excessive temperatures however, a specific value could not be obtained in the sheeted dikes. Electrical resistivity measurements predict porosity changes which largely mirror the observed drop in measured permeability over layers 2A and 2B, (Fig. 34). If porosity changes from the electrical measurements are extrapolated into the dikes to guess at a value for the permeability for that layer, predictions are for very low values; near that of unfractured granite (10-20 m squared) (Becker et al. 1983).

Though we could reach only simplistic conclusions about the spectral changes within the dikes, we can conclusively state that the dikes are at least "acoustically" permeable. That is, energy (e.g., from Normal Modes) is being attenuated in zones of fracturing within the

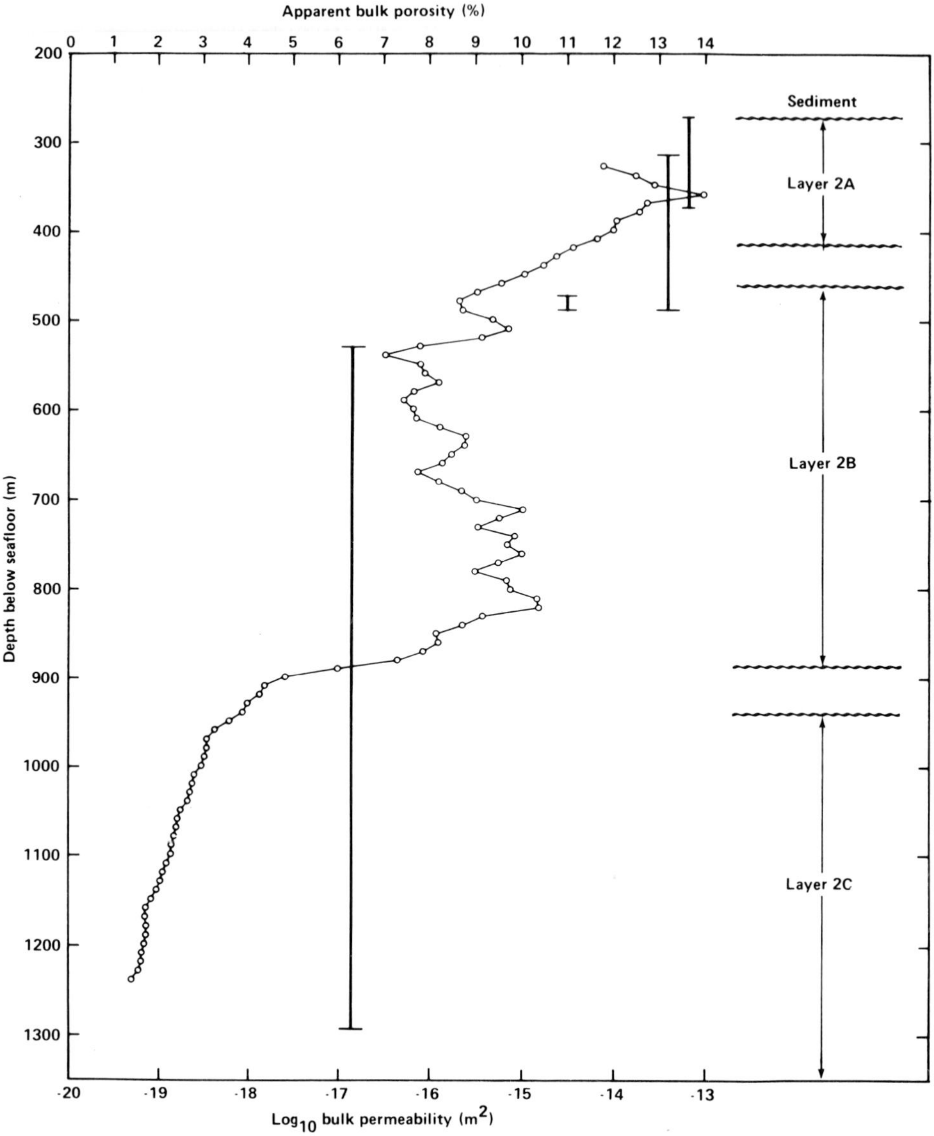


Fig. 34. Apparent bulk porosity from long-spacing electrical resistivity experiment run at hole 504B. Plotted on the same graph are permeability measurements in same well. The two scales are not related, but the relationship between the two data trends is obvious (Becker et al. 1983). The resistivity curve would predict very low permeability in the dikes at the bottom of the hole, but the acoustic permeability of this interval is still high, implying hydraulic permeability several orders-of-magnitude greater than 10-20 m² (see text)

dikes. The fact that 30 to 40 cm wavelength headwave energy is also heavily attenuated within zones of fracturing in the dikes suggests that such zones are likely to be hydraulically permeable as well, because intersecting "vents" are required if anti-resonances are to occur. We predict that the dikes are several orders of magnitude more permeable than unfractured granite, which would not attenuate and would have no vents.

In summary, geophysical logging in oceanic crust is only just entering an era where quantitative statements can be made about virtually every important physical property from in situ measurements made in deep boreholes. We have presented methods for the quantitative calculation of hydroxyl-mineral content of the formation, for use of the borehole-televiewer-fracture log and full-waveform-sonic-log spectral changes for lithology identification, fracture characterization, and acoustic permeability determination, and for the use of integrated geophysical log cross-correlation to determine overall geology of Deep Sea Drilling Project boreholes. We look forward to the day when many more than two crustal holes have been logged.

*Acknowledgments.* We thank the officers, crews and scientific parties of Glomar Challenger legs 82 and 83, without whose diligence and hard work these logs could not have been successfully completed. Schlumberger, Ltd., kindly provided computer and analyst time for this project. We particularly wish to thank David Leslie and Olive Liu from performing the spectral processing used in this paper, and Bill Menke for writing the FFT spectral analyses programs used in this paper. They contributed invaluable scientific insight as well. Dan Moos, Dave Egolf, Kier Becker, Bill Murphy, Ken Winkler, and two anonymous reviewers greatly improved the content of this paper. None of this work could have been completed without the support and encouragement of Mark Zoback. This work was supported by Office of Naval Research contract TO-0098, scope HH, and National Science Foundation grant OCE 81-10919.

## References

Alt JC, Laverne C (1984, in press) Alteration of the oceanic crust: mineralogy and processes in DSDP hole 504B. Init Reports, Deep Sea Drilling Project, Leg 83, US Gov Printing Office

Anderson RN, Honnerez J, Becker K (1982) DSDP Hole 504B, the first reference section over 1 km through layer 2 of the oceanic crust. Nature (Lond) 300: 589

Anderson RN, Zoback MD (1983) Permeability, underpressures and convection in the oceanic crust near the Costa Rica Rift, Eastern Equatorial Pacific. J Geophys Res 87: 2860

Anderson RN, Zoback MD, Hickman SH, Newmark R (1984, in press) In Situ permeability versus depth in the oceanic crust at DSDP Hole 504B. Init Reports, Leg 83, Deep Sea Drilling Project, US Gov Printing Office

Becker K, Von Herzen RP, Francis TJG et al. (1982) In situ electrical resistivity and bulk porosity of the oceanic crust, Costa Rica Rift, Nature (Lond) 300: 594-598

Becker K, Langseth MG, Von Herzen RP, Anderson RN (1983) Deep crustal geothermal measurements, Hole 504B, Costa Rica Rift. J Geophys Res 88: 3447

Bougault H, Cande S et al. (1982) Chemical inhomogeneities in the mantle: results from DSDP Leg 82. Nature (Lond)

Egolf DP, Tree DR, Feth LL (1978) Mathematical predictions of electroacoustic frequency response of in situ hearing aids. J Acoust Soc Am 63: 264

Hill IA (1983) Downhole physical property measurements in Atlantic ocean crust from wireline logging results. EOS 64: 314
Houtz R, Ewing J (1976) Upper crustal structure as a function of plate age. J Geophys Res 81: 2490-2498
Khatchikian A (1983) Log evaluation of oil-bearing igneous rocks, World Oil, December 79
Menke W (1983) Frequency response from scattering of seismic energy by multilayered media. Geophys J R Astron Soc
Moos D, Zoback MD (1983) In Situ studies of velocity in fractured crystalline rocks. J Geophys Res 80: 2345
Newmark R, Zoback MD, Moos D, Anderson RN (1984) Sonic and ultrasonic logging of Hole 504B and its implications for structure, porosity, and stress in the upper 1 km of the oceanic crust. Init Reports Deep Sea Drilling Project, Leg 83, US Gov Printing Office
Paillet FL (1981) Predicting the frequency content of acoustic waves in boreholes, 22nd Annu Logging Symp. Trans Soc Prof Well Log Analysts, Mexico City, p 26
Zoback MD, Anderson RN (1982) The implications of fracture distribution, structure, and stratigraphy from borehole televiewer imagery of the oceanic crust on the Costa Rica Rift. Nature (Lond) 295: 375

# RESEARCH AND DEVELOPMENT ACTIVITIES IN GEOTHERMAL DRILLING, COMPLETION, AND LOGGING

John Finger[1]

Sandia National Laboratories manages the Geothermal Drilling and Completion Program for the US Department of Energy. The primary purpose of this program is to expand access to the geothermal resource by reducing the cost of drilling and completing geothermal wells. This cost is typically two to five times that for an oil well of comparable depth, and well cost is usually about half the total capital investment for a geothermally-powered plant to generate electricity.

Geothermal drilling is plagued by a combination of problems not shared by the rest of the drilling industry. Formations are hard, abrasive, fractured, and frequently underpressured, which results in slow drilling and lost circulation. The fluids produced are often corrosive and high in dissolved solids. Completions are difficult, with casing failure a major cost in many geothermal wells. Aggravating this catalog of complications is the fact that the geothermal industry is so small compared to the oil and gas industry that oil tool manufacturers and service companies have little incentive to do research on geothermal problems. Thus, if we accept the premise that development of this alternative energy source is in the national interest, reduction of geothermal well cost is a legitimate expenditure of taxpayer's money.

Our principal goal in this program is the development of an advanced drilling system that can drastically reduce well costs. Since this is a large task, and we do not have the resources to attack the whole problem, we have chosen projects that would be part of an advanced system but would also have stand-alone value as successful research and development. Since a great variety of projects is available, we have conducted systems analyses to select work that is appropriate to the goal, scaled to our resources, and likely to have a high payoff.

Our program is defined in five basic elements:

- Rock Penetration Mechanics
- Fluid Technology
- Borehole Mechanics
- Diagnostics
- Permeability Enhancement

Each of these elements will be briefly summarized with examples of typical projects in each area.

[1]Div. 6241, Geothermal Technology Development, Sandia National Laboratories, Albuquerque, NM 87185, USA

Observation of the Continental Crust through Drilling I (ed. by C.B. Raleigh)

## Rock Penetration Mechanics

This element includes all the aspects of actually making a hole in the rock. In general, the work has been directed toward analytical and experimental investigation of rock breakage mechanisms and drill bit behavior. Some examples are:

- Erosion drilling — A series of contracts with Hydronautics, Inc has led to the analytic description and experimental verification of cavitating and pulsating water jets. These jets produce a periodic train of ring vortices which can impinge on a wellbore surface. Jets of this type are interesting for several potential uses - direct rock breakage, augmentation of mechanical rock breakage, and improved bottom hole cleaning. A roller cone bit with pulsating jet nozzles has been commercially produced and shown to have performance superior to the same bit with conventional nozzles.
- Rock/cutter interaction — A finite element model of a drag bit cutter moving through rock has been developed and compared to laboratory experiment. The model gives forces on the cutter, stresses in the cutter, and fracture patterns in the rock. Qualitatively, the agreement with experiment has been good, although the calculated loads are lower than actual loads. All cutting tests have been done at atmospheric pressure, but model predictions should be better when there is borehole or confining pressure to suppress tensile cracks. A series of pressurized cutting tests to investigate cutter forces and rock damage patterns is imminent.
- Drill String Dynamics — A finite element model of a drill string - from the bit/rock interface to the surface - is under construction. This model incorporates user-definable formation, bit geometry, and drill string configuration. Knowledge of the dynamic behavior of the drill string is important when considering hole deviation, bit performance, and the environment to be imposed on potentially delicate downhole equipment. The elements of this model will be validated by lab experiment; parts of this have already been done. This project is being supported jointly by Sandia plus four oil and service companies.

## Fluid Technology

This element deals primarily with drilling fluid; its properties and its flow in the borehole.

- Aqueous Foam — Because many geothermal formations are underpressured, the use of low density drilling fluids is either advantageous or necessary. Compared to air, aqueous foam has higher carrying capacity and lower air flow rate requirements. Early foam work investigated drainage times of various surfactants at high temperature; current work is focussed on heat transfer properties and rheology of flowing foam.
- Clay-Based Drilling Fluids — Because some wells will be drilled with mud, we have measured properties of clay-based drilling fluids at high temperature, including the effects of several salts and hydroxides on viscosity. Current work includes a look at in-situ clay-to-cement conversion and the formulation of a high temperature bentonite mud.

- Bit Hydraulics — We operate a Sandia-designed test stand that can be used to visualize flow across the face of a rotating bit. Identification of stagnant points in the flow is important when considering bottom hole cleaning, bit balling, and convective cooling of the cutters. The facility can also be used to study the effect of changing nozzle and junk slot placement.

## Borehole Mechanics

This element includes the factors which affect the integrity of the borehole. Past work involved such projects as casing stress analysis and a water jet casing de-scaler, but current activity is concentrated on problems of lost circulation. A Sandia-designed Lost Circulation Test Facility (LCTF) has been used to study the dynamic behavior of lost circulation material in a simulated borehole. The LCTF is a 1000 psi (6.89 MPa), 400°F (204°C) flow loop which accurately simulates the movement of particles past, as well as into, a borehole crack. Well over 100 tests have been done and the data used to establish criteria for crack bridging. Development of an analytic model complementary to these experiments is under way.

## Diagnostics

The diagnostics element involves the techniques used for getting information out of the wellbore. Earlier work was concentrated on the development of a family of high temperature electronic components that would operate continuously at 527°F (275°C) and could be used either for wireline logging or measurement while drilling (MWD). Other recent or current work includes:

- Lost Circulation Zone Mapping Tool — The low temperature acoustic televiewer commercially available from Simplec was extensively redesigned to operate for 5 hours at 527°F (275°C) with the electronics in a dewar. The tool has been field tested in hot dry rock and commercial geothermal wells; it has provided information about fractures (location, density, orientation) as well as about cracks and corrosion in casing. The latest design iteration for the tool includes temperature logging capabilities.
- Wellbore Inertial Navigation System (WINS) — Based on Sandia weapons guidance technology, this tool gives a continuous survey of wellbore location. The important innovation was in software; signal processing uses Kalman filtering with periodic updates to give a position accuracy of one foot (0.3 m) per 1000 feet (305 m) of well depth. This tool has been field tested, but is not packaged to operate at high temperature.
- Cement Bond Log — Casing failure is a major geothermal problem and much of it is due to poor cement jobs. The bond log tool uses acoustic data to evaluate the bond between the casing and the cement. A dewar is not used here, the tool is designed to operate continuously at 482°F (250°C). The design is complete and has been bench tested.

## Permeability Enhancement

Increasing production from an existing well is the focus of this element. Assessment of natural and man-made fracture systems as well as actual stimulation techniques are being pursued.

- Fracture Assessment – Since most of the fluid in a geothermal well is produced through fractures, it is important to know both the natural fracture system around a well and how that system may have been altered after treatment. Fractures that intersect the wellbore can be detected by the televiewer described above, but that tool does not give significant definition beyond the wall. To detect non-intersecting fractures, we have tested a 30 MHz downhole radar with directional transmitter and receiver. This tool has detected a brine-filled fracture 35 feet (10.7 m) away in granite; current effort is aimed at improving data interpretation and antenna directionality.
- Modified Pressure Pulse – If a wellbore is pressurized to fracture the formation, the pressure pulse rise time will control the way that the cracks are formed. If the rise time is long, as in hydraulic fracturing, a pair of cracks perpendicular to the least principal stress will form. If these cracks are parallel to the "natural" fracture system, production won't be significantly increased. If the rise time is very short, as in explosive fracturing, the rock near the wellbore may deform plastically, creating a "stress cage" that will actually inhibit production. There is an intermediate rise time, however, that will drive cracks in many directions. It is then more likely that the wellbore fractures will intersect the natural fractures and connect them for increased production. The rise time is tailored by adjusting the mixture of propellants that creates the pulse. The concept has been field tested and verified - a series of experiments to address specific geothermal problems will be done in late 1984.
- Modelling – The modified pressure pulse fracturing has been modelled in a finite element code so that performance can be predicted when key variables such as acoustic wave speed, rock strength, and in-situ stress are altered.

This outline of activities is not all-inclusive and project descriptions are necessarily minimal. Further information on these and other projects can be found in the References, which also include names and telephone numbers of the people involved in specific tasks.

*Acknowledgment.* This work was performed under the auspices of the U.S. Dept. of Energy Contract DE-AC04-76DP00789.

## References

Kelsey JR (ed; 1981) Geothermal technology development program. Annu progress report October 1980 - September 1981 SAND81-2124

Kelsey JR (ed; 1982) Geothermal technology development program. Annu progress report October 1981 - September 1982 SAND82-2690

Kelsey JR (ed; 1983) Geothermal technology development program. Annu progress report October 1982 - September 1983 SAND84-1028

Varnado SG (ed; 1980) Geothermal drilling and completion technology development program. Annu progress report October 1979 - September 1980 SAND81-2179

# VERTICAL ELECTROMAGNETIC PROFILING (VEMP)

R. J. Lytle†1

## Introduction

A question of great significance is, "What are the properties of the ground?" This is of importance in locating natural resources (such as mineral and oil deposits) and in fundamental studies (such as scientific continental drilling interests). Much prior work has addressed surface and borehole measurements for economically driven natural resource applications. Geophysical measurements to ascertain fundamental properties of the earth have seen less attention. The availability of one or more deep boreholes thus opens new scientific vistas.

Assuming one or more deep boreholes are available for scientific studies, a key question is, "What data can be acquired which may yield results of fundamental importance?" It is, of course, recognized that the data acquired is particular to the general region surrounding the borehole(s). Given the objectives envisioned for the scientific continental boreholes, it is useful to attain as much data as possible. Both physical results which stand alone and physical results which provide complementary results are useful. A suite of physical observables is desirable. A partial list of physical observables might include the chemical, seismic, electromagnetic, and mechanical properties, among others.

Some data should be sought that has relatively high resolution but relatively little global sensing capability. Some data should also be sought that has relatively little resolution but has relatively high global sensing capability.

## Electromagnetic Probing Considerations

The electromagnetic properties of the ground are one set of physical observables that can satisfy the need for providing stand-alone data, complementary data, a locally confined high resolution capability and a global sensing capability. This paper addresses the utility of electromagnetic measurements performed in a deep borehole.

Because the major focus has been on economic objectives rather than fundamental scientific objectives, electromagnetic means of sensing the ground have been largely concentrated on surface-based measurements and single-borehole measurements of earth properties. Probing from these surface and borehole locations has been concerned with detecting economic features of interest.

---

[1]Electronics Engineering Department, Lawrence Livermore National Laboratory, University of California, P.O. Box 808, Livermore, CA 94550, USA

Observation of the Continental Crust through Drilling I (ed. by C.B. Raleigh)

Those features that are detectable from the surface may or may not be detectable from a borehole, and vice versa. The detectability of features is partially dictated by their proximity in skin depths relative to the surface or the borehole. Since the spatial extent of effective probing is partially governed by the electromagnetic skin depth, which is a function of frequency, it is sometimes wise to use a suite of frequencies in performing remote probing. This can be done using either pulses (time domain techniques) or a set of frequencies (frequency domain techniques).

Due to equipment or operational considerations, it may not be possible or useful to employ the same probing means in the borehole and at the ground surface. For example, this might mean that different frequency ranges, antenna types and antenna orientations might be used on the surface and in the borehole. Since a variety of operational scenarios can be considered, modeling should be performed with the objective of designing an experiment, that attains the most useful data at the least expense. This is not always as straightforward as it may appear, especially for objectives related to scientific continental drilling. One reason for the complexity of the problem is that although it is conceptually relatively easy to envision a set of data being collected and analyzed, the necessary equipment may not exist or it may not be designed to survive in the possible adverse environment in which it has to operate.

Prior work (Ref. 1-28) has been performed with the objective of profiling the electrical or electromagnetic properties of the ground. This is true with regard to both the variation with depth and with lateral location. From a historical perspective, the word electrical probing has commonly been differentiated from the word electromagnetic probing. Typically the designation electrical probing has referred to probing where conduction currents dominate displacement currents. This regime is also sometimes called the low frequency regime. The designation electromagnetic probing has commonly referred to probing where displacement currents dominate conduction currents. This regime is also sometimes called the high frequency regime. In a technical sense, the word electromagnetic actually encompasses the complete spectrum of frequencies for electromagnetic energy. This spectrum of frequencies is somewhat facetiously described as covering from dc to daylight, i.e., from zero frequency to frequencies much higher than are commonly used in probing the ground to depths greater than a few millimeters. When talking of ground probing measurements that may include either high or low frequency energy, the most general designator is therefore electromagnetic probing.

The seismic community has been very successful in developing a technology for probing the ground in the vicinity of a borehole. The seismic measurements which are performed are made at the earth surface (e.g., surface reflection seismic probing), in the borehole (e.g., conventional seismic inter-borehole and intra-borehole logging) and between locations near the earth surface and locations in the borehole (i.e., surface-to-borehole seismic measurements). In the way in which the data is collected and analyzed, the attainable resolution of features in the vicinity of the borehole is superior to that obtainable from data collected just at the earth surface or just from within the borehole. This seismic technology is designated as vertical seismic profiling (VSP).

In recognition of the utility of combining these separate means of collecting data to attain a superior overall result, it is prudent to consider the technology of vertical electromagnetic profiling. By this is meant that a wide range of frequencies (either in time or frequency domain form) are used to perform electromagnetic probing from locations at the earth surface (e.g., magnetotelluric measurements), from within the borehole (e.g., intra-borehole and inter-borehole resistivity logging) and from locations at the earth surface to locations within the borehole (surface-to-borehole measurements). This technology is designated as vertical electromagnetic profiling and is given the acronym of VEMP, in analogy to the acronym VSP (Fig. 1).

VEMP has seen limited application to date. The opportunity exists for more extensive application of VEMP. Before this will occur, fundamental scientific studies may need to be performed which address critical aspects of the overall problem. Philosophically, a fundamental question is, "What sources and measurements will yield the most useful data on the lateral and depth variations in ground properties?" Although it is sometimes thought that the answers to this question are intuitively obvious, a reevaluation is often warranted considering advances in technology. Such advances may be occurring, or have already occurred, although they may not have been applied to VEMP. In other cases, the significant advantages that a technology breakthrough (e.g., in instrumentation capability or data analysis) would merit a research and development effort to achieve this goal.

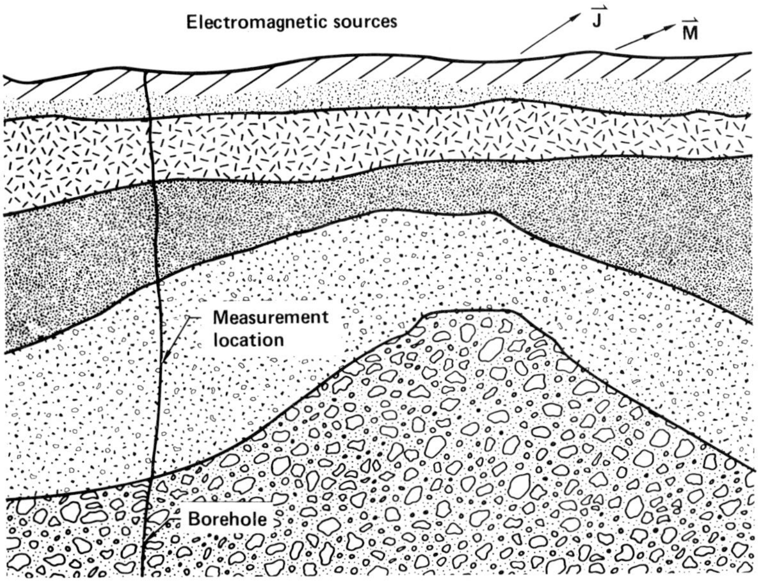


Fig. 1. Vertical electromagnetic profiling (VEMP) consists of using electromagnetic sources (electric currents $\vec{J}$ and magnetic currents $\vec{M}$) and receivers at the ground surface and in one or more boreholes in the ground. From this data, interpretations are made as to lateral and depth variations in electromagnetic constitutive properties in the ground

This paper briefly addresses a number of the subparts of which VEMP is comprised. For brevity, primarily mathematical modeling aspects of VEMP are addressed here. Other aspects such as laboratory measurements on cores (such measurements can provide key physical relationships which may yield more definitive diagnostic information than is presently attained) and design of instruments that will survive and reliably function, even in adverse environmental conditions, are not addressed here.

## Mathematical Modeling

The problems which are studied using mathematical modeling are the forward problem (i.e., calculation of responses given specific situations with known source/receiver/ground types, excitations, locations of sources/receivers/anomalies) and the inverse problem (i.e., providing estimates of the locations, extent and constitutive properties of the ground and any anomalous features using measurement results from known source/receiver/excitation types and locations).

Both the forward and inverse problem procedures are essential in adequately developing a proper experimental design. Performing a number of such experimental designs can help define the advantages and disadvantages of the various probing methods. This knowledge is helpful in deciding which methods are best suited to particular probing objectives and scenarios. The knowledge is also helpful in determining the limits of detectability, both from a signal to noise standpoint as well as from anomalous feature "invisibility".

A pictorial depiction of the VEMP process is given in Figure 1. Electromagnetic sources are shown nearby the earth surface. These sources can be electric dipoles, magnetic dipoles or special cases of these basic source types (e.g., plane wave excitations). The electromagnetic sources could also be located in the borehole. Measurement locations could also be near the earth surface and in other boreholes (not shown).

The types of excitation can be frequency domain or time domain signals. Dependent upon the frequency content of the excitation, electromagnetic conduction currents (e.g., such as excited by four probe resistivity methods), and displacement currents (e.g., such as excited by wave propagation phenomena as reflection, transmission, diffraction and refraction) can be excited in the ground. These currents can be altered in passing through the ground. The way in which they are altered is sometimes diagnostic of the anomalous features which caused the alteration. Examples of anomalies that have been conventionally modeled are a layered ground, two-dimensional anomalies (e.g., anomalous cylinders of arbitrary shape) and three-dimensional anomalies (e.g., spheres, plates, discs), among others, including even rather complex variations in shape and constitutive parameter variation.

Electromagnetic probing with source and receiver in the borehole is a subset of vertical electromagnetic profiling. It is an exploration tool which can be used when the depth and resolution capabilities of ground based electromagnetic probing systems are known to be inadequate. It can also be used as a check on the validity of interpretations obtained using exclusively ground-based system data.

Primarily because of the economic interest in mineral exploration, there has been a surge of activity in developing and validating such borehole electromagnetic systems for relatively shallow depths (i.e., not the depths being considered for scientific continental boreholes). Detailed scale model, theoretical and field studies have established that if a borehole passes in the vicinity of an electrically conductive body, the borehole-based system will detect the presence of the body and return information about the size, orientation and conductivity-thickness product for the body.

Advantages in using vertical electromagnetic profiling include: providing information in regions where cores are lost; identifying lithologic changes that may not be readily apparent in visual inspection of cores; providing information on features which may be remote from the borehole and which may be useful in planning future activities in the area; providing information regarding the presence, extent and constitutive properties of geophysical features remote from or intersecting the borehole and providing electromagnetic data for the ground that is of higher local accuracy than surface-based data is; thus, it can be used to calibrate and thereby improve interpretations of surface-based data.

In performing vertical electromagnetic profiling, it is commonly advocated that a pulse excited large horizontal transmitting loop antenna near the top of the borehole provides the greatest depth of penetration and greatest extent of probing remote from the borehole. These statements are made as: (1) frequency domain excitations can have problems in accurately determining the secondary field in the presence of a large primary field, (2) small loop transmitting antennas can excite relatively small close by conducting anomalies more than larger further removed bodies of greater interest, and (3) it is commonly easier to use a receiving antenna in the borehole and a transmitting antenna at the ground surface.

These points are valid points. In certain circumstances, however, other operational considerations may dictate that an alternative system is preferable. Each probing situation should be evaluated based upon the knowledge available for local site conditions, experimental objectives and instrumentation suitability. Computer modeling can be used to assess expected results. As an example, one type of computer modeling study is shown in Figs. 2-6.

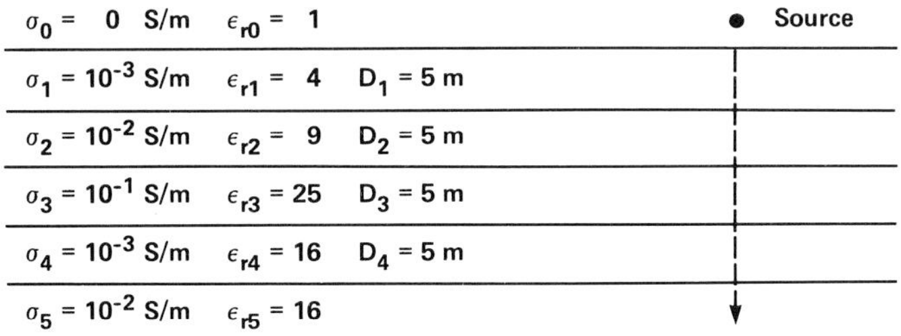


Fig. 2. The high frequency model (f=10 MHz) has horizontal magnetic and electric current sources

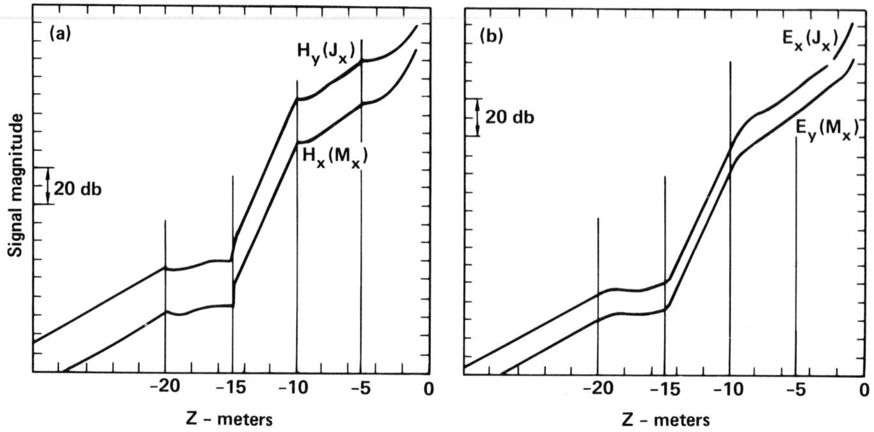


Fig. 3a,b. The electromagnetic fields excited by electric and magnetic dipoles are somewhat similar in shape. Both show near field differences and wave interference effects near interfaces

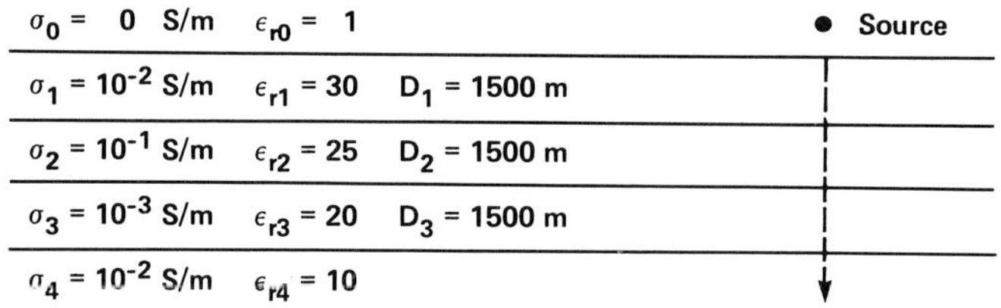


Fig. 4. The low frequency model (f=5 Hz) has horizontal magnetic and electric current sources

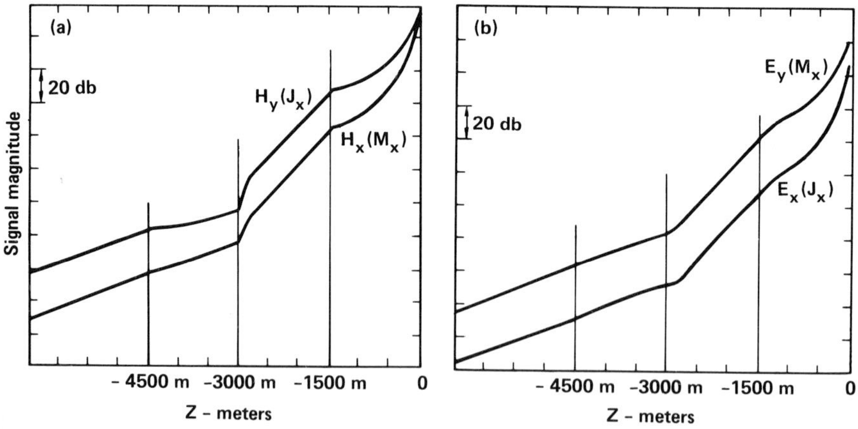


Fig. 5a,b. The electromagnetic fields excited by electric and magnetic dipoles are somewhat similar in shape. Both show near field differences and wave interference effects near interfaces

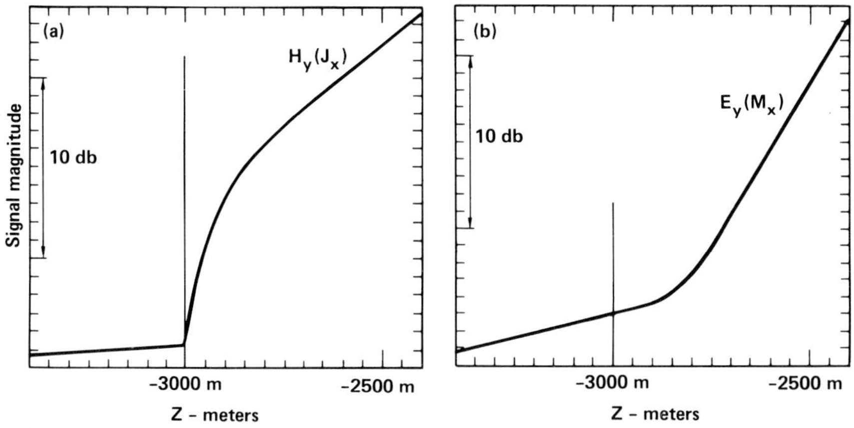


Fig. 6a,b. The electromagnetic fields excited by plane wave excitations due to various sources show wave interference effects near interfaces

Near surface probing results are shown in Fig. 3 for the profile and source conditions depicted in Fig. 2. This situation involves a horizontal electric Hertzian dipole source (J) and a horizontal magnetic Hertzian dipole source (M). For the high frequency excitation considered here, the electric and magnetic fields that are excited are similar in shape. Key features in the results are that the two sources excite similar field variations, the fields have different near field structure and interference effects (due to superposition of upgoing and downgoing waves) are apparent in the field behavior near changes in media constitutive properties. These results are expected for high frequency probing. It is important to recognize that similar observations can apply for lower frequency probing to much greater depths.

Figure 5 shows the field response within the ground structure of Fig. 4 for low frequency horizontal electric and magnetic dipole Hertzian sources at the ground surface. Once again, the differences in near field variations for the two different sources are seen – the general features of the two fields are similar in shape and the presence of interference effects (due to the superposition of upgoing and downgoing waves) is evident in the vicinity of where there are changes in constitutive properties within the ground. The extent of these interference regions can be significant.

To further illustrate this interference interaction zone extent, more results are shown for the ground situation of Fig. 4. In the case results shown in Fig. 6, the excitation is a plane wave normally incident upon the ground. These results clearly indicate the need for care if interpreting the local low frequency field decay in terms of just a downgoing wave alone. Unless the measurement location is sufficiently far removed from any significant changes in media properties, errors in interpretation for the local ground properties can occur.

It is also possible to encounter situations where sideways going waves can be excited due to lateral variations in constitutive properties in the ground (e.g., see Fig. 7). As with the cases shown above for upgoing and downgoing waves, the extent of influence for the sideways going waves is dictated by the frequency of excitation and the changes in constitutive properties in the ground. Extensive modeling studies

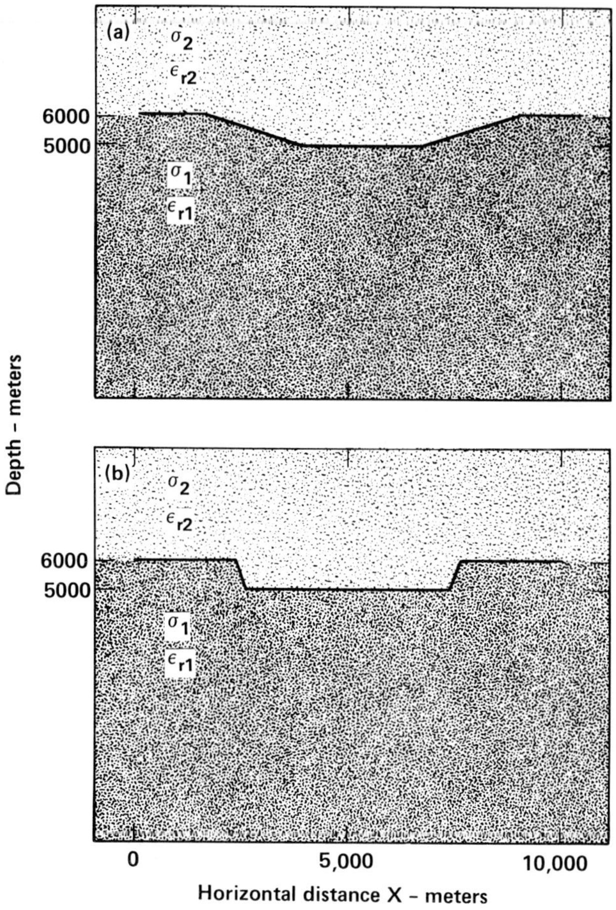


Fig. 7a,b. A borehole passing through the ground may or may not intersect all features of interest. Some means of remotely detecting the lateral variation in constitutive properties remote from the borehole is need. VEMP may provide useful data by sampling sideways-going energy

have been performed assessing the utility of such sidegoing waves in interpreting vertical seismic profiling results. The sideways moving electromagnetic energy has been studied to a lesser extent.

To a large extent, this disparity in study has been because of the perceived larger benefits attainable in using the sideways moving seismic energy as compared to the sideways moving electromagnetic energy. Before this perception is cast as reality, further modeling work is needed on the uses of and attainable information present in the sideways moving electromagnetic energy. It is not difficult to conceive of situations where lateral changes in electromagnetic constitutive properties occur but lateral changes in seismic constitutive properties do not occur. In such cases, the geophysical features that need to be detected would be invisible to VSP but might be readily detectable to VEMP.

A problem of importance in interpreting VEMP data is the effect of the ground contacting region (e.g., the borehole fluid, probe position within the borehole, near surface effects). Recent work has addressed these problems. Further work is needed to remove these effects from the data.

## Final Remarks

This overview has indicated the following points:

1. Near field influences affect the apparent field decay, thus caution is needed in interpreting VEMP data.
2. Media contrasts (i.e., changes in electrical constitutive parameters) also affect the apparent field decay. In general, both near field and media contrast effects are present and need to be accounted.
3. Interference effects occur for both high and low frequencies. The impact of the two features mentioned in items 2 and 3 above are evidenced in the field behavior.
4. Effects due to lateral inhomogeneities remote from the borehole may be detectable in the borehole but may not be detectable at the earth surface.
5. A number of specific solutions to inverse problems exist. Examples include methods for inferring the depth and constitutive properties of remote planar features as well as methods for determining the location, diameter and constitutive properties of a spherical anomaly remote from a borehole.
6. General solution procedures for solving the inverse problem for lateral inhomogeneities of general structure need to be developed.
7. Usable techniques exist and have been used for generating solutions to forward modeling problems. This is true for planar layered grounds as well as for 2-D and 3-D anomalous features in a homogeneous ground or in a layered ground. Computer costs become appreciable as the complexity of the ground structure increases.

To have VEMP become a major contributor to scientific continental drilling interests, advances are needed in experimental design, borehole instrumentation which works in adverse environments, and data processing procedures incorporating information acquired within a deep borehole.

*Acknowledgments.* The author has benefited by discussions with his Lawrence Livermore National Laboratory colleagues Ed Laine, Bill Daily, Clyde Dease, and Edna Didwall. Any errors are those of the author. Work performed under the auspices of the U.S. Department of Energy by the Lawrence Livermore National Laboratory under Contract W-7405-Eng-48.

## References

Barnett CT (1984) "Simple Inversion of Time-Domain Electromagnetic Data". Geophys 49 (7): 925-933

Boyd GW, Wiles CJ (1984) "The Newmont Drill-Hole EMP System – Examples from Eastern Australia". Geophys 49 (7): 949-956

Cagniard L (1953) "Basic theory of the magneto-telluric method of geophysical prospecting". Geophys 18: 605-635

Clarke AR, Salt DJ (1951) "The investigation of earth resistivities in the vicinity of a diamond drill hole". Geophys 16: 659-665
Daniels JJ (1977) "Three-dimensional resistivity and induced - polarization modeling using buried electrodes". Geophys 42 (5): 1006-1019
Daniels JJ (1978) "Interpretation of buried electrode resistivity data using a layered earth model". Geophys 43 (5): 988-1001
Daniels JJ, Dyck AH (1984) "Borehole resistivity and electromagnetic methods applied to mineral exploration". IEEE Geosci Remote Sensing GE-22 (1): 80-87
Devaney AJ (1984) "Geophysical diffraction tomography". IEEE Trans Geosci Remote Sensing GE-22 (1): 3-13
Dyck AV, West GF (1984) "The role of simple computer models in interpretations of wide-band, drill-hole electromagnetic surveys in mineral exploration". Geophys 49 (7): 957-980
Eaton PA, Hohmann GW (1984) "The influence of a conductive host on two-dimension borehole transient electromagnetic responses". Geophys 49 (7): 861-869
International Symposium on Observation of the Continental Crust Through Drilling, May 20-25 (1984) Tarrytown, New York, sponsored by the US Department of Energy, the US Geological Survey and the National Science Foundation, Organizing Committee Chairman Dr. Barry Raleigh, Lamont-Doherty Geological Observatory, Columbia Univ
Kennett BLN (1972) "Seismic waves in laterally inhomogeneous media". Geophys JR Astr Soc 27: 301-325
Lytle RJ (1984) "The influence of the ground contacting system on cross-borehole and borehole-to-surface electrical measurements". (submitted to IEEE Trans Geosci Remote Sensing, LLNL report UCRL-90932, June)
Lytle RJ, Lager DL (1976a) "Theory relating to remote electromagnetic probing of nonuniform thickness coal seam". Radio Sci 2 (5): 465-476
Lytle RJ, Lager DL (1976b) "Using the natural-frequency concept in remote probing of the earth". Radio Sci 11 (3): 199-209
Nabighian MN, Oppliger GL, Edwards RN, Lo BBH, Cheesman SJ (1984) Cross-hole Magnetometric Resistivity (MMR). Geophys 49 (8): 1313-1326
Ogilvy RD (1983a) "Application of down-hole pulse electromagnetic surveys for off-hole mineral exploration". Trans Inst Mining Metallurg 92: B148-B153
Ogilvy RD (1983b) "Transient electromagnetic prospecting technique applied to mineral exploration in the UK". Trans Inst Mining Metallurg 92: B132-B157
Ogilvy RD (1984) "Down-hole IP surveys applied to off-hole mineral explorations - some design considerations". Geoexplor 22 (1): 59-74
Oristaglio ML, Hohmann GW (1984) "Diffusion of Electromagnetic Fields into a Two-Dimensional Earth: A Finite-Difference Approach". Geophys 49 (7): 870-894
Seigel HO (1952) "Ore body size determination in electrical prospecting". Geophys 17 (4) 907-914
Telford WM, Geldart LP, Sheriff RE, Keys DA (1976) Applied geophysics, Cambridge Univ Press
Wait JR (1951) "A conducting sphere in a time varying magnetic field". Geophys 16 (4): 666-672
Wait JR (1984) "General formulation of the induction logging problem for concentric layers about the borehole". IEEE Trans Geosci Remote Sensing GE-22 (1): 34-42
Wannamaker PE, Hohmann GW, SanFilipo WA (1984) "Electromagnetic modeling of three-dimensional bodies in layered earths using integral equations". Geophys 49 (1): 60-74

Wannamaker PE, Hohmann GW, Ward SH "Magnetotelluric responses of three-dimensional bodies in layered earths". (submitted to Geophys)
Ward SH (1959) "Unique determination of conductivity, susceptibility, size, and depth in multifrequency electromagnetic exploration". Geophys 24 (3): 531-546
Worthington MH, Kuckes A, Oristaglio M (1981) "A borehole induction procedure for investigating electrical conductivity structure within the broad vicinity of a hole". Geophys 46 (1): 65-67

# RECENT RESULTS FROM SCIENTIFIC DRILLING

# RESULTS OF DEEP DRILLING IN AUSTRIA

Hermann F. Spoerker[1]

Austria is a small country in the heart of Europe and most foreigners have problems in not calling it Australia; even if they heard of it before, they think of Johann Strauss waltzes and the world-famous Sacher-Torte. Very few – even in the oil business – know that the deepest wildcats in the search for hydrocarbons worldwide, outside of the USA, have been drilled here. Oil and gas exploration in Austria started early this century in 1909. Galicia, at that time part of the Austro-Hungarian Empire, ranked third in world oil production. In 1934 the first commercial oil discovery was made in Austria, within the borders as we know them today. By 1943 the annual production passed the 1 million tons per year mark and in 1956 Austria produced more than 3.6 million tons of crude oil. Because of the intensive exploration programs during the 1940's and 1950's our company, the OeMV Aktiengesellschaft, was forced to drill deeper and deeper.

The first well in Austria to reach a depth of 3000 m was the Palterndorf 1 in 1957, the 4000 m mark was passed in 1961 in the Schönfeld 1 well.

The era of ultra deep exploration began in Austria in 1966 when OeMV (the Austrian State Oil Company) spudded the first 6000-m well: the Schönkirchen T 32, some 35 km northeast of Vienna. In this area the Matzen Oil Field, at that time the largest oil field in Western Europe, was discovered in 1949 (Fig. 1).

Below the tertiary oil and gas accumulations a new oil discovery was made in 1962 in the top part of the basement: this field was called Schönkirchen Deep.

Following the geological concept that one should drill in a proven oil and gas province deeper and deeper until one has definite indications that no more hydrocarbon accumulations can be expected, the OeMV spudded the before-mentioned ultra deep test: Schönkirchen T 32 (Fig. 2). Drilling operations progressed according to plan and something exploration people can only dream of happened: the very first ultra deep test discovered another gas field: Schönkirchen Ultra Deep. As can be seen on the geological cross-section of this area, the discovery was made in a dolomite body of the highly tectonically disturbed stratas of the Mesozoic Basement. The test showed sour gas, containing approx. 13% $CO_2$ and 2% $H_2S$. Due to the fact that the reservoir was in a steep dipping dolomite body the reservoir pressure gradient on top was about 1.4 atm - 10 m.

[1]Director Exploration and Production, OeMV Aktiengesellschaft, Otto-Wagner-Platz 5, 1090 Vienna, Austria

Observation of the Continental Crust through Drilling I (ed. by C.B. Raleigh)

Fig. 1. Vienna Basin

This field was developed by other 6000 m wells and in 1977 an even deeper gas discovery was made in the wildcat Gänserndorf ÜT1 which reached a T.D. of 6344 m.

These deep discoveries stimulated the exploration efforts of the OeMV and already in 1975 it was decided to order one of the biggest land drilling rigs available at that time to be able to reach to depths of 8000 and 9000 m. This rig, an IDECO H 3000 (Fig. 3), was rigged up on

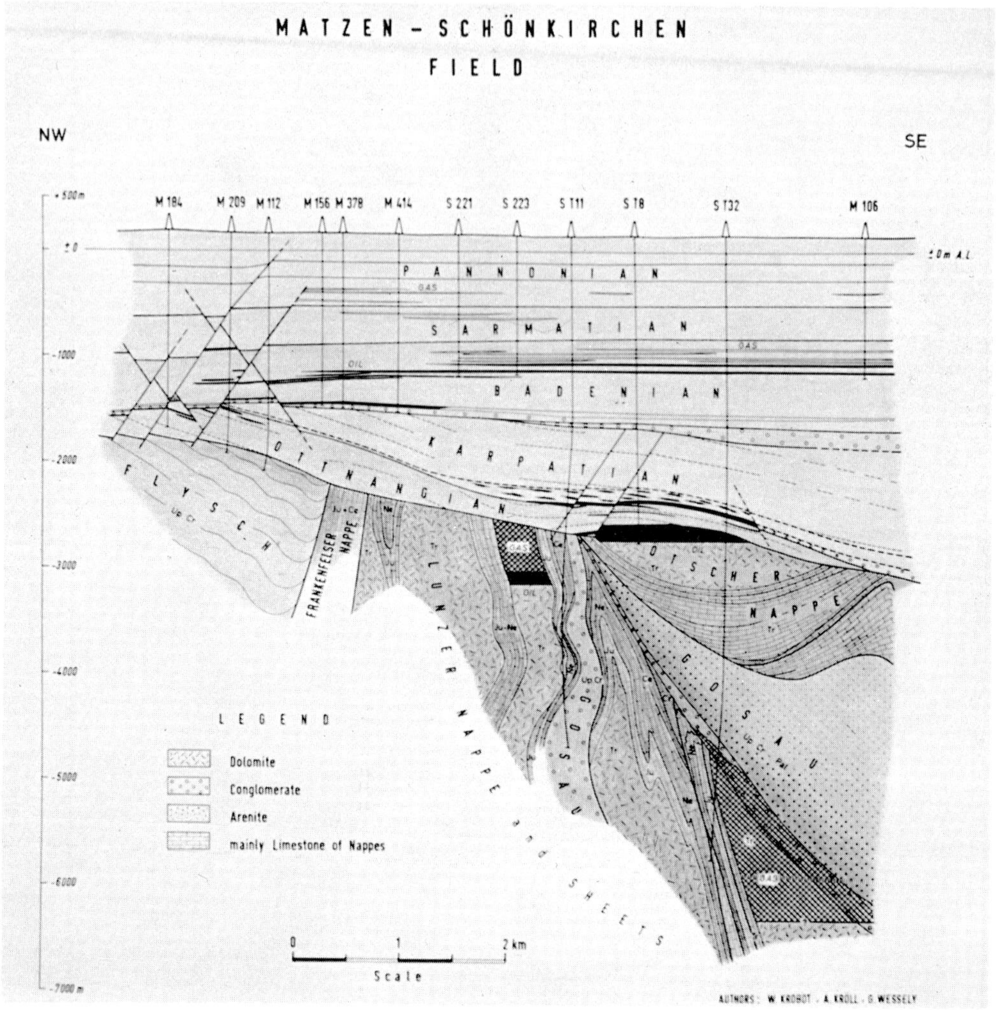


Fig. 2. Schönkirchen, geological cross-section

a location called Zistersdorf ÜT1, about 5 km east of the village of Zistersdorf, the place where the first commercial oil discovery had been made in 1934.

This well was planned to reach, for the first time in Austria and Western Europe, the record depth of 7000 m. The rig is an IDECO H 3000 model and was manufactured in the United States:

The single drum drawworks has a maximum line pull of 61.6 tons and the drum is grooved for 1 5/8" wireline (Fig. 3). Under normal conditions 4 turbocharged Caterpillar V 16 Diesel Engines deliver up to 1035 HP each through torque convertors to the compound. From the compound the two 1600 HP EMSCO single acting triplex pumps are driven. If all the power of the four V 16 diesel engines is needed to drive the two slush pumps – for example when drilling large diameter holes in the upper section of a deep well – an extra diesel engine, a turbocharged Cater-

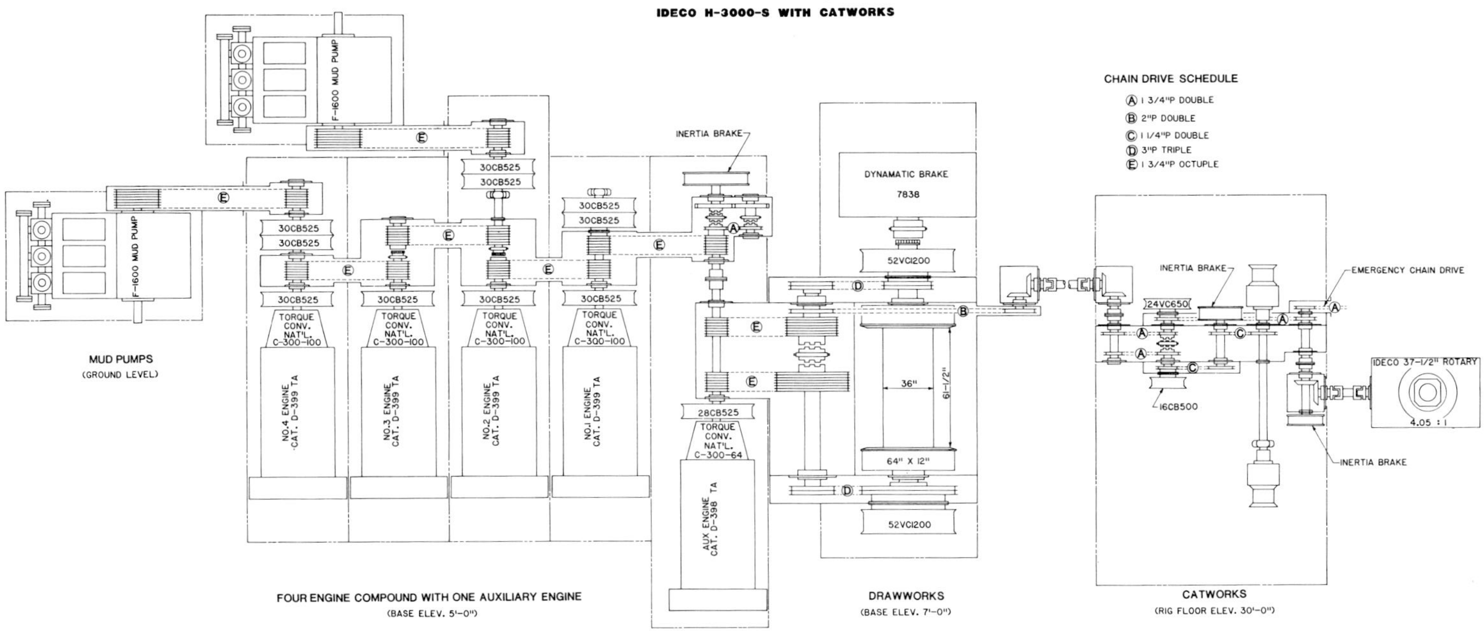


Fig. 3. IDECO H 3000, mechanical layout

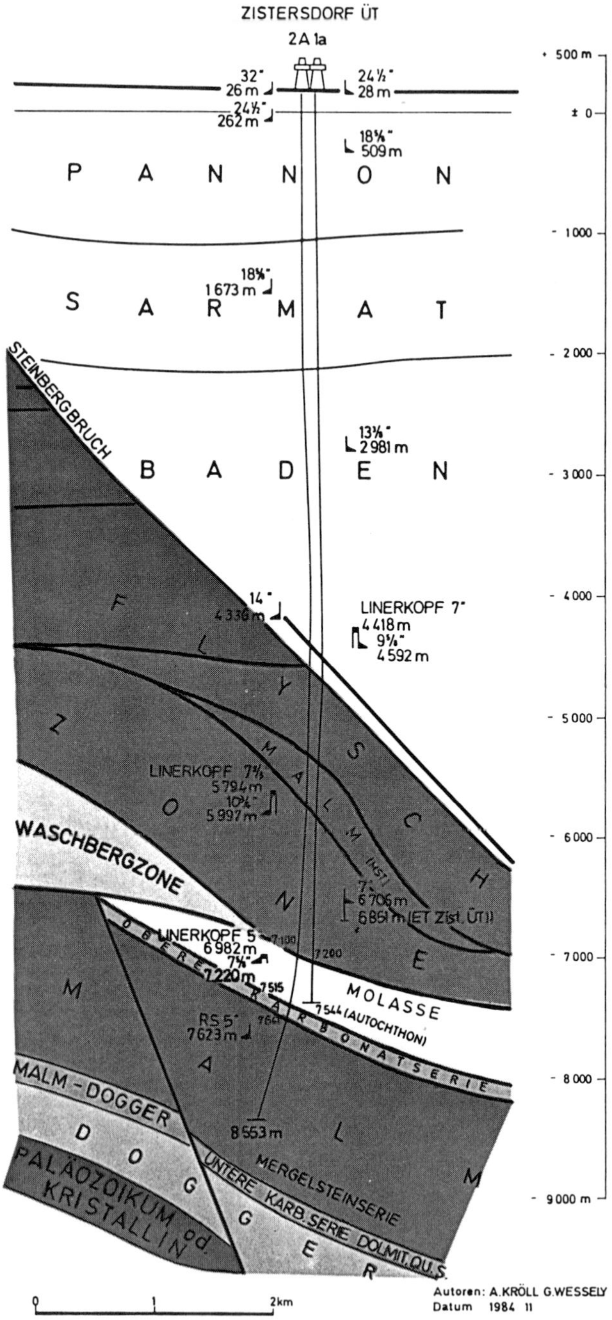


Fig. 4. Zistersdorf ÜT 1 – Geological cross section

pillar V 12 can be rigged up to drive the rotary table through the drawworks and the catworks. The mast in use is a Lee C. Moore 142' x 32' base Cantilever type and has a 30' raised floor. The rated static hook-load capacity of this mast with 14 lines strung to the travelling block is 1,555.000 pounds per API standard 4E. According to Austrian standards the maximum hook load is 655 metric tons.

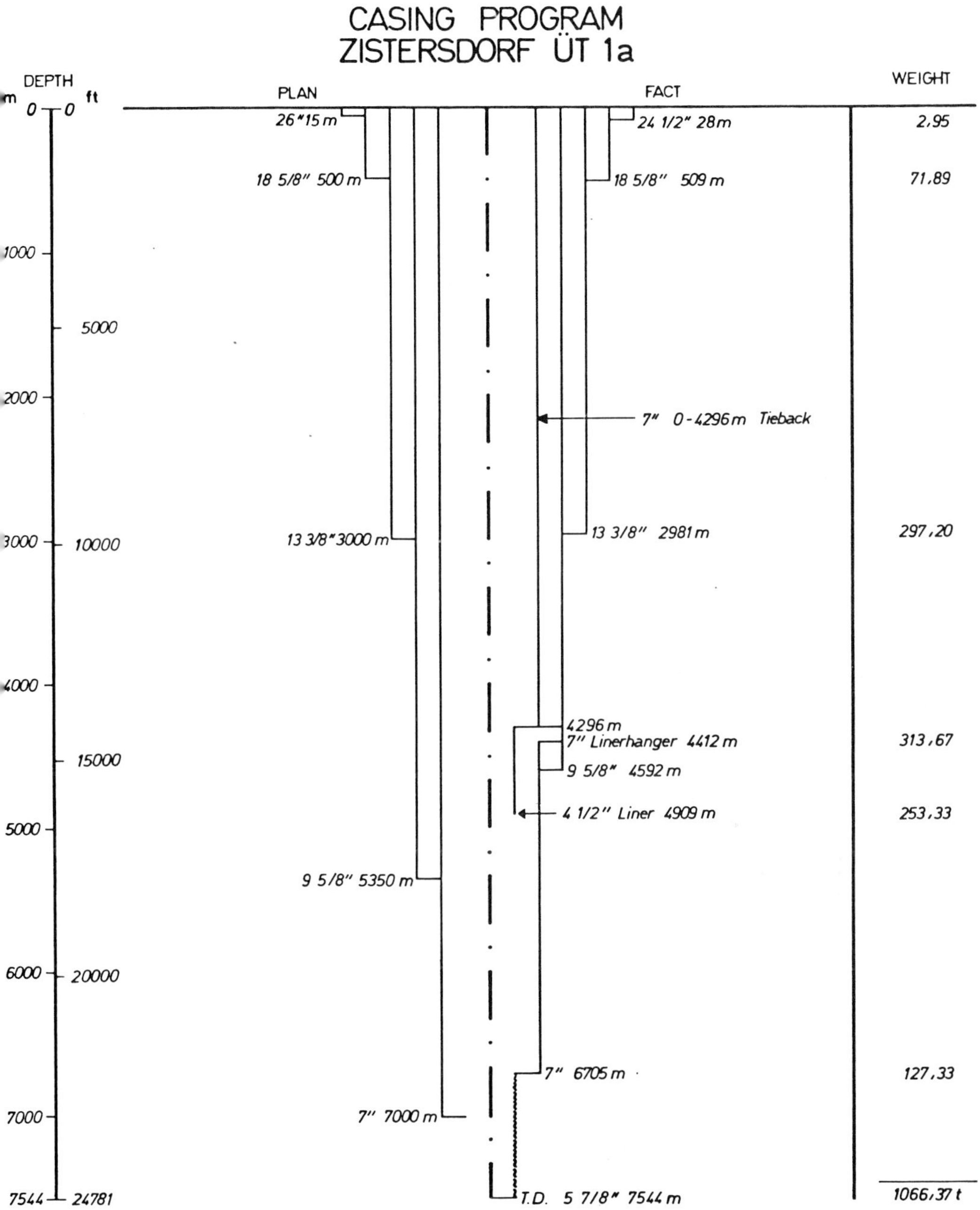


Fig. 5. Zistersdorf ÜT 1a – casing program

The substructure is 30' high and gives enough room to mount the necessary blowout preventer stacks on the wellhead.

Drilling on the Zistersdorf ÜT 1a location began on November 2nd, 1977. Figure 4 shows the geological cross-section. The casing program was as follows (Fig. 5):

| | | Plan | | Fact |
|---|---|---|---|---|
| 18 5/8" | | 500 m | | 509 m |
| 13 3/8" | | 3000 m | | 2981 m |
| 9 5/8" | | 5300 m | | 4592 m |
| 7" | T.D. | 7000 m | | 6705 m |
| | | | T.D. | 7544 m |

No problems were encountered at first and the 18 5/8" casing and the 13 3/8" string were set according to plan. When drilling the 12 1/4" hole, which was planned to go to a depth of 5300 m to the top of the Flysch formation, severe gas cutting of the drilling mud at about 4500 m required an increase of the mud weight. As it was not possible to increase the mud weight above 2.0 kg $l^{-1}$ because of the fracture gradient at the 13 3/8" casing shoe at 2981 m, 9 5/8" casing had to be set at 4592 m.

After running and cementing the 9 5/8" casing string, the water base mud in use to drill to the depth was changed to an oil base system. We knew from previous wells that using a water base mud would create many problems when drilling the Flysch formation, such as poor hole stability; and that in the deep part of the hole we would have to use an oil base fluid anyway because of the anticipated high bottom hole temperatures. An open hole drill stem test at a depth of 4670 m showed gas with a pressure gradient of 1.84 atm/10 m and from then on we had to increase the mud weight to 2.22 - 2.24 kg $l^{-1}$ and carry that mud weight until total depth.

When drilling deeper the geological information showed that the target formation Paleozoic or crystalline rocks would not be reached before 7500 m. As we had to set 9 5/8" casing prematurely at approx. 4600 m and we knew from previous experience that one should not plan, after setting a 7" liner, to drill more than 500 to 600 m of 5 7/8" or 6" hole, we were forced to try to drill the 8 1/2" hole as deep as possible before setting the 7" drilling liner.

Beside the fact that we had to carry a very high mud weight and thus had to be extremely cautious during round-trips, we experienced no problems in drilling the 8 1/2" hole and finally reached a depth of 6851 m, that means that we already had an 8 1/2" open hole section of more than 2200 m, which is rather unusual under conditions such as we had. Thus we decided to run the drilling liner to secure the bore hole. To be able to achieve a maximum internal diameter of the liner to continue drilling with a bit diameter of at least 5 7/8", we had to use a very high quality steel for the liner tubes. We had some field reports from other oil companies about the use of V 150 casing and as we had no indications of $H_2S$ in the hole and were using an oil base drilling mud as an extra precaution, we had no objections to the use of high strength casing.

Early in 1980 we ran this 7" liner. Steel quality for tube and coupling was V 150 and the thread configuration Mannesmann BDS (BDS = Buttress Double Seal) (Fig. 6). The length of the liner was about 2500 m to cover the open hole from 6851 m, with a 200 m overlap into the 9 5/8" casing, to 4350 m. The linershoe was at a depth of 6400 m when the hole was circulated before continuing to lower the 7" string to the bottom of the hole. No unusual observations were made during circulation, but

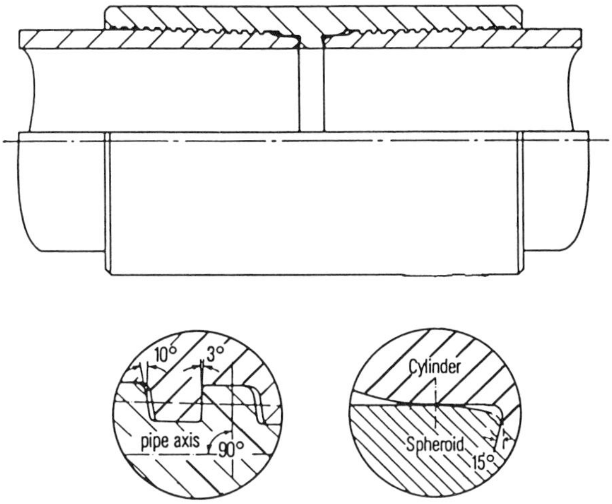


▲
Fig. 6. Mannesmann BDS (Buttress Double Seal) is externally threaded on both ends of a non-upset pipe. Single lengths are joined by an internally threaded coupling. Coupling is of standard and special clearance (reduced OD) design. The thread profile is original buttress. Metal-to-metal seals are provided at the pin end

Fig. 7. Typical fracture V-150 coupling ▶

when the kelly was set aside a vibration was noticed in the substructure, and when the string was picked up the weight indicator showed a loss in total weight of approx. 80 tons.

On its emergence for the hole, it was found that a coupling of the 7" liner had failed. When trying to fish out the lost part of the liner, nine identical failures occurred (Fig. 7). Finally the 8 1/2" hole was cleaned down to a depth of 5024 m and the lowest part of the liner was found to be intact and could not be pulled; unfortunately the 7" liner shoe was stuck about 150 m off bottom of the 8 1/2" hole.

All the failures of the 7" V 150 couplings were alike, and at first, because of the nature of the failures, $H_2S$ was blamed for their occurrence. To be able to continue drilling operations, a 7" V 150 liner with P 110 oversize couplings was run into the well joined with the remainder of the 7" liner at a depth of 5024 m and cemented.

Finally the top of the 7" liner was successfully squeezed. It should be noted that during all these time-consuming operations the oil base drilling fluid performed perfectly and it was possible, after running the "tie back liner", to break circulation without any problems after 52 days.

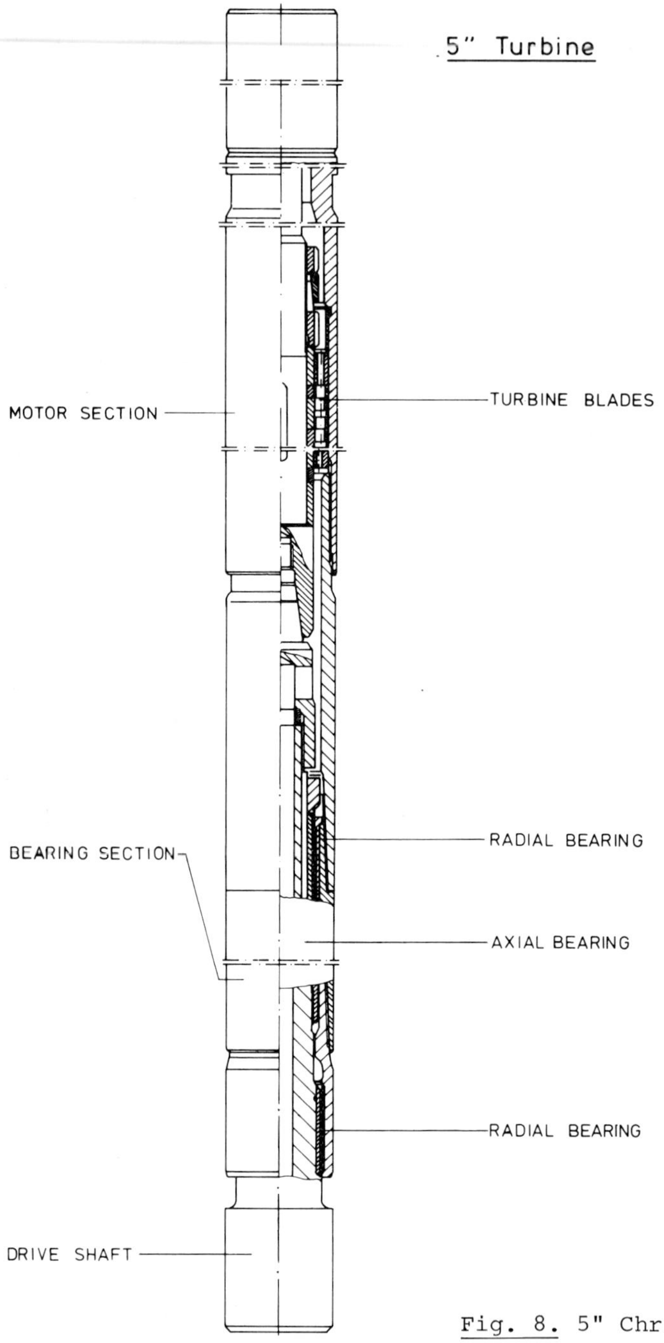


Fig. 8. 5" Christensen turbine

When drilling ahead we were facing the problem of having about 150 m of 8 1/2" hole below the 7" liner and we were now using a 5 7/8" bit, 4 3/4" drill collars and 3 1/2" drill pipe. Unfortunately, exactly what we had feared most happened – we had a failure of a 4 3/4" drill

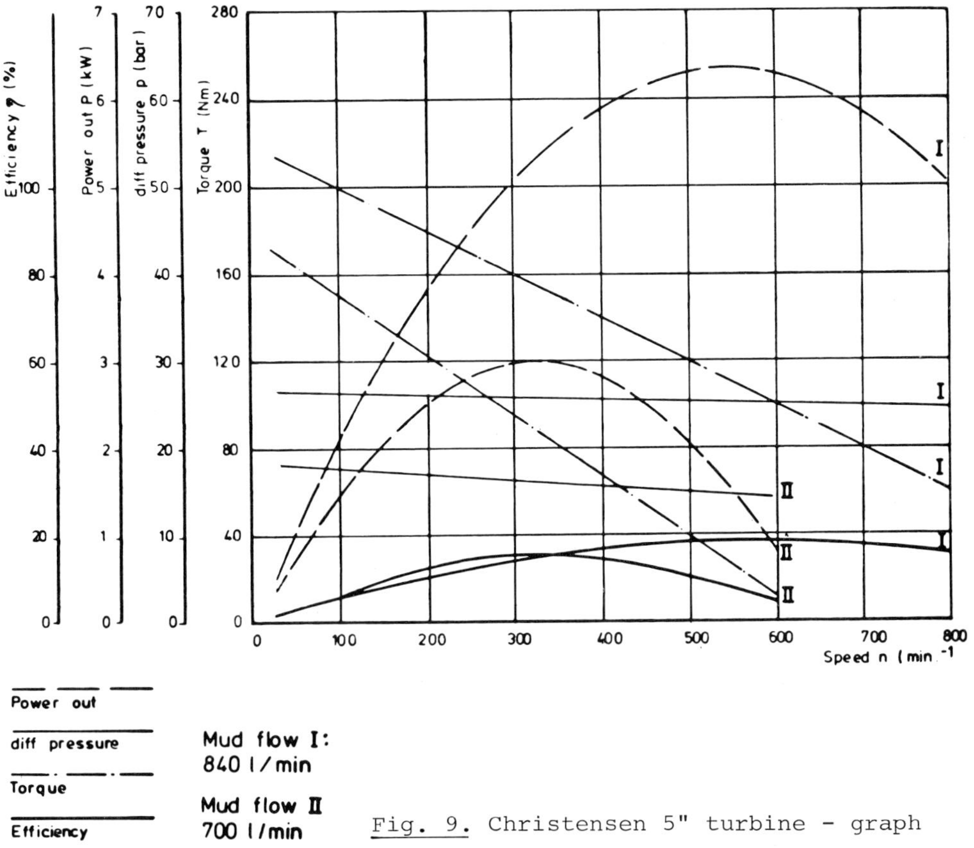


Fig. 9. Christensen 5" turbine - graph

collar connection and were unable to catch the top of the fish in the 8 1/2" hole because the maximum outside diameter of the fishing tools to run were limited by the internal diameter of the 7" liner.

Unable to recover the fish (the top was at 6781 m) we were facing the problem of sidetracking, with an oilbase fluid of 2.22 kg $l^{-1}$ in the hole and a bottom hole temperature of about 165°C.

We finally decided to try it: the 8 1/2" hole was cleaned by using an underreamer and a cement plug was set. The first trial, using a standard 5" turbo drill, failed because of the high bottom hole temperature, swelling of the bearing rubber parts prevented the rotation of the drive shaft. It was also not possible to use a Moineau motor (Dynadrill) for this operation because the polymerized rubber stator would have been destroyed by the high well bore temperature after a short run.

OeMV contacted Christensen Diamond Co. and they manufactured a 5" turbodrill with no rubber or polymere parts within an extremely short time (Figs. 8, 9). With this downhole motor we were able to perform a successfull side-tracking job.

Here are the parameters of this outstanding job, as far as we could investigate the deepest sidetracking worldwide:

- 2.2 kb $l^{-1}$ oil base mud
- bottom hole temperature: 165°C
- down hole motor: 5" turbodrill, special design of Christensen Diamond Co.
- bit: 5 27/32 SST diamond bit
- flow rate: 700 $l\ min^{-1}$
- circulation pressure (stand pipe): 400 bar on bottom; 390 bar off bottom

The turbodrill ran for 23.5 h and drilled 9.5 m. The kick-off depth was 6734 m and no wear could be noticed.

After this sidetracking job, drilling continued with no problems other than that the bit was pulled from and run into the open hole with circulation, the mud weight had to remain high not only because of continuous gas cutting but also to keep the hole open in the Flysch formation.

On January 16th, 1980, while drilling ahead at 7544 m, with the hole configuration as we had it (the OeMV had decided to deepen the 5 7/8" hole as far as possible with the hope of reaching a geological marker which would either justify another well or rule out further ultra deep exploration in this area), light mud losses occurred. At first it was thought that the hole might breathe – something we had already noticed before – that the well would take some mud and flow it back later; but this time the gain increased rapidly and soon we had to close the blowout preventer and put the well on the choke. But we had to find out that the mud we pumped into the drill string was lost to the formation and, at the same time, the formation produced gas in ever-increasing quantities. About 12 h after the first kick, the well produced gas at a rate of more than 1 million $m^3$ per day. This gas contained 98% $CH_4$ and about 2% $CO_2$. No higher hydrocarbon components could be detected (Table 1).

Table 1. Zistersdorf ÜT 1a – gas analysis

| | | |
|---|---|---|
| $CO_2$ | (Vol %) | 1,89 |
| $CH_4$ | | 97,67 |
| $C_2H_6$ | | 0,21 |
| $C_3H_8$ | | 0,02 |
| $iC_4H_{10}$ | | traces |

After some time (approx. 3 days) the open hole bridged and the gas flow ceased. As soon as possible the annulus was filled with mud, but it was too late, as we shall see later.

During the flaring of the gas we pumped some drilling mud into the drill string every hour, just to keep the string open to the bit. After about 24 h we noticed an increase in pumping pressure and finally found that the string had become plugged.

After the flow from the annulus had stopped and there was no pressure on the drill string, we ran a wire line caliper and localized a bridge inside the 5" drill pipe at about 900 m. This meant that the only possibility of controlling the deep gas flow below the bridge in the 5 7/8" bore hole around the 3 1/2" drill string was to drill out the bridges

Fig. 10. Otis snubbing unit

inside the drill string, try to get contact with the drill bit, perforate the drill collars (or shoot off the diamond bit) and squeeze the gas zone.

As a consequence of this decision, a B 747 Cargo load of special snubbing equipment was flown from Oklahoma City and Houston to Vienna. Six 15,000 psi blowout preventers and the snubbing unit were mounted on the 5" drill string (Fig. 10). Then a combination macaroni string – 1", 1 1/4" and 2 1/16" – was used to wash down inside the 5" drill string. To be able to circulate the mud through the macaroni string, 20,000 psi Halliburton frac units had to be brought to the drill site.

Shortly after starting this operation the first bridge inside the 5" drill string had been drilled up, and circulation to the 9 5/8" annulus was observed. There was a leak in the 5" drill string and later, by

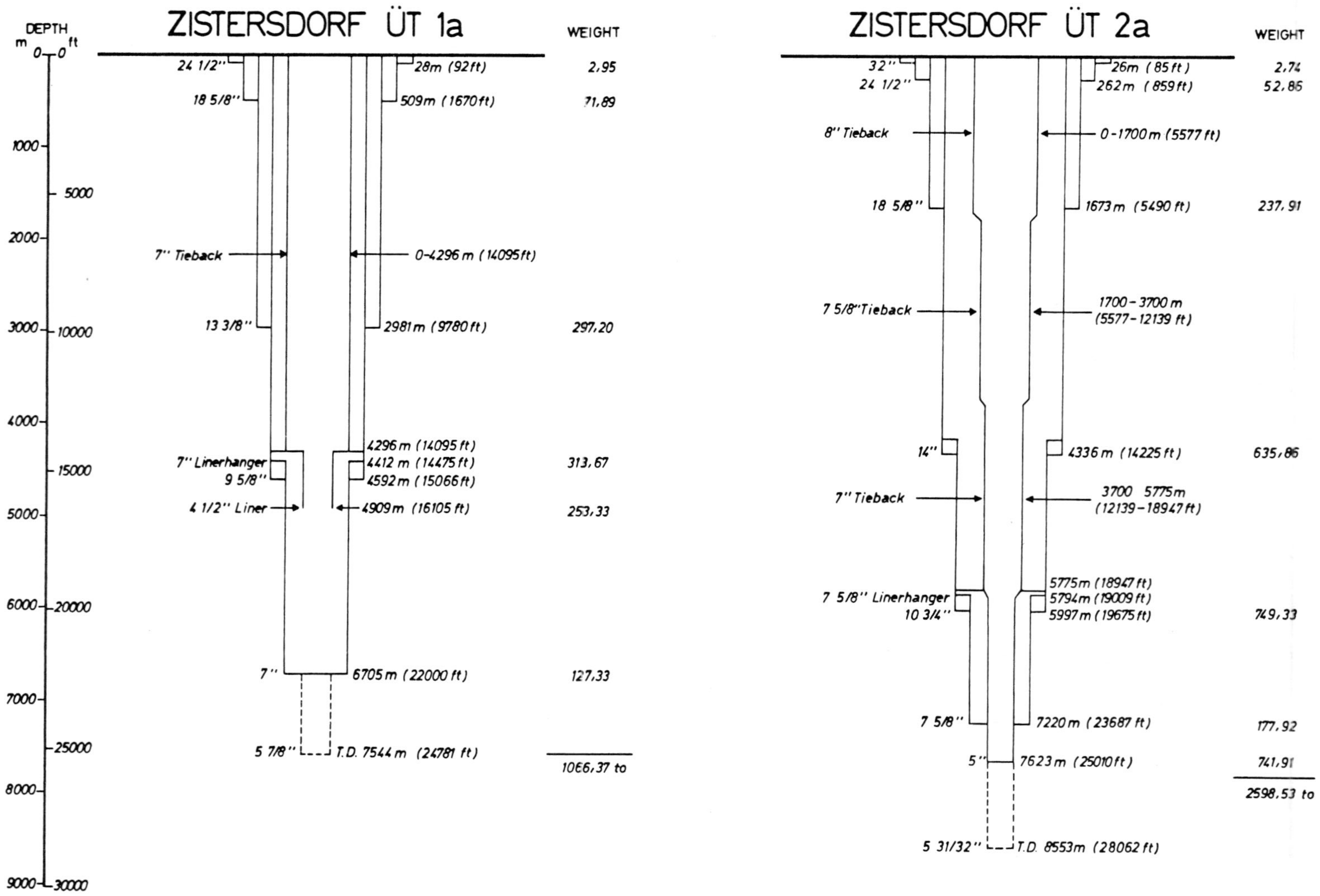


Fig. 11. Casing programs, Zistersdorf ÜT 1a/Zistersdorf ÜT 2A

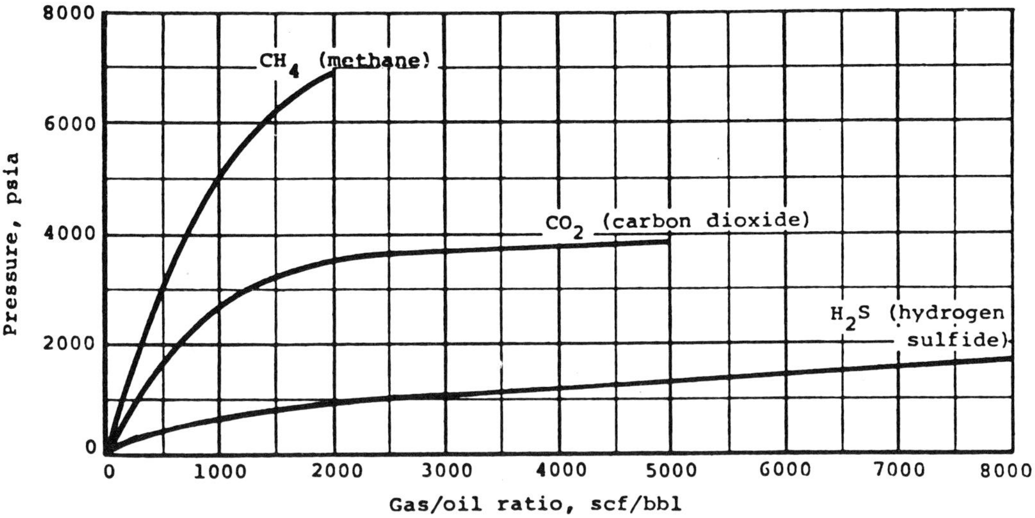


Fig. 12. Solubility curves for $H_2S$, $CO_2$, and $CH_4$ in no. 2 Diesel oil at 250°F

back off and pulling the upper part of the drill string, we found a tool joint wash out at 4104 m.

When we tried to run the intact 5" string again to screw it into the box of the 5" string left in the hole, we could not do so because we found later that the 9 5/8" casing had collapsed at a depth of 4352 m, only a few meters below the crossover from 5" to 3 1/2" drill pipe (4340 m). With difficult and time-consuming operations we were able to make contact with the 3 1/2" portion of the drill string and again made a 3 1/2" × 5" drill pipe connection to the surface.

Then we started again with the snubbing operation to reach the diamond bit inside the drill string.

Many more problems occurred – more or less only because of failures of thread connections of the macaroni string – we even had to fly in another full macaroni string. But we finally had to give up when again a coupling failure in the 2 1/16" tubing occurred on top of the string. The tiny milling tool inside the 3 1/2" × 5" drill string was at 6634 m at that moment. Now stand by stand, joint by joint, we had to fish out the macaroni string and the drill string. Seeing no possibility to reach the gas zone at the bottom of the hole (7544 m) we finally decided to plug the well inside the 7" liner. This was done by setting two bridge plugs at 4915 and 4910 m. Then a 4 1/2" × 7" casing string was run to 4800 m and cemented. A production test on perforations made between 4660 m and 4670 m showed the same high-pressure low-volume gas influx as noticed during the open hole test, 30 months earlier.

The results of the Zistersdorf ÜT 1a well were unsatisfactory in quantity, but very interesting and encouraging in quality.

We had spent a lot of money to see natural gas flowing from a depth never seen in production before in Europe, or even outside the USA. We flared – more or less during a blowout-situation – quantities of gas, which would allow us to hope that we scratched a potential reser-

Fig. 13. 30,000 psi christmas tree

voir. Finally we came up with more pro's than con's about the ultra deep exploration in this region.

Although the total costs of the Zistersdorf ÜT 1a well were in excess of 25 million US$, the OeMV's management decided to go ahead and drill a second well – a replacement well – to test the potential of the discovery.

The rig was moved about 150 m to the north of the old location and on March 6th, 1981, the replacment well "Zistersdorf ÜT 2a" was spudded. Because of the experience of no. 1 well (Fig. 11) the second well has a modified casing program and also takes into consideration that 2% $CO_2$ under this partial pressure makes the encountered gas a sour gas.

The casing program of the no. 2 well and all the procedures are planned to be well on the safe side:

| | | | | |
|---|---|---|---|---|
| 24 1/2" | 250 m | 10 3/4" | | 5700 m |
| 18 5/8" | 1650 m | 7 5/8" | | 7250 m |
| 14" | 4350 m | 5" | TD | 8000 m |

<u>Fig. 14.</u> 30,000 psi blow out preventer stack

The problem with the use of oilbase drilling fluid in potential gas zones (Fig. 12) will be overcome by the use of a new developed water base mud, available for temperature and pressure conditions as they are relevant in the deep formations of the Zistersdorf area. There was no doubt that we would need sour gas surface equipment for a pressure rating between 25,000 and 30,000 psi and special consideration must be given to completion techniques. Expensive high pressure sour gas equipment had to be ordered because it was not available on a rental basis worldwide and did not exist in Europe at all (Figs. 13, 14). Table 2 gives some specifications for the casing used in the Zistersdorf ÜT 2A well.

No special problems were encountered while drilling this well and it was terminated at a depth of 8553 m. As the lower part was found unproductive, the 5" liner was set only to 7623 m (Fig. 15). Figure 16 shows the mud weight used and temperature recorded. Special logging tools were used and all necessary logs for the evaluation of the well could be run satisfactorily (Table 3). For safety reasons a complete second mud mixing, storing and pumping facility was set up, but not used.

Table 2. Casing specification for Zistersdorf ÜT 2A Well

| Casing OD | | Wall mm | Connection | Grade | Pressure Burst | Collapse |
|---|---|---|---|---|---|---|
| 24 1/2" | | 13.5 | B | J 55 | 141 | 34 |
| 18 5/8" | | 12.32 | B | N 80 | 251 | 60 |
| 14" | | 17.6 | 14" BDS-S | 130 | 777 | 409 |
| 10 3/4" | | 20.15 | 10 1/4" MUST-M | 125 | 1113 | 1065 |
| 7 5/8" | Liner | 19.6 | 7" MUST-M | MW 125 | 1604 | 1561 |
| 8" | | 31.81 | 7 5/8" BDS-S | C 95 | 1844 | 1727 |
| 7 5/8" | Tieback | 28.17 | 7 5/8" BDS-S | C 95 | 1714 | 1626 |
| 7" | | 20.8 | 7" BDS | P 110 | 1600 | 1570 |
| 5" | | 14.7 | 5" BDS-S | MW 125 | 1896 | 1768 |
| 5" | Liner | 16.5 | 5" BDS-S | MW 125 | 2123 | 1944 |
| 5" | | 14.7 | 5" BDS-S | AF 22/130 | 1973 | 1832 |

Table 3. Schlumberger logs planned and recorded below 7000 m

Zistersdorf ÜT 1A

6700 - 7050 m

mud: Oil faze
max. temp. 350°F
max. pressure 23000 psi

1. Run: CDR-CAL-GR
2. Run: FDC
3. Run: CNL

(with standard tools)

Zistersdorf ÜT 2A

7200 - 7600 m

mud: XP-20 HT
max. temp. 392°F
max. pressure 21000 psi

1. Run: DLL-GR
2. Run: FDC
3. Run: CNL-GR
4. Run: HDT
5. Run: BHC

(with standard tools)

7600 - 8550 m

mud: XP-20 HT
max. temp. 442°F
max. pressure 27200 psi

1. Run: BHC-GR
2. Run: DLL-GR
3. Run: HDT

(with HEL tools going down)

with upgraded standard tools could only be logged from 8250 m up

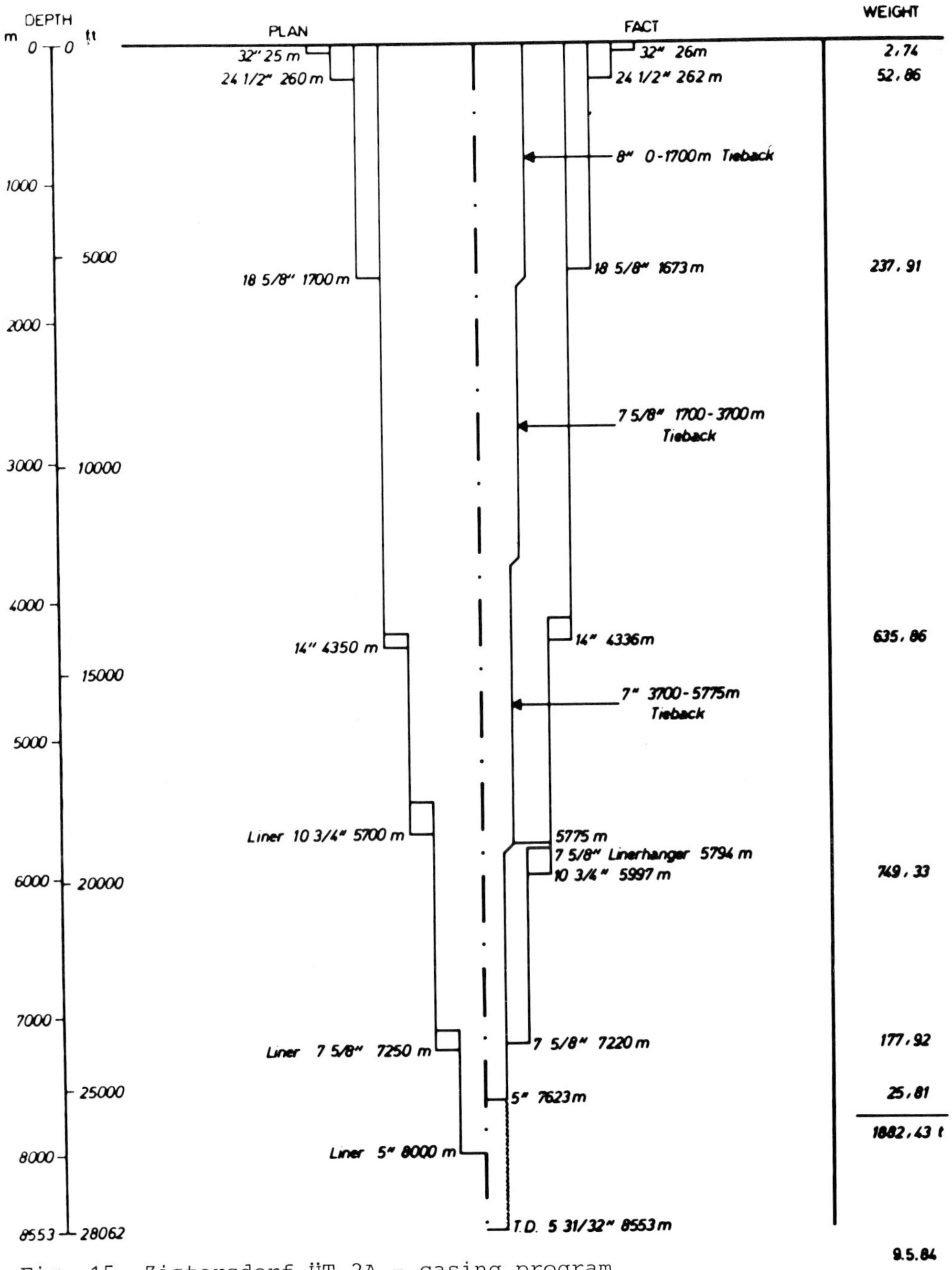


Fig. 15. Zistersdorf ÜT 2A - casing program

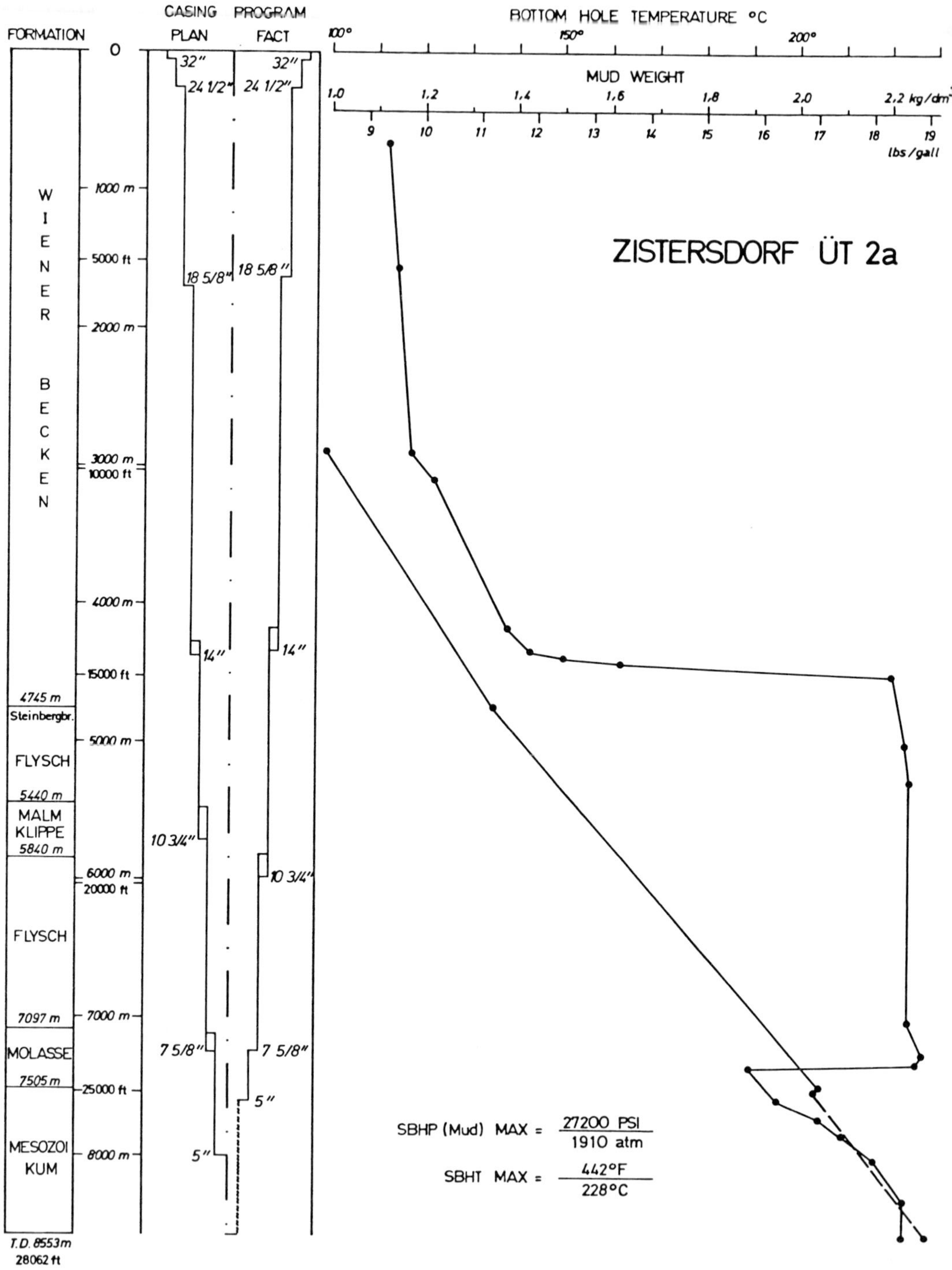


Fig. 16. Zistersdorf ÜT 2A - mud weight and temperature vs. depth

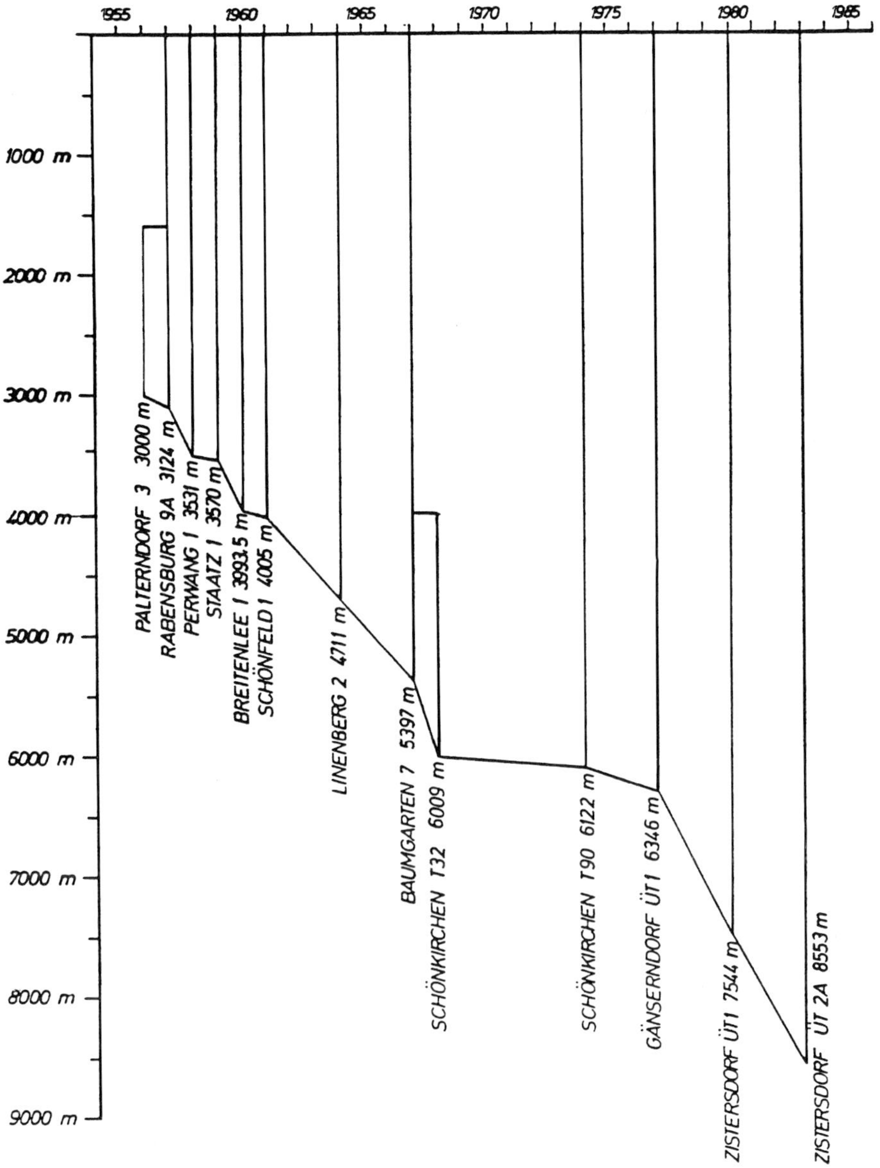


Fig. 17. Drilling depth records in Austria

Table 4. Wells deeper than 6000 m

| | | Spudded | End of drilling | Final depth |
|---|---|---|---|---|
| Schönkirchen | T 32 | 22. Nov. 1966 | 5. Dec. 1967 | 6009 m |
| Schönkirchen | T 90 | 20. May 1973 | 28. May 1975 | 6122 m |
| Gänserndorf | ÜT 1 | 27. Oct. 1975 | 15. Dec. 1977 | 6346 m |
| Berndorf | 1 | 21. April 1978 | 6. June 1979 | 6028 m |
| Prottes | ÜT 2 | 28. April 1981 | 14. Aug. 1982 | 6043 m |
| Zistersdorf | ÜT 1 | 2. Nov. 1977 | 29. Oct. 1980 | 7544 m |
| Zistersdorf | ÜT 2 | 6. March 1981 | 12. June 1983 | 8553 m |
| Aderklaa | Ultra 1 | 12. July 1982 | Present depth | 6119 m |
| Maustrenk | ÜT 1 | 14. Sept. 1983 | Present depth | 6285 m |

Since OeMV started the ultra deep exploration program in 1966, nine wells have been drilled to the present to T.D. below 6000 m. One of them reached 7544 m and the record well 8553 m (Table 4). The progress in the OeMV's ultra deep drilling is shown in Figure 17.

The problems of drilling and completing such ultra deep, high pressure, sour gas wells are very complex. Today only very few companies, or stated more correctly few individuals, are confronted with and working on these problems. It is my thorough belief that they should communicate very frequently and freely to the benefit of all of us because ultra deep exploration and production is very expensive – and today we in Western Europe are no longer competitors in gas production! Let us help each other to find and produce as many hydrocarbons as possible to our mutual benefit.

I hope that my presentation has shown that the OeMV is very active in the field of deep exploration and that at the present time we are confident to be able to drill and complete wells to more than 8000 m under extreme conditions, such as high pressure and temperature. Under normal conditions we think that reaching depths of 10,000 m should be well within our technical capability.

# DEEP DRILLING TECHNOLOGY FOR HOT CRYSTALLINE ROCK

John C. Rowley[1]

## Introduction

This paper recounts the experience of deep drilling into hot crystalline rock by the Fenton Hill HDR project (Fig. 1) to illustrate that such operations for scientific purposes are indeed possible. The discussion is intended to provide general guidance for project planning for future scientific drilling operations. The discussion also supports the thesis that experience from past ambitious drilling projects can aid in the design of the strategy of drilling and coring programs and help reduce costs and risks of such efforts in order to maximize the likelihood of success and scientific yield. Drilling hardware, operations, problems, and costs are described to give information to those

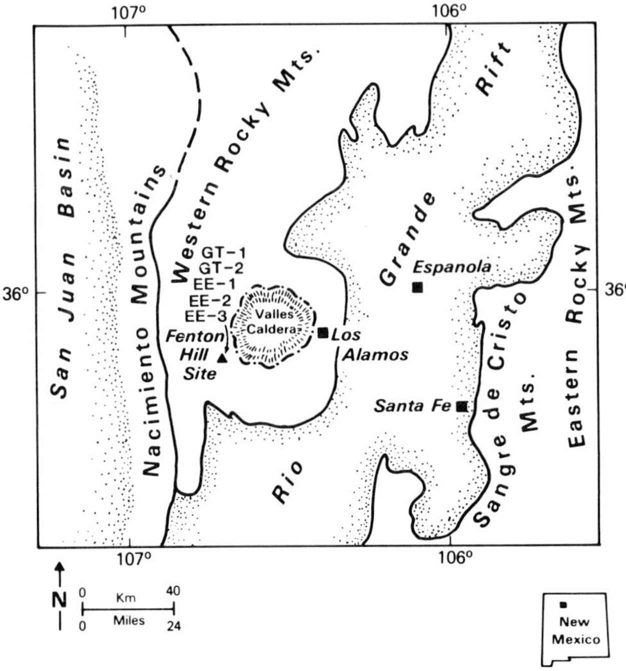


Fig. 1. Location map of the Fenton Hill, New Mexico, Hot Dry Rock Geothermal Site, 32 km west of the Los Alamos National Laboratory on the western flank of the Valles Caldera. Two deep borehole pairs (GT-2 and EE-1; EE-2 and EE-3) penetrate to 3 km and 4.4 km depth in the hot, crystalline crust

[1]Earth and Space Sciences Division, MS D462, Los Alamos National Laboratory, Los Alamos, NM 87545, USA

Observation of the Continental Crust through Drilling I (ed. by C.B. Raleigh)

interested in future deep, scientific drilling projects in the continental crust. The focus is on deep (>3km) large diameter [31.1 cm (12-1/2 in.) to 22.2 cm (8-3/4 in.)] boreholes with spot coring of selected sections. The experience related here is in contrast to the relatively shallow, small diameter, continuous scientific coring operations such as those conducted to a depth of about 2 km by the Iceland Research Drilling Project (Fridleifsson et al. 1982)

The high formation temperatures and very hard, often fractured, abrasive rocks penetrated at Fenton Hill are potentially to be encountered in the deeper portions of deep (~10 km) boreholes in areas of typical or average geothermal gradients (e.g., 25°C $km^{-1}$), but especially in future projects planned to perform deep sampling into geothermal anomalies. These types of projects have been discussed as part of the Thermal Regimes effort of the Continental Scientific Drilling Program (CSDP). Potential sites include the Valles Caldera, NM; Long Valley, CA; Yellowstone National Park, WY; and the Salton Sea, CA. Details of thermal regime drilling objectives are recorded in the National Research Council, Continental Scientific Drilling Program (1979) and Goff and Waters (1980). Drilling, coring, borehole measurements, and sampling for scientific purposes in such hostile environments can be unexpectedly costly unless detailed planning and supervision of the drilling and coring operations by those experienced in similar operations is carefully performed.

Projects contemplating core and fluid sampling operations under these more severe conditions, especially higher temperature, will benefit from the technology improvements outlined here. The special coring hardware and drilling procedures successfully developed for the Deep Sea Drilling Project and summarized by Larson et al. (1981) for deep ocean sediment sampling are illustrative of those developments proposed here. The value of specially designed drilling equipment and procedures specifically for such scientific drilling projects cannot be overemphasized. The high-temperature rated, high-performance turbodrills designed and fielded for the HDR geothermal project (Neudecker and Rowley 1982) are an example of the type of advanced technology development potentially required for successful achievement of directional control of deep crustal boreholes. A second example, of wider interest to CSDP, is the hybrid core bit that was needed to achieve adequate core quality, reliability, and recovery in the deep crystalline rock at Fenton Hill (Pettitt et al. 1980). Finally, it should be noted that most drilling technology advances and case histories of potential interest to the scientific community are reported in the engineering literature and are usually familiar only to a very small and specialized segment of the drilling and related industries. Thus, Oliver (1982) quite correctly stresses the availability of rig capability to perhaps 15 km (50,000 ft) depths. It is likely that harder rocks, scientific coring and sampling requirements, and high temperatures in drilling of potential thermal regime targets will necessitate significant shifts in drilling strategy away from that used in the petroleum industry procedures, which primarily focus an optimum penetration rate for commercial exploration and development. In addition, new technological developments away from the main stream of commercial drilling activity are difficult to transfer within the petroleum industry (Knight 1983). Knowledgeable and experienced personnel will play a most important role in the appropriate planning, supervising and on-line problem solving necessary to successfully conduct and field CSDP deep sampling projects. Costs can escalate rapidly and projects may fail if effective planning and experienced supervision are neglected.

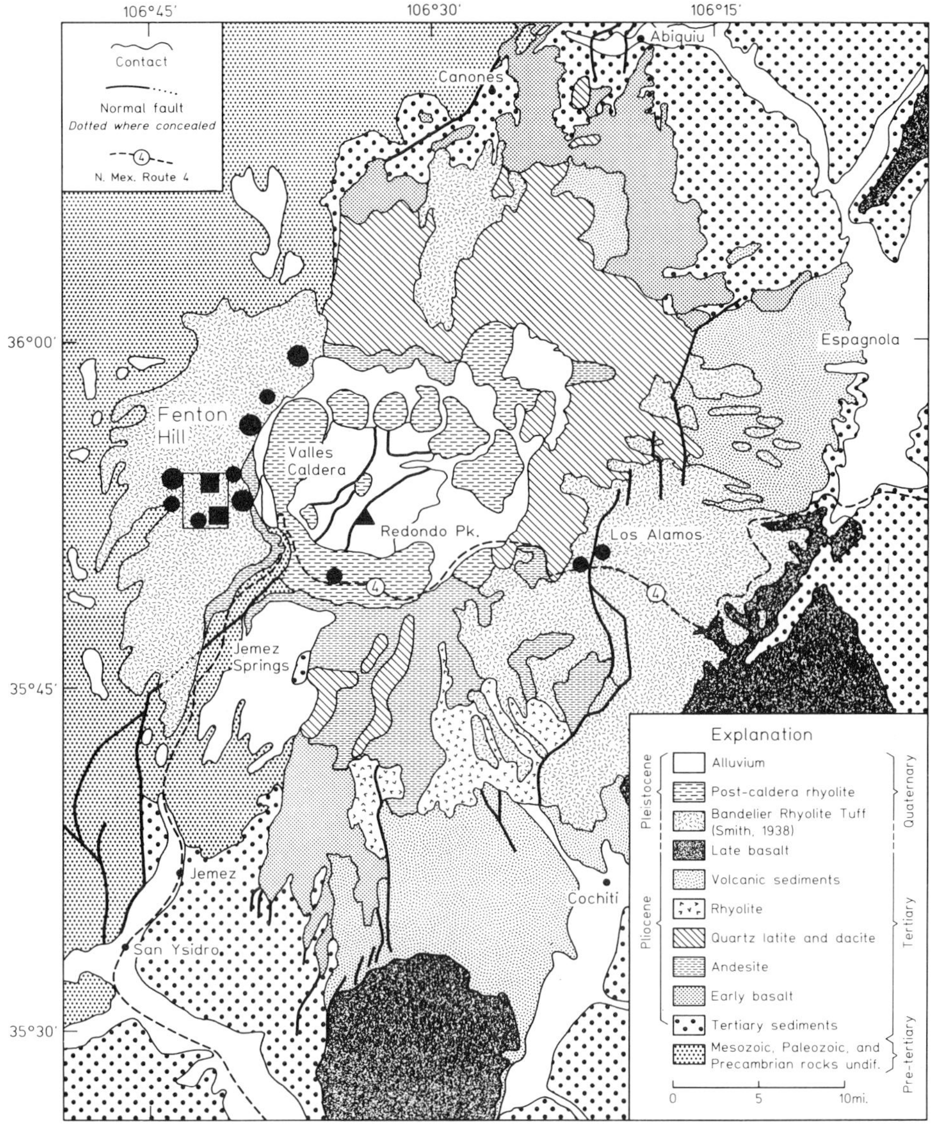


Fig. 2. Geologic map of Jemez Mountains (Bailey and Smith 1978, M.C. Smith 1983) indicating exploration drilling. *Small* and *large circles* locate shallow and deeper heat flow holes. *Squares* are crystalline basement rock test holes GT-1 and GT-2

## 1. Drill Site Location

The Fenton Hill site is located in Sandoval County, New Mexico, approximately 32 km west of the Los Alamos National Laboratory (Fig. 1). Fenton Hill is in the Jemez Mountains (Fig. 2) and lies about 1 km outside the western topographic rim of the Valles Caldera in a region termed the Jemez Plateau. Laughlin (1981), and Laughlin et al. (1983) review the selection and the regional setting of the site. The brief summary here from Laughlin et al. (1983) is intended as background for the drilling technology discussions that follow:

"The Jemez Mountains lie on the boundary between the Colorado Plateau and the Rio Grande rift in north-central New Mexico. Volcanism occurred over the interval from roughly 10 to 0.1 m.y. ago (Bailey et al. 1969) with major caldera collapses occurring at 1.4 and 1.1 m.y. ago to form the Toledo and Valles calderas. Rhyolite domes were extruded along ring faults of the Valles Caldera and they surround a large resurgent structural dome, Redono Peak. The Redono Peak area, which is highly faulted, is the site of a conventional (hydrothermal) geothermal prospect currently under investigation by Union Geothermal Company of New Mexico.

The plateaus that surround the Valles and Toledo calderas are capped by ash-flow sheets of Bandelier Tuff erupted during formation of the two calderas. Beneath the tuff in descending order are older volcanic rocks of the Jemez Mountains (Tschicoma and Paliza Canyon Formations), the Miocene Abiquiu Formation (Abiquiu Tuff of Smith 1938), the Permian Abo Formation (red beds), the Carboniferous Magdalena Group (mainly limestone), and Precambrian metamorphic and igneous rocks. The Precambrian surface lies at a depth of about 730 m at Fenton Hill. Precambrian rocks are exposed at the surface a few kilometers south and west of Fenton Hill.

The Valles Caldera region and associated subjacent magma reservoirs provide an ideal setting for development of both hydrothermal and HDR (geothermal) systems. Only the presence or absence of secondary permeability distinguishes these two systems.

During early stages in the selection of the Fenton Hill site, eleven holes were drilled to determine the heat flow around the periphery of the Valles Caldera (Potter 1973; Reiter et al. 1976). The four deepest holes, which reached depths of 152 to 229 m and so are believed to yield the most reliable data, indicate heat flow values of 92 to 247 mW $m^{-2}$. The drill hole yielding the 92 mW $m^{-2}$ value was radially farther from the caldera, and probably sampled the regional heat flow."

## 2. Geologic Section

The locations of the shallow heat flow holes are shown in Figure 2. Geologic features of interest to drilling technology are summarized in the stratigraphic section of Figure 3 based upon results from subsequent deep drilling of boreholes, designated as GT-1, GT-2, EE-1, EE-2, and EE-3 in Figure 1.

The geologic section at Fenton Hill consists of approximately 152 m (500 ft) of welded volcanic tuff overlying 244 m (800 ft) of Permian red-beds (shales and sandstones), 335 m (1100 ft) of Pennsylvanian limestone, and below 732 m (2400 ft) Precambrian crystalline basement of extremely low permeability and porosity to 4.5 km (14,000 ft). The lithology represented in Figure 3 was established by spot cores and cuttings, and generally without the benefit of geophysical logging in the deeper hotter sections. The temperature at the top of the Precambrian formations is 100°C (212°F), and the geothermal conductive gradient in this crystalline rock increases from 55°C $km^{-1}$ (30°F/1000 ft) to 89°C $km^{-1}$ (49°F/1000 ft) at 4.7 km (15,000 ft) depth (Fig. 3). The temperature at the bottom of the deepest borehole (EE-2) is 320°C (608°F).

The Permian red beds and the Pennsylvanian clays, shales, and limestones are sensitive to a fresh water based drilling fluid and contain

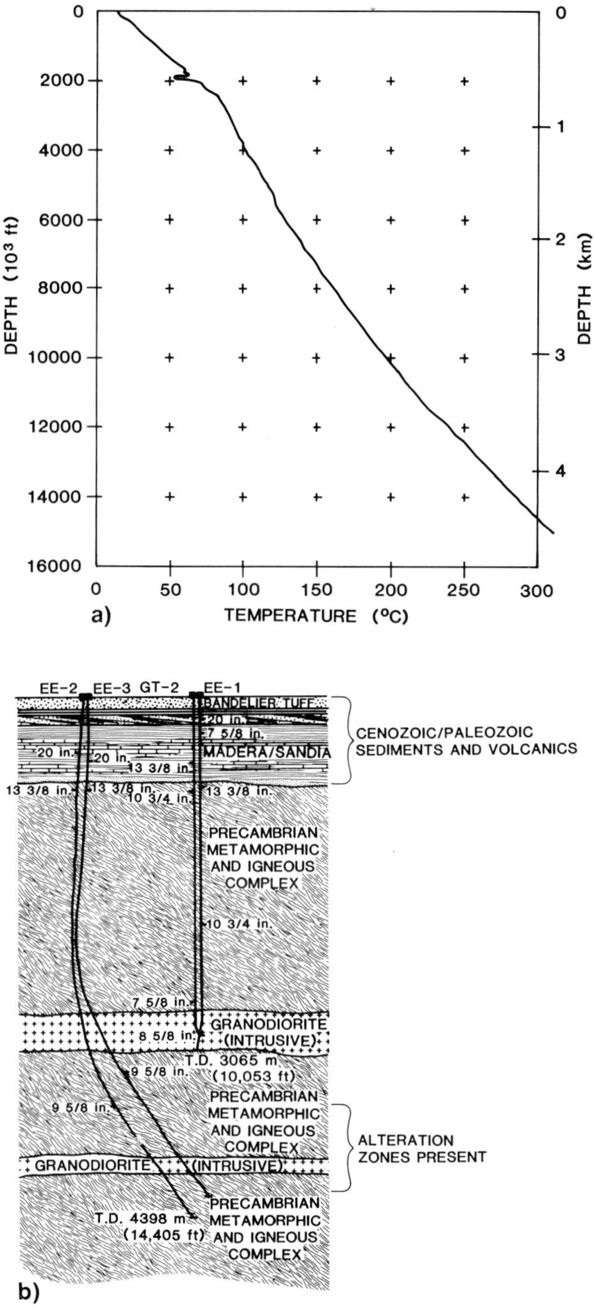


Fig. 3. (a) Temperature profile and (b) generalized geologic section at Fenton Hill, with well configurations indicated. *Triangular flags* indicate casing depths (set points) and diameters as shown

several highly permeable zones that can cause lost circulation of drilling fluid. Several units in the Permian and Pennsylvanian formations in this section will also swell and become unstable causing sloughing of the wellbore (Nuckols et al. 1981). In addition, a cavernous zone in the Pennsylvanian limestone just above the crystalline basement rock at about 800 m (2400 ft) caused many extremely difficult drilling and casing problems related primarily to severe lost circulation (Laughlin et al. 1983).

The Precambrian crystalline basement has high intact compressive strengths, low permeability and is very abrasive. Much of its gneiss with no discernible large-scale horizontal bedding. Several large intervals of granodiorite were encountered within the genesis, and a number of narrow fractured zones that contained chlorite, pyrite, and considerable amounts of calcite. Much chemical alteration has occurred in these zones. They evidently are shear zones or locally shattered zones sealed or mineralized by the forceful late injection of fluids and volatiles into the host rock (Laughlin et al. 1983). Variations in drilling performance and unexpected directional-drilling problems further indicated the presence of faults or altered zones in the wellbores.

Drilling the first slim (small diameter) core hole, GT-1, was an essential step in developing the scientific and engineering aspects of the site exploration strategy. The GT-1 borehole also served to delineate the potential drilling problems and provided a means to forecast for the larger diameter holes to be drilled through the section above the Precambrian rocks. The GT-1 corehole also discovered a shallow water-bearing unit used to supply drilling fluid for the later wells. Analysis of the crystalline rock core from the bottom of GT-1 established an initial baseline for the drilling properties of this unit. Such a slim diameter, shallow, core hole should be considered as an essential element of a deep CSDP hole, especially in a frontier area where no previous drilling records are available. The effort could usually be justified on the basis of the drilling-related lithologic data alone.

## The Hot Dry Rock (HDR) Project

Natural geothermal reservoirs can form where water has contacted rocks heated by shallow heat sources within the crust. In a few places, large circulating convection systems exist within porous or fractured reservoirs. These natural hydrothermal systems, often capped by an impervious layer, are characterized by high heat flux at the surface and are sometimes associated with leakage of fluids forming hot springs, geysers, and fumaroles. These natural hydrothermal reservoirs are rare, but may be excellent energy resources when properly exploited. Worldwide development of such high-grade hydrothermal reservoirs, greater than 180°C at depth, is at an installed electrical capacity of about 3200 MW(e) (DiPippo 1983) in 13 nations, from about 3000 drilled production wells. Exploration activities have located 147 hydrothermal reservoirs located in 43 countries. The current worldwide growth rate of installed geothermal generating capacity is approximately 15% per year.

Pursuit of research and development projects for innovative geothermal energy extraction systems has been motivated by the recognition that many conventional hydrothermal exploration and development wells do not produce fluids. Furthermore, the ratio of dry to productive wells in hydrothermal exploration averages somewhat greater than 3 to 1. The majority of the dry holes are hot, but penetrate rocks that lack sufficient fluids and natural permeability. Such hot, dry holes can be excellent prospects for stimulation (Campbell et al. 1981) or for possible development as hot dry rock energy extraction reservoir systems. The latent energy contained in the HDR resource is immense, and its success-

Fig. 4. Photograph of Fenton Hill site (1982) showing work towers over GT-2, EE-1, and EE-2 boreholes. Such towers can provide for post-drilling logging, and sampling operations. A workover rig is set up over the EE-3 well bore

ful development offers one promising alternative to supply a part of the world's increasing energy demands. Several HDR geothermal projects have been initiated around the world, with those in the Unites States, England, Germany, France, and Japan being the most advanced (Heiken and Goff 1983).

Production of heat energy by the HDR technique involves creating of in situ permeability in hot rock formations by hydraulic fracturing followed by injection and circulation of water. A two-well HDR reservoir system (GT-2 and EE-1), was created in the hot crystalline rock at Fenton Hill. Water was circulatd and heat energy extracted to demonstrate the basic feasibility of the concept.

The HDR project was initiated at the Fenton Hill experimental site in 1974. The first two deep boreholes (GT-2 and EE-1, see Fig. 4 and Table 1) were drilled to about 3 km (10,000 ft) depth, at a bottom hole temperature of about 200°C (400°F). This well pair was fractured in 1975 - 1977, to create the first reservoir. Results of extensive ciruclation and heat extraction experiments during 1978 - 1980 (Murphy et al. 1981, Dash et al. 1983) produced power at a 3 to 5 MW (thermal) level for about 1 year and motivated the drilling of the deeper well pair, EE-2 and EE-3, in an attempt to develop a larger and hotter HDR geothermal reservoir.

Table 1. Summary descriptions of the Fenton Hill deep boreholes

| Designation | Date drilling completed | Drilled depth (m) | Bottom-hole temperature (°C) | Purpose/Results | |
|---|---|---|---|---|---|
| GT-1[a] | 6/1972 | 785 | 100.4 | Small diameter exploration, 143 m Precambrian core, drilling data, volcanic and sedimentary section | |
| GT-2 | 10/1974 | 2928 | 197 | Test conductive gradient, core, and test equipment | Developed first HDR reservoir system (Murphy et al. 1981) |
| EE-1[b] | 10/1975 | 3065 | 200 | Energy producing, directional drilling | |
| EE-2 | 5/1980 | 4660 | 320 | Injection, core below 3 km | Drilled as prototype HDR well pair, inclined 35° to vertical current fracturing operations |
| EE-3 | 8/1981 | 4400 | 280 | Production | |

[a]Granite test hole no. 1; [b]Energy extraction hole no. 1.

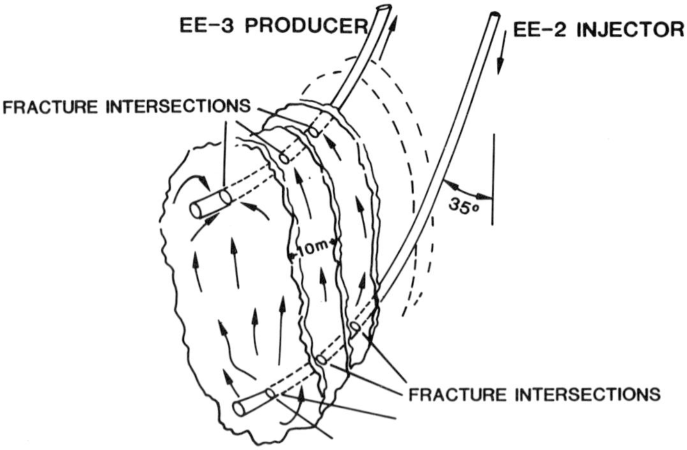


Fig. 5. Schematic of HDR parallel fracture system reservoir to be created in open-hole, 35° inclined sections of EE-2 and EE-3 wellbores

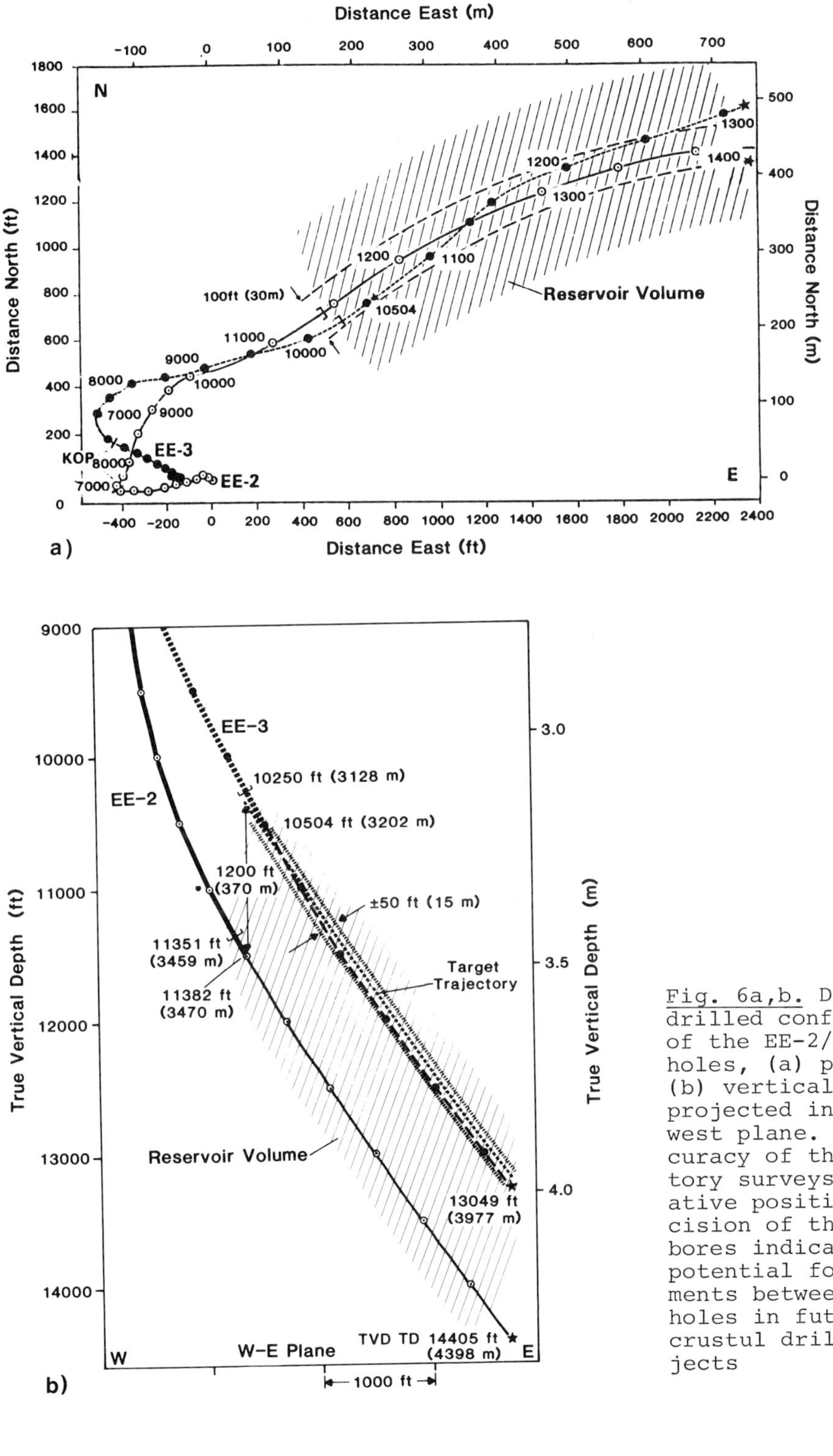


Fig. 6a,b. Directional drilled configuration of the EE-2/EE-3 boreholes, (a) plan view, (b) vertical section projected into east-west plane. The accuracy of the trajectory surveys and relative position precision of the wellbores indicates the potential for measurements between boreholes in future deep crustul drilling projects

(a) EE-2 DRILLING HISTORY

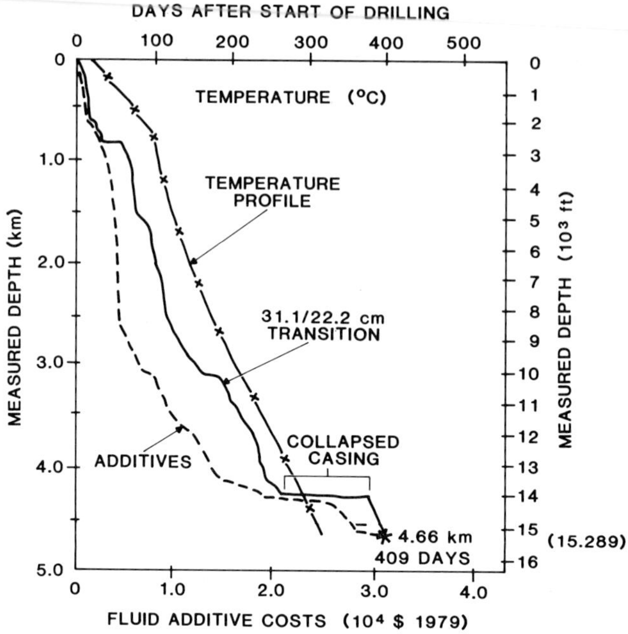


(b) EE-3 DRILLING HISTORY

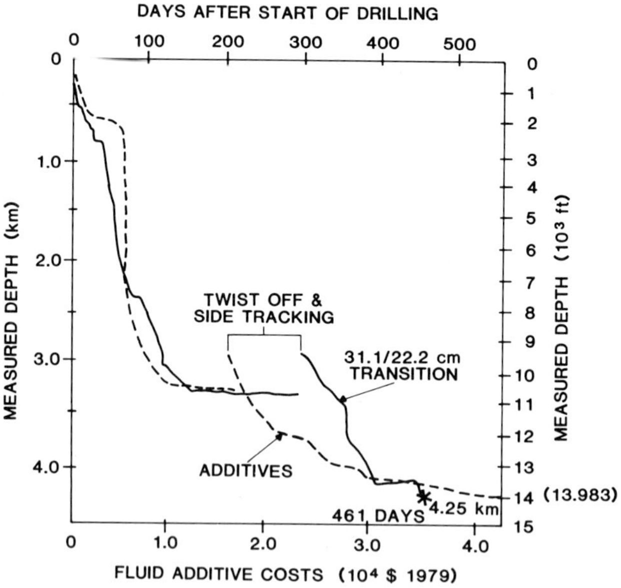


Fig. 7a,b. Drilling history, solid curve, of (a) EE-2 and (b) EE-3 boreholes. Cost of circulating fluid chemical additives *(dashed curve)* is also shown (converted to 1979 dollars). A major mechanical problem is indicated for each borehole. The temperature profile at Fenton Hill is indicated in (a) and the depths shown are measured along the wellbore

This paper concentrates on the drilling experience from EE-2 and EE-3. These wellbores are inclined at 35° from the vertical and configured so that parallel fracture could be made between the wellbores to create a HDR geothermal reservoir (Fig. 5). The drilling of EE-2 and EE-3 was finished in August 1981 (Helmick et al. 1982; Rowley and Carden 1982). The boreholes were drilled with open hole reservoir sections at 22.2 cm (8-3/4in.) drilled diameters. The EE-2 injection well was drilled to a measured depth of 4.66 km (15,298 ft) and had a bottom hole static temperature of 320°C (608°F). EE-3, the production well, was drilled to a measured depth of 4.25 km (13,933 ft) and had a bottom hole temperature of 280°C (580°F). High-precision directional drilling was used to position EE-3 at 370 ± 15 m (1200 ± 50 ft) above EE-2 and within lateral location tolerances of ±30 m (±100 ft). This was done to maximize the possibilities for connection during post-drilling fracturing operations. Figure 6 depicts the directionally drilled wellbore trajectories for the EE-2/EE-3 well pair, and Figure 7 summarizes the drilling histories for these two boreholes.

## HDR Experience Relevant to Scientific Drilling in the Continental Crust

Scientific objectives of a deep scientific hole in the continental crust should be considered as early as possible to allow time for the development of equipment and drilling procedures that are specifically designed for the anticipated downhole conditions. Fenton Hill experience indicates that three principles should dominate this development phase:

- *Do not* assume that the drilling industry or another project has already developed a solution to a specific problem.
- *Do* seek multiple and alternative solutions with in-depth backup. Duplication of potential solutions will considerably reduce risk.
- *Do* expect special problems caused by high temperature. *Elevated* temperature performance for conventional drilling equipment and related services generally is limited to 150°C, but services to 200°C can be obtained in a few instances.

### 1. Core Bits

Coring technology and hardware is an area of special importance to CSDP projects and illustrates the above principles very well. The original drilling plan for the first deep borehole, granite test No. 2, GT-2, in the crystalline rock at Fenton Hill called for continuous coring using the Deep Sea Drilling Program (DSDP) core bit. The DSDP had succeeded in developing a successful rotary coring technology (Larson et al. 1981) for the sea floor sediments. The variety of coring tools that resulted were designed to obtain quality cores from the unique conditions and lithologies represented by the deep sea sediments. These tools include: (1) a piston corer for unconsolidated near sea bed surface, oozes and sediments, (2) a rotary driven, extended retractable core barrel for layered sections of highly variable hardness, and (3) a wireline deployed core barrel, coupled to a rotary driven four roller cone bit with tungsten carbide insert (TCI) cutting structure (Fig. 8). This core bit was designed to provide very long bit life through imposition of low loads on the roller cone bearings and long cutting structure life for both the core trimming and drilling functions. These design features resulted in long bit life and therefore long intervals could be cored without tripping the bit, and provided high quality core. The annulus of the borehole drilled by this bit might be viewed as

Fig. 8. Four-cone, tungsten carbide insert core bit developed by Deep Sea Drilling Project for sea bed sedimentary rock coring. Bit has 25.1 cm (9-7/8-in.) diameter and cuts a 7.0 cm (2-7/8-in.) diameter core

Fig. 9. Hybrid core bit with both four-cone and polycrystalline diamond compact (PDC) drag cutting structures. Developed for the Los Alamos HDR Geothermal Project for coring of deep crystalline rocks. (a) Before use, (b) after first trial run (Table 2). Bit has 20.0 cm (7-7/8 in.) diameter and cuts a 7.5 cm (3-in.) core

Table 2. Comparison of diamond core recovery and initial test of hybrid core bit.
[During 24.5 cm (9-5/8 in.) redrill of GT-2 borehole]

| Date (1977) | Cored interval depth [m (ft)] | Recovery (%) | Bit type[a] | Core bit size | Remarks |
|---|---|---|---|---|---|
| May 9 | 2799.3 | 0 | D | 24.5-cm o.d. by 11.5-cm i.d. (9-5/8 by 4-1/2 in.) | Junk basket run first, but no reaming bit run. Experienced difficulty getting past kick-off point at 2482.3 m (8144 ft). Reamed to bottom with coring bit; bit badly worn on gauge (outside) surfaces. Probably no core cut |
| May 29 | 2672.8-2674.0 | 52 | D | 21.6-cm o.d. by 11.4-cm i.d. (8-1/2 by 4-1/2 in.) | *Biotite Granodiorite.* Previous drilling was 9 m (30 ft) of straight hole cut with 24.5-cm (9-5/8 in.) bit; hit fracture zone at 2672.8 m (8769 ft). Core showed three fracture planes, two still sealed with calcite and one reopened |
| June 9 | 2710.9-2712.1 | 0 | D | 21.6-cm o.d. by 11.4-cm i.d. (8-1/2 by 4-1/2 in.) | Previous 18.3 m (60 ft) reamed with rerun 24.5-cm (9-5/8-in.) bit, plus 4.6 m (15 ft) new hole. Supposedly cored 1.2 m (4 ft) in 4 h. Core bit showed extreme wear on cutting surface, deep into matrix metal. Diagnosis: loose pieces of rock rolling around under bit, no core cut |
| June 15 | 2712.7 (8900) | 0 | D | 24.5-cm o.d. by 14.0-cm i.d. (9-5/8 by 5-1/2 in.) | Reamed hole to bottom with new 24.5-cm (9-5/8-in.) bit and junk basket. Could only get core bit down to within 12.8 m (42 ft) of bottom. Gauge diamonds of core bit completely worn off |
| June 18 | 2712.7-2713.3 (8900-8902) | 0 | D | 21.6-cm o.d. by 14.0-cm i.d. (8-1/2 by 5-1/2 in.) | Core bit was worn in same pattern as June 9; same diagnosis |
| June 19 | 2713.3-2714-9 (8902-8907) | 89 | H | 20.0-cm o.d. by 7.5-cm i.d. (7-7/8 by 3 in.) | *Biotite Granodiorite.* Penetration rate 0.6 m/h (2 ft/h). Bit was retrieved with three studs broken off and chisel teeth worn smooth. Core contained several distinct fracture surfaces, with possible evidence of pumping flow |

[a]D - Conventional diamond.
H - Hybrid, combined polycrystalline diamond compact (PDC) and chisel-tooth, tungsten carbide insert, roller cone.

excessive with 25.1 cm (9-7/8-in.) outer bit diameter needed to cut a 7 cm (2-7/8-in.) diameter core. However, this bit design provided a very effective solution and a single bit could often be used to core the entire sedimentary section before it was dulled by a short penetration into the basalt basement below. Initial trials at Fenton Hill, however, showed rather poor core recovery, slow penetration rates, and short bit life for the DSDP four-cone core bit. The primary causes were the hardness and extreme abrasiveness of the crystalline rock which rapidly wore the core-trimming TCI buttons. This caused the core diameter to gradually increase and to jam in the core catcher at the entrance of the core barrel. This jamming resulted in breakage of the core into 5 - 10 cm pieces, which were ground and abraded into near-spheres. Also the TCI buttons on the roller cones that perform well in sea bed sediments, wore rapidly in the deep GT-2 borehole. A solution to the core problem was sought using conventional diamond core bits, but recovery was often very poor (Table 2). Finally, during the GT-2 drilling a solution was devised by a bit manufacture who used a hybrid four roller cone and drag bit (Fig. 9). The drag cutters were fashioned from polycrystalline diamond compact (PDC) cutting elements (Herrick 1978). The core trimming was thereafter successful because of the low wear rate of these PDC drag cutters. Good core recovery and reasonable penetration rates of approximately 0.6 m $h^{-1}$ (2.0 ft $h^{-1}$) were achieved.

Although this hybrid core bit was an adequate solution for the HDR project, which only required spot coring through crystalline rock at the 3 to 5 km depth interval, it seems clear that improvements in core bits and coring systems will be needed if deep, extensive or continuous coring of crystalline rocks are required for CSDP projects. It is judged that, even when optimized, a limit on bit life, penetration rate, and core quality will be reached for this type of hybrid core bit if a rotary drive is used at great depths.

A suggested development approach in coring technology research to solve this problem would involve use of a hollow shaft turbodrill such as that initially proposed by the Moho project for coring sea floor basement rocks. Section 4, Fig. 16, indicates a conceptual design for such a coring system. The hollow shaft turbodrill would allow wireline core barrel retrieval. The core bit and core trimmer should be developed based on the same philosophy as used for the DSDP tool designs. Long bit life can be enhanced through designs that use multiple cones (for low bearing loads) that are combined with drag cutters, and include heavy gauge and cutting structure wear protection. Optimizations to increase bit life, and thereby reduce tripping of the drill string, and improve core quality should dominate the development. The turbodrill would provide the high bit torque necessary to drive the PDC drag cutters. A higher rotating speed of the turbodrill relative to conventional rotary drives will enhance penetration rate and core quality through better bit dynamics, and improved bottom hole and drill string mechanics. In addition, the fatigue life of the drill string will be much longer, a distinct advantage when long cored intervals are planned. Another design feature to be considered is to allow the opportunity to replace or redress the core cutters with each core barrel run.

Turbodrill technology as applied to deep drilling operations in crystalline rocks was furthered by the HDR project when several deep directional drilling operations were conducted below 3 km and at temperatures exceeding 260°C (500°F). The successful development and use of such a high-temperature rated, high-torque, low rational speed turbodrill (Neudecker and Rowley 1982) could form the basis for design and development of hollow shaft turbodrills for CSDP coring application. These high-temperature rated turbodrills could uniquely be used for hole straightening, sidetracking, and directional drilling operations for CSDP projects; especially in deep hot boreholes in the harder crystalline rocks.

Rezanov (1981) describes deep scientific drilling in the Soviet Union. From a literature study, and a visit to one deep drill site (Rowley and Watkins 1977), it appears that the deeper geologic and stratigraphic studies in the Soviet Union rely heavily on cuttings, and that a deep coring technology has not yet been developed. A core bit with 21.4 cm (8-1/2-in.) outer diameter and 5.0 cm (2-in.) core diameter is mentioned, but apparently used only for the shallower parts of the boreholes. Ioanesyan et al. (1981) discuss a hollow shaft Soviet turbodrill used with high pressure drag bits, but no mention was made of use with coring tools. Deep continuous coring is discussed as a needed and desirable goal. These very deep (greater than 5 to 7 km) scientific boreholes in the Soviet Union are being drilled by both rotary and turbodrill but most of the Soviet oil and gas exploration and production drilling is accomplished nearly exclusively with turbodrills. Indeed, the Soviet petroleum industry personnel indicated that the deep drilling program was looked upon by them as a drilling technology development effort, and was directed toward the future when deeper petroleum production drilling would be more heavily dependent on drilling in deeper, harder sedimentary rocks. It is certainly true, as experience with the deep HDR boreholes has demonstrated, that the high temperatures, high dynamic forces and fatigue stresses imposed on all downhole hardware, represent a very good test bed for finding weakness in drilling hardware, and for developing improved drilling hardware and techniques.

The Soviets also report:

- Development of a high torque, low rotational speed turbodrill suitable for crystalline rock drilling.
- That the 21.4 cm (8-1/2-in.) roller cone bit size is optimum for bit life and penetration rate in crystalline rock.
- Difficulty in stabilizing turbodrill assemblies to maintain a straight hole.

These technological developments and observations by the Soviets for deep drilling are in accord with Fenton Hill experience. The significance of the possibility of long, deep, open-hole sections has many practical advantages for the planning and drilling aspects of CSDP projects, and especially for the potential scientific yield. The use of a protective or sacrificial casing string (see Section "Casing Design") is worth serious consideration. This approach could be very valuable if long open-hole sections are to be rotary drilled or cored because wear of permanent casing strings and drilling of the shallow larger diameter holes can be very costly and possibly present high risks. The Fenton Hill experience confirms the long-term stability of open hole crystalline rock sections. The EE-1 and GT-2 boreholes have remained stable since 1975 from about 800 m to about 3 km. Also the EE-2 and EE-3 wellbores are open hole below the production 23.5 cm (9-5/8-in.) casing point with about 900 m of open hole below as shown in Figure 3.

## 2. Rock Bits

This section summarizes from the HDR project deep crystalline rock drilling experiences in bit performance and selection that should be considered in any deep CSDP drilling plan and represent areas where serious and costly drilling problems could arise. Technical reports by Helmick et al. (1982), and Rowley and Carden (1982) give details.

The drilling of the deeper borehole pair EE-2/EE-3, summarized in Table 1, started with the spudding of the deeper borehole, EE-2, on April 3, 1979. This deeper well was drilled to a measured depth of 4.66 km (15,299 ft); a true vertical depth of 4398 m (14,405 ft). The bottom open hole section was drilled to 22.2 cm (8-3/4-in.) diameter and inclined at 35° from the vertical by directional drilling. The drilling time was 409 days. The second, shallower borehole, EE-3, was drilled to a measured depth of 4.25 km (13,933 ft); or a true vertical depth of 3977 m (13,049 ft). EE-3 was drilled parallel to EE-2, as shown in Figure 6, at a vertical depth of 370 m (1200 ft) above EE-2. EE-3 took 461 days to drill and complete. The drilling history vs. depth plots for the two boreholes are depicted in Figure 7. This type of presentation is used throughout the drilling industry, and is a valuable way to summarize a drilling program or plan. As noted on the figure, both wells suffered major mechanical problems; EE-2 a collapsed intermediate (34.0 cm, 13-3/8-in.) casing string and EE-3 a severe drill string twist off and subsequent sidetracking operation.

The proper choice of drill bits is an important aspect of deep drilling in hard and abrasive crystalline rocks. Attention to early downhole bit evaluation and tests can optimize drill bit performance in a given lithology and avoid serious problems that otherwise could lead to lengthy remedial (fishing) operations. In this and the following sections it is assumed that initial deep CSDP projects will rely mainly on rotary drilling with spot coring. The major consideration in bit selection in deep drilling is bit life, with rate of penetration a secondary concern. This is because of the length of time required for round trips in deep holes greater than 3 km.

Deep rotary drilling experience at Fenton Hill through over 7.6 km (24,000 ft) of hot crystalline Precambrian rock has led to the following conclusions about drill bits:

1. Bit life is the major factor in bit selection.
2. Drilling rates in crystalline rock are uniform with depth, ranging roughly from 3 to 10 m/h (10 to 30 ft/h).
3. Drilling rate is primarily dependent upon bit cutting structure, tooth configuration, and the force on the bit.
4. Outer diameter gauge wear is the major bit problem.
5. Sealed bit bearings are temperature limited; open bearing bits are required above about 200°C (400°F).
6. Bit bearing life decreases with increasing temperature (i.e. depth).
7. Only tungsten carbide insert (TCI) cutting structures were found effective for granite drilling.

Severe drilling conditions in hard crystalline rock will rapidly destroy poor quality rock bits. Even minor manufacturing defects lead to rapid bit failures, which might not have occurred in less severe drilling conditions. For example, each new bit run must ream the previous hole diameter slightly because of hole taper due to gauge wear and hole diameter reduction from the previous bit run. Also, bits with weak leg structures readily "pinch", lose cutting structure teeth and cone tips, and torque-up quickly. Bit configurations based on blast hole type rock bits proved superior for reaming and drilling granitic and gneissic rocks.

Fig. 10. Special 22.2 cm (8-3/4-in.) diameter rock bit used for drilling in deep, hot crustal crystalline rocks. Design features tungsten carbide insert (TCI) and gauge pad wear protection; and open, air-cooled type bearings

Based upon these considerations, a special geothermal three-cone rock bit was used to great advantage and almost exclusively to drill the deep, 22.2 cm (8-3/4-in.) diameter sections of the EE-2 and EE-3 boreholes. This bit (see Fig. 10), is a nonsealed, air circulation, TCI bit. Several special design features that account for this superior performance are:

- harder and more abrasive resistant carbide TCI buttons were placed on the gauge row of the cones,
- hard facing for extra wear resistance was used on the leg skirt-tails,
- wear pad build-up and carbide inserts were included on the shanks of the legs,
- use of nonsealed bearings, and
- additional bearing clearances allowed for increased fluid circulation.

These features increased bit life and reduced the number of trips to an extent that more than offset the modest (10%) additional cost of the bit. These bits routinely provided bit runs greater than 60 m (200 ft), with average bit life greater than 90 m (300 ft). These long runs were made in the 35° inclined sections of the holes where abrasive wear is especially severe.

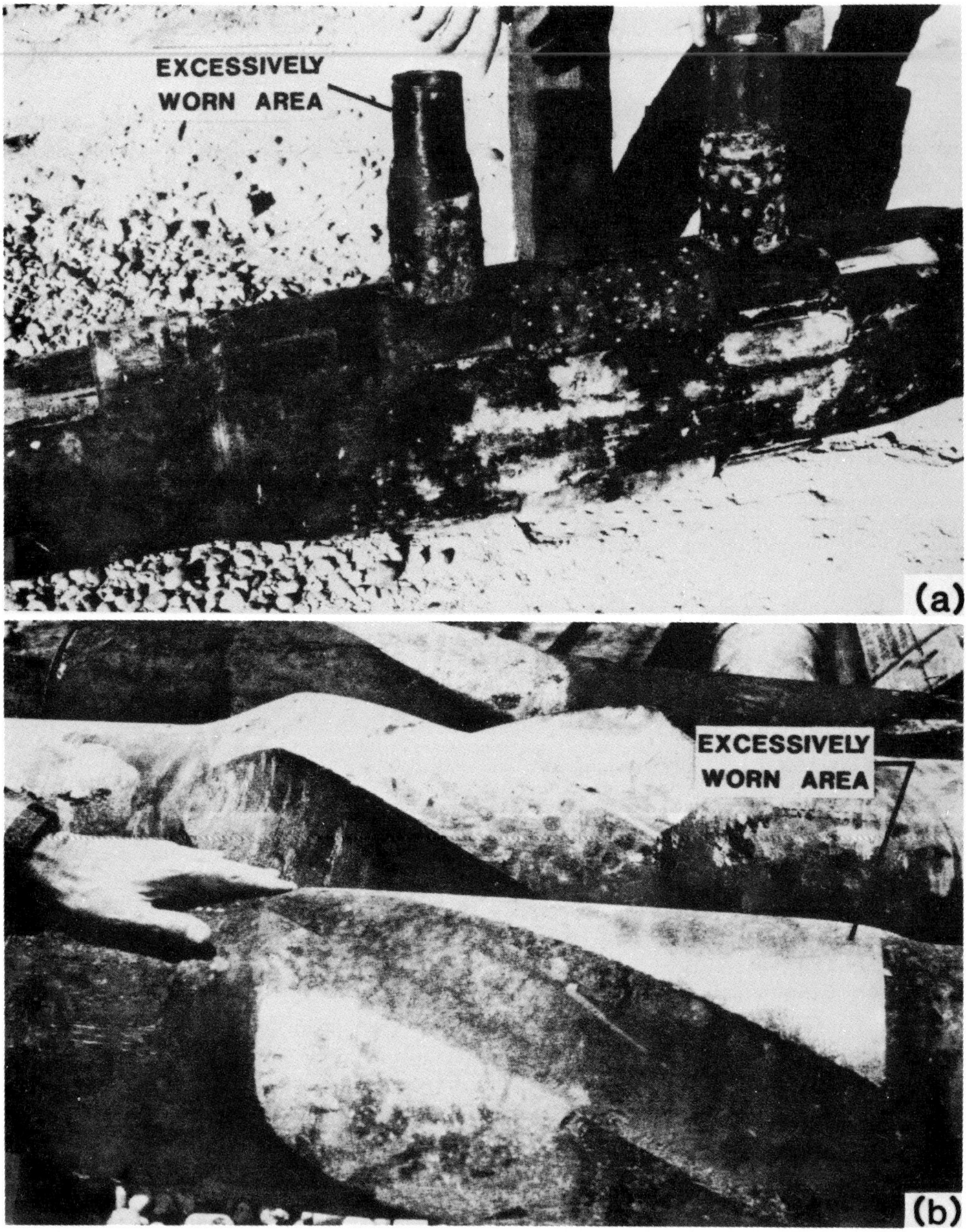


Fig. 11. (a) Near bit roller reamer and (b) blade stabilizer after extreme abrasive wear. The roller reamer and stabilizer configuration are preferred in abrasive hard rocks

## 3. Bottom Hole Assembly (BHA) and Drill String

The selections of a proper bottom hole assembly (BHA) and drill string are equally important for hard, abrasive crystalline rocks. The following factors are most important:

- A near-bit roller reamer (Fig. 11) should be placed immediately above the bit to aid with the undersize hole gauge problem caused by bit gauge wear. Blade type stabilizers (Fig. 11) wear far too rapidly to be used effectively.
- Drill string diameter and strength should be chosen with stress fatigue conditions as a primary guide. Generally, heavy wall drill strings with diameters equal to or greater than 12.5 cm (5 in.) will be necessary to accommodate the fatigue associated with extended hours of deep rotary drilling and coring. It should be noted that the larger inside diameters provided by these larger diameter drill strings will accomodate wireline core barrel operations and offers potential for through-core-bit logging.
- A very thorough and rigorous drill string inspection program is also necessary.
- The bottom hole drilling configuration should be a well stabilized, strong pendulum-effect assembly used with modest bit forces. This will provide for straight hole drilling; an extremely important objective in very hard formations. Crooked holes contribute to drill string wear and stress cycle fatigue that result in downhole failures. Further, it is extremely difficult to straighten, or ream-out, a crooked hole in crystalline rocks, therefore *it is far more important to drill a straight hole than to drill at a high penetration rate.*
- For very deep holes, greater than 5 to 7 km depths, a specially designed, tapered diameter drill string will be required.

The importance of high dynamic forces on BHA, fatigue life, and the influence of BHA design has recently been stressed by an analytical and experimental study reported by Darring (1984). These results relate especially to drilling or coring of harder rocks, and the problems avoiding resonance conditions in the BHA collars that are driven by bit rotational impact forces.

## 4. Drilling Fluid System

Choice of suitable drilling fluid systems for deep CSDP boreholes will be dictated by many complex and possibly conflicting factors. The primary and essential function of the circulating fluid is to clean the drill cuttings from the hole. It may also be necessary to stabilize the upper formations, or those deeper zones, found to be water susceptible. This is accomplished by increases in fluid density and viscosity and additions of chemicals. The density and viscosity enhancements are usually accomplished by adding various clays (muds). It is often necessary to add chemicals to neutralize reactions between the formations and the fluid, in order to prevent excessive wash out of or squeezing into the hole. In the deeper, hotter portion of the boreholes cooling of bottom-hole hardware is an important consideration. Chemical additions for corrosion inhibition are necessary, especially in hot holes. These circulating fluid requirements may be in conflict with the scientific requirements to minimize the contamination of borehole, cores, cuttings, and formation fluid samples. To partially counter the contamination problem it is possible to consider the use of appropriate tracers injected into the circulating fluids to monitor the extent and nature of the invasions and contamination.

Many crystalline rocks are water-insensitive, and are mechanically stable and therefore a so-called "clear water" circulating fluid system can be considered. Such a circulating fluid system is relatively clean,

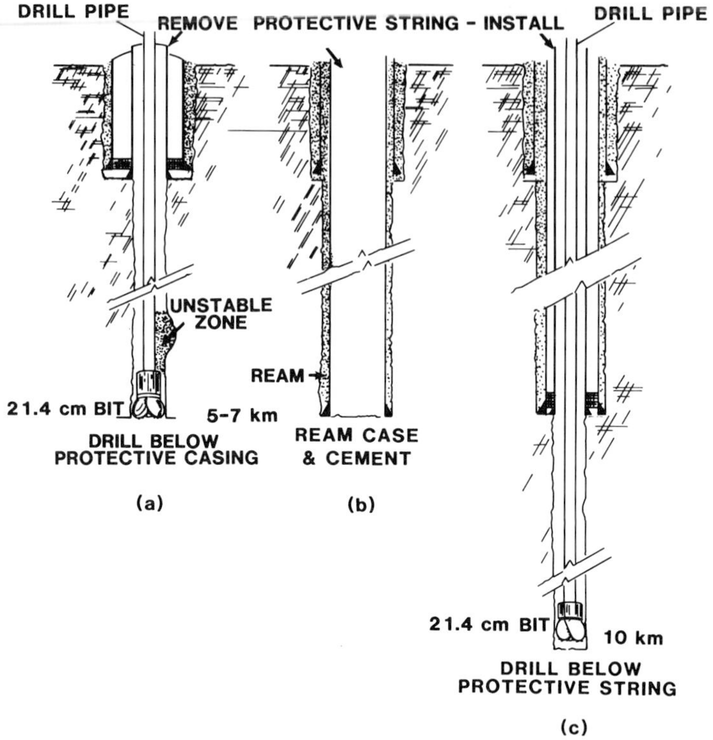


Fig. 12a-c. Sketches of the Soviet advance open hole drilling strategy for ultra-deep boreholes in stable formation. (a) Initial borehole with 24.5-cm (9-5/8-in.) diameter uncemented protective casing string set and drilling below with turbodrill and 21.4 cm (8-1/2-in.) bit. (b) When wellbore sloughing is encountered remove protective string, ream at 39.4 cm (15-1/2-in.), and set and cement-in a 32.5 cm (12-3/4-in.) casing. (c) Install a 24.5-cm (9-5/8-in.) protective casing and continue drilling below. This technique is apparently applicable to most crystalline rocks, and can be a possible general approach to deep crustal drilling. Fenton Hill crystalline rock borehole stability would appear to be sufficient to apply the Soviet method

but it has a relatively low viscosity and lacks efficient carrying capacity for the cuttings. This leads to poor bottom hole cleaning of chips, so that the cuttings are reground to a fine powder, and results in inefficient drilling. Surface inspection of fine cuttings is more difficult and can reduce the scientific quality of this evaluation. The fine cuttings increase abrasive wear of downhole hardware. Higher fluid pumping rates, that is higher fluid velocities, and periodic sweeps with discrete slugs ("pills") of high viscosity mud can be used to clean the hole. A second major disadvantage of the clear water fluid system is that only slightly lubricity of the wellbore is provided. That is, high friction between drillstring, bottom-hole assembly and the borehole wall can develop. Formations of low permeability cause little or no mud cake to form on the borehole walls and lubricity will be reduced further. Low lubricity leads to both high drag fources with increased risks of drill string failure (twistoffs), stuck drillstring, and increased abrasive wear of downhole hardware. A complex mud system will not always be a drilling advantage, or a clear water system may be dictated by the scientific goals. In that circumstance, lubricity agents must also be added; a selection will be based on the chemical contamination restrictions of the project. Most mud systems have temperature limits, and tend to degrade at high temperatures. In addition, if circulation is halted (for example to change a dulled bit) muds will harden in the hot wellbore sufficiently so that they must be drilled out. If fracture permeability is encountered mud loss can be very severe and, with temperature hardening the mud will plug deep into the fractures and possibly restrict desired fluid sampling from fracture zones. Fluid-sensitive or highly permeable shallower zones, perhaps sedimentary or metamorphic rocks, might best be sampled with a separate small diameter (slim), shallow core hole, as outlined in the next Section. This approach has the advantages of providing good core quality for these shallow zones, and the detailed stratigraphy needed to effectively plan the subsequent, deeper, much larger diameter, holes that can then be drilled with no core requirements and therefore with an optimum circulating fluid program that is designed to mitigate hole stability and lost circulating fluid problems.

## 5. Casing Design

A casing design for a deep CSDP borehole will usually be quite different than that for a deep oil and gas well. This is due mainly to: (1) different rock types expected, (2) stable formations can be anticipated, and (3) larger diameter bits will be required to yield long bit life in the deeper, hard rock sections of the holes. One design option might be to consider an open hole method practiced by the Soviets. This element of the Soviet "ultra deep" drilling technology that might potentially be very important in future U.S. deeper boreholes is the procedure termed by the Soviet drillers as the "advance open borehole drilling method". It is illustrated in Figure 12. The more usual casing design that involves cementing a sequence of progressively smaller casing diameters and drilling below with successively smaller bits, is replaced by a casing program that drills at an initial large diameter of 92.0 cm (36-1/4-in.) with a cemented-in 72.0 cm (28-1/2-in.) casing, Figure 12a. A 24.5 cm (9-5/8-in.) diameter protective, or sacrificial, casing string is run in but *not cemented*. Drilling with a turbodrill below this protective casing string proceeds with a 21.4 cm (8-1/2-in.) diameter bit and the hole remains in an open condition to a depth and for as long a duration as the borehole wall remains reasonably stable. When hole caving or sloughing occurs the drilling assembly is pulled from the hole and the protecting 24.5 cm (9-5/8-in.) casing string is retrieved. The hole is reamed to a larger diameter below the unstable zone, or to the bottom of the previous drilling. Typically a 39.4 cm (15-1/2-in.) diameter drill bit (or reamer) size is used to open the hole diameter and clean out the borehole. Then a deep cemented-in casing of 32.5 cm (12-3/4-in.) diameter is installed (Fig. 12b). Drilling operations continue (Fig. 12c), following the installation in the reamed and cased hole again of the protective or sacrificial 24.5 cm (9-5/8-in.) casing that is *not cemented* in place. Drilling below proceeds with the 21.4 cm (8-1/2-in.) bit. The Soviet reports stress the obvious advantages of such a procedure of long open hole intervals for geophysical logging. Also they have reported heavy casing wear of the protective strings from just the tripping in and out of the very deep holes. This technique was first used in the Kola peninsula borehole (abbreviated as UD-3; ultradeep No. 3) successfully to a 10,780 m (35,367 ft) depth in 1981.

A second scheme is illustrated in Figure 13, a possible design for a thermal regime CSDP project. A sequence of successfully deeper holes provides data for the casing design for the next deeper core hole,

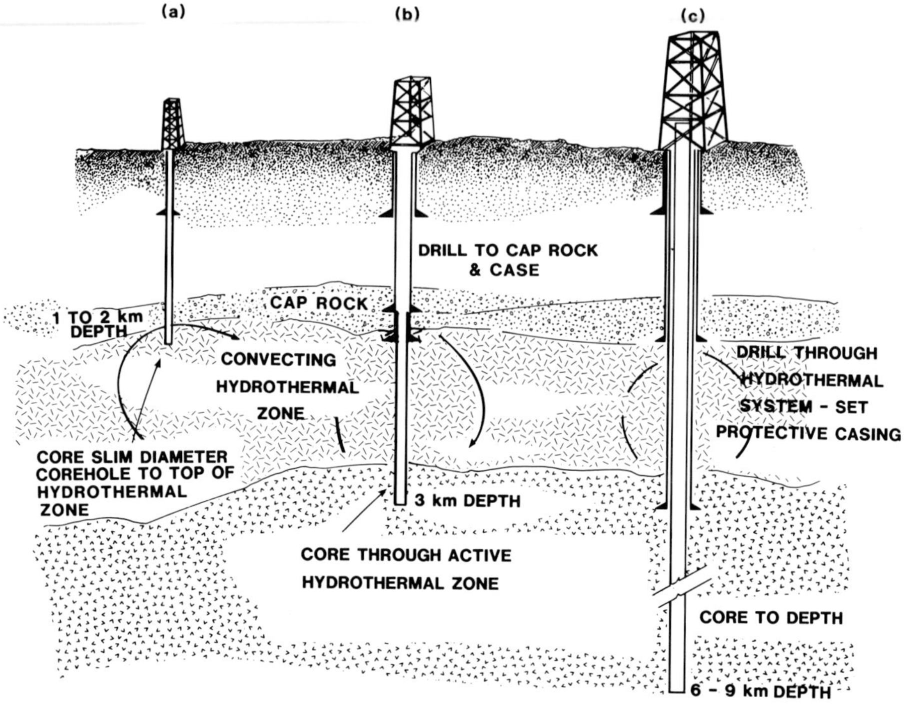


Fig. 13a-c. A recommended drilling strategy for CSDP thermal regimes investigations of hydrothermal convection system. (a) Slim hole core to top of hydrothermal reservoir zones of high permeability, (b) drill second borehole through shallow formations previously cored and set casing into the caprock, and core through the permeable zones, possibly using a dual string drill rig, to a depth of about 3 km. (c) The third deep borehole is drilled through the hydrothermal reservoir and a heavy casing is set; possibly use two strings. Then coring proceeds below to the 6 - 9 km depths

while likely assuring higher quality cores, logging, and sampling of the upper part of the section. This approach provides the opportunity to design a sufficiently durable and stable final casing program so that wear and collapse conditions can be projected and accounted for. The major wear will come from the many round trips into the deep holes and the long rotating hours caused by the slow penetration rates experienced in hard rocks for both coring and drilling. Uphole casing collapse can result from wear, poor cementing due to lost circulation zones, and exposure to high temperature fluids produced from deep formations. Geologic knowledge from the earlier shallower core holes and knowledgable drilling and casing designs can avoid or help mitigate these problems. A collapsed casing can require lengthy remedial operations (fishing) and repair, or in a worst case, the hole may be lost. This second example is in reality a brief discussion of a general strategy for CSDP-thermal regimes deep drilling. First, it must be appreciated that there is a necessity to design the drilling, coring, and sampling program in response to the scientific requirements of a particular thermal regimes project. This is illustrated here by an example project that requires continuous coring through and below an active hydrothermal system to a depth of about 8 km. This example strategy is intended to maximize the recovery and quality of core, provide appropriate fluid samples, and preserve open-hole sections for geophysical measurements. CSDP-thermal regimes drilling and coring planning must focus on the specific site scientific goals and objectives of the particular CSDP project. These are expected to concentrate on coring and sampling. Therefore the drilling strategy to optimize the scientific yield will also be very site specific. The strategy proposed to achieve these goals suggests a sequence of progressively deeper and larger diameter core holes.

The sketches in Figure 13 illustrate one possible approach intended to optimize the scientific yield. It is assumed that the scientific objectives require essentially continuous core, fluid samples and open-hole logging. A three-stage, drilling and coring program is proposed:

1. Slim hole core to the top of the hydrothermal system (Fig. 13a), shown at about 1 km depth.
2. Use the lithology from this first stage to plan a second, larger diameter hole designed to core through the hydrothermal system, and below the depth reached by the first slim core hole. This second section is expected to be primarily in permeable fractured rock formations; problems of lost circulation of drilling fluids, borehole instability, and high temperatures will dominate this stage. A dual string, gas lift core drilling system (Dareing and Kelsey 1981) is suggested (Fig. 13b) as one method of minimizing lost circulation problems. A depth of 3 km is presumed to just penetrate into the unfractured crystalline formation below the active hydrothermal convection system.
3. Drill a third, larger diameter hole through the hydrothermal system. The knowledge from the first and second holes is used to design the upper portion of this third hole. It is a straight drilling section, with no coring. An extra heavy casing is set and cemented a few hundred meters into the crystalline rock. A second, protective, or sacrificial casing is set inside this primary casing string (Fig. 13c). Coring below is assumed to be into completely unknown formations. The heavy casing program is suggested because core drilling to a depth of 8 km and perhaps temperature of $\sim$500°C are anticipated. Very slow coring rate (many rotating hours) and the time needed for solution of downhole problems must be anticipated to ensure good recovery of high quality core. A 21.6 cm (8-1/2-in.) diameter bit is suggested for this deepest core hole to assure long bit life.

All three holes would be sited in close proximity at the surface so that drilling surprises would be minimized.

This brief exposition illustrates the types and range of drilling strategy, hole designs, and casing schemes that might be necessary to conduct a CSDP deep drilling project. Other CSDP projects will require different drilling and sampling strategies. However, it is probably always wise to consider an exploratory, slim core hole to 1.5 or 2 km depths as an initial step in any unknown or wildcat area.

## 6. Fishing

As a final topic from the experience gained from the Fenton Hill drilling, it is necessary to discuss the subject of problem solving operations in deep boreholes. These remedial operations are usually referred to as "fishing" in the drilling industry. The important consideration is that, next to avoiding problems, effective and efficient downhole

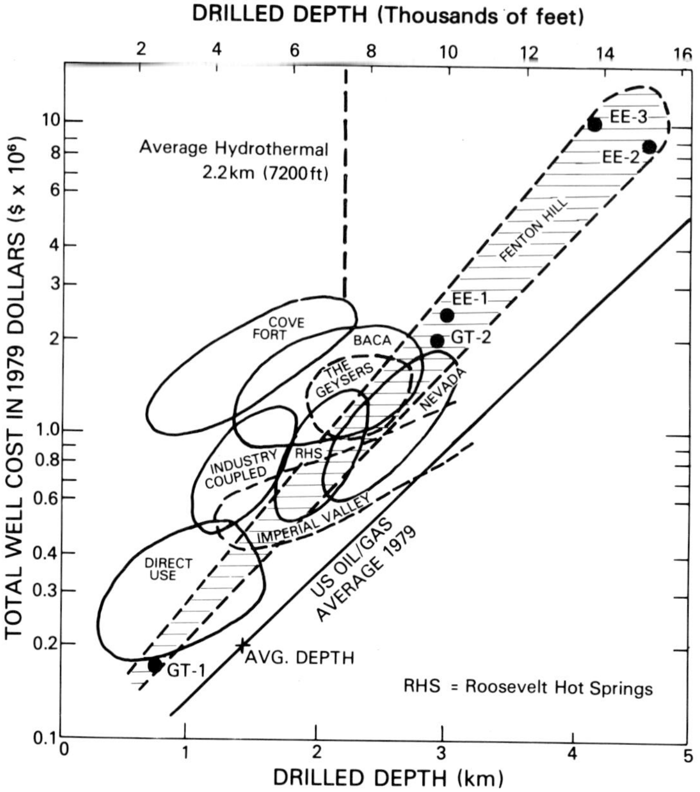


Fig. 14. Historical drilling cost comparisons for hydrothermal, HDR, and oil and gas wells (after Carson and Lin 1981). Costs normalized to value of 1979 dollars through use of a 17% escalation factor

problem solving is extremely important. Experienced personnel and the proper equipment and tools are the essential ingredients to successful removal of broken hardware from the borehole. It should be recognized that for the hot hole conditions to be encountered in thermal regimes CSDP projects, many fishing tools are temperature limited, or have temperature degradable components. Examples are the often needed wireline conductor cables and high explosive charges used to unscrew (backoff) downhole threaded connections. One slight advantage can be cited for the very hard deeper formations anticipated in many CSDP projects, it is generally easier to fish objects (junk) from holes in the harder formations because there is less likelihood for the junk to become jammed or pushed out into the rocks, as is often the case in softer formations. Failure to remove (fish) junk from the hole can have significant impact on a drilling project because then the decision must be made whether to abandon the hole or to drill around the junk (sidetrack the hole) from above the unretrieved objects and thus bypass the junk in order to continue drilling. Sidetracking in crystalline rock has proven to be very difficult (Pettitt and Carden 1981) and when completed, the crooked hole section produced introduces additional drillstring drag forces and increased stress levels. These hole conditions will add to the drilling risks in the deeper parts of the hole. Especially important is the additional fatigue induced during deep rotary drilling or coring.

Both EE-2 and EE-3 were problem wells, and plagued with lengthy fishing operations as shown on Figure 7. EE-2 suffered a collapsed intermediate casing string and 34.5% of the 409 drilling days were needed to clear the problem. The EE-3 borehole suffered from a large number (26) of drillstring failures (twistoffs). One especially bad twistoff required sidetracking. These associated fishing and sidetracking problems occupied nearly 50% of the 461 drilling days.

## 7. Drilling Costs

The usual method of deriving detailed drilling cost estimates is to sum up all the individual cost element estimates of a specific detailed drilling plan and borehole design. For a hole in the 7 to 9 km depth range, this is a considerable task. Such a design, engineering optimization, and cost analysis effort might require one to two years of time, and this effort should be allowed for in any CSDP planning and scheduling process. A quick order-of-magnitude cost estimate can be gained from historical data. An example of this type of data is shown in Figure 14. Drilling costs are plotted on a log scale against well depth for intermediate diameter holes between 15.0 cm (6-in.) and <30.5 cm (12-in.). These are cost data for geothermal production and exploration well drilling and are refered to 1979 dollars, with an escalation factor of about 17% per year used through 1981. For reference and comparison the average cost trends for U.S. oil and gas wells are also plotted, as are the costs of drilling the five Fenton Hill boreholes, adjusted to the value of 1979 dollars. As an important observation, the increased cost effects of hard, fractured rocks and higher temperatures experienced in geothermal and HDR drilling are evident, when compared to the oil and gas drilling experiences at the same depths. The cost depth trends are seen to be exponential with depth. Clearly scientific boreholes drilled to 9 - 10 km depths will be dear.

## Recommended Drilling Technology Improvements

This discussion is an outline of needed drilling technology improvements, with special emphasis on solutions to problems associated with deep, high temperature drilling, coring, and sampling in the continental crust. These needs are drawn from direct rig-floor, field experience attained during the Los Alamos HDR drilling operations and from interactions with drillers from Italy, France, England, the Federal Republic of Germany, Japan, and Canada, as well as many U.S. geothermal drilling operators. Experience was also gained through collaboration with the private sector to develop high temperature downhole drilling tools such as hydraulic drilling jars, core bits, shock absorbers, and turbodrills. In addition, involvement with development field trials and of advanced rock melting drilling systems (Rowley 1974) provided a unique perspective on problems of high temperature drilling. Generally, these brief technology descriptions proceed from the simpler components needing improvements to longer term, more complex hardware items. Discussions of advanced drilling methods or techniques (Maurer 1980) that might be required for coring and sampling in deep formations at temperatures in excess of 600°C are not included.

This outline of needed improvements is organized into five major categories for the purpose of focusing the presentation. The topics selected are drill pipe and bottom-hole assemblies, coring systems, circulating fluids, fishing, and special tools and instruments.

## 1. Drill Pipe and Bottom-Hole Assemblies (BHA)

*Pipe Dope.* A special high temperature thread lubricant (pipe dope) is needed when temperatures exceed 300°C. Adequate lubrication must be provided to prevent galling, seisure, and tool joint (thread) make-up downhole. A positive seal must be maintained to prevent drilling fluid leaks and washouts. Sulfur must be excluded from the lubricant because of associated corrosion problems. A two-component system appears to be required: (a) an anti-galling metallurgical treatment or coating on the threads, and (b) a high-temperature "grease" that is in turn, composed of two components; one that will assure proper flow characteristics at ambient temperatures and a high-temperature portion that will remain in place as a seal and lubricating film. An essential property of the thread dope or compound is that it must be capable of breakout (being unscrewed) after extended stress cycles at high temperatures.

*Drill Collars and Drill Pipe Design.* Special alloy collars with tailored dimensions, and connection designs will be needed. The inside diameter of the BHA and drillpipe should be selected to be at least 10 cm (4 in.) so that wireline core barrels can be run and logging can be performed within the drill string. This will require special connection design to increase high temperature strength and fatigue life. The alloy should be selected to provide enhanced high-temperature hardness (abrasion resistance), corrosion resistance, and extended fatigue life. Development of a special drill pipe is recommended with a minimum 10 cm (4-in.) inside diameter for the reason given above. It will be necessary to design a special drill string for the extreme depths (>5 to 7 km), with larger diameter and thicker walled sections located at the top of the hole.

*Drill Pipe and BHA Inspection Strategy and Equipment.* It is suggested that two drill pipe strings be used in the coring of very deep formations, i.e., greater than 5 km. Avoidance of drill pipe failures in CSDP drilling projects is very important, and therefore frequent and detailed inspection will be necessary. One effective strategy is to inspect one string while the second is in use. A second element of an effective strategy should be to develop a special dedicated set of inspection equipment that will provide sensitive and high reliability detection of flaws and allow rejection of flawed sections (joints) of the drill string.

*Drill Pipe Identification System.* A technique to uniquely and reliably identify each length (joint) of drill pipe is needed. The system should include rig floor identification reading scanners with on-line talley, drill string stress computations, and accumulated stress fatigue cycle evaluations. A schematic sketch of such a system is presented in Figure 15.

*Heavy Weight Drill Pipe Joints.* To improve BHA design and downhole drilling dynamics, and to reduce drag forces during rotary drilling and coring, a special set of heavy wall drill pipe joints should be provided to replace upper section of the drill collars in the BHA. Special alloy selection considerations should be similar to those suggested above for drill strings and collars.

*Hardbanding.* Adequate high-temperature hardbanding methods should be developed to protect the drill pipe connections and collars from excessively high abrasive wear rates. This is especially important in crystalline rock.

*Float Valves.* Both rock bit drilling and coring operations in high-temperature boreholes require the use of reliable check valves in the

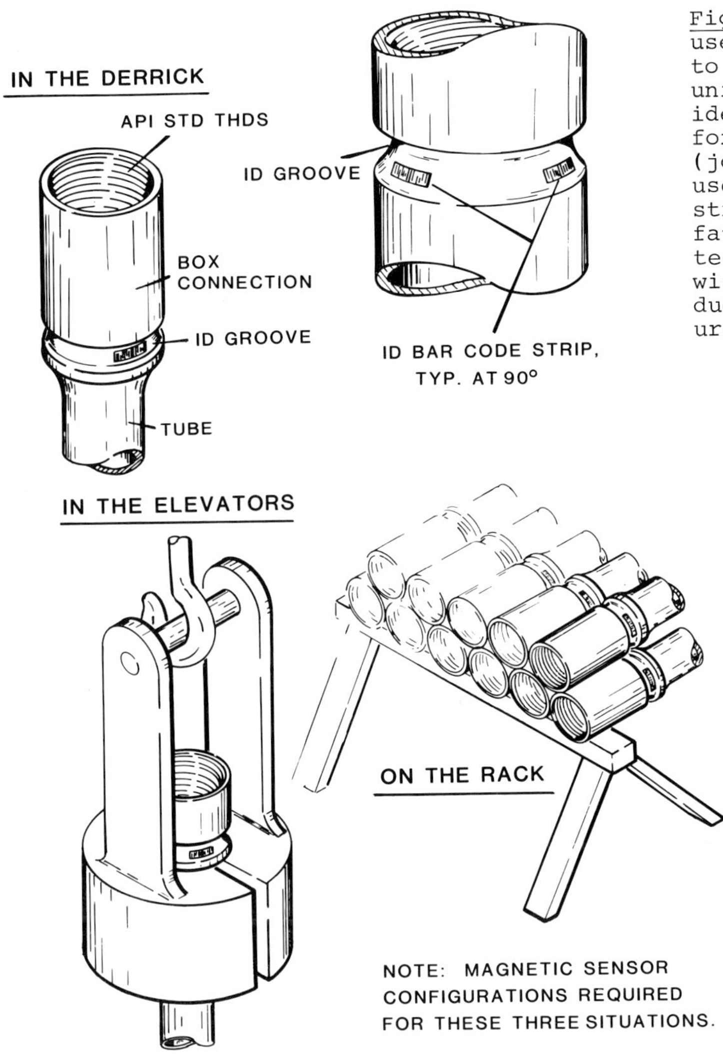


Fig. 15. Schematic of use of bar code strips to provide reliable, unique, and automatic identification system for drill pipe sections (joints). Code can be used to calculate stress and estimate fatigue life, and integrate inspection data with the result of reduced drill pipe failures

BHA above the bit. Otherwise cuttings and well fluids are prone to backflow up inside the drill string and tend to plug the components of the bottom-hole assembly. The problem is especially severe for wireline coring operations that tend to swab the drill pipe when the barrel is retrieved, and for use with downhole motors or other components that have tight clearances. Temperature upgrading of float valves is recommended.

*Dual String Drill Pipe System.* Drilling and coring of active hydrothermal systems, with high permeability formations, usually involve severe lost drilling fluid circulation. Fractured rock units with extremely high permeabilities are commonplace and subhydrostatic conditions are often encountered. Solutions for these difficult conditions can sometimes be achieved by use of concentric, dual drill strings as indicated by Dareing and Kelsey (1981). Design and availability of such a dual string system that can be used to at least 3 km depths, with wireline core retrieval [10-cm (4-in.) I.D.] capability for the inner string, is recommended.

*Reamers and Stabilizers.* In hard, abrasive rock, drilling or coring bottom-hole assemblies (BHA) will require use of roller stabilizers in place of blade stabilizers that wear too rapidly. A near-bit, roller reamer should be run with all drilling and coring assemblies. Excessively rapid bit guage wear results in drilling of a tapered hole diameter. A tapered hole in hard rock can cause the BHA to become friction jammed in the hole. Extended reaming of the borehole with each new bit will be minimized by use of the near-bit reamer. Use of roller reamers is also recommended, because blade-type tools wear too rapidly and this loss of diameter can cause the BHA to lose directional control effectiveness. Drilling of a crooked hole can result, and this is an especially severe problem in hard rock.

*Drill Bit Selection.* In the drilling of hard, crystalline rock, bit selection should primarily emphasize bit life. Development of sealed bearing, high-temperature bits would be an important and distinct improvement. Blast hole type bits, with heavy-section leg structures are suggested, to avoid leg bending and cone pinching. Heavy guage wear protection is essential to retard bit guage wear rates. Bit cutting structures must emphasize tungsten carbide insert (TCI) button selection that provides wear resistance on guage rows and high impact resistance for the insert cutters on the cones. Air cooled type bits, designed for high-temperature geothermal drilling, can be used to advantage with clear water or high-temperature mud fluid circulation systems. Bit selection should emphasize bit life for deep drilling in crystalline rock. Penetration rate is a secondary factor in bit selection for hard, crystalline rock. Penetration rate depends primarily upon the details of the cutting structure design, for these rock types.

## 2. Coring Systems

*Rotary Diamond Coring.* Technology and equipment exists for slim hole coring to depths of 1.5 to 2.0 km depths. Efforts to develop and extend these techniques and equipment to beyond 3 km depths are recommended with emphasis on development of higher temperature diamond bit capabilities.

*Downhole Motor Driven Systems.* One drilling method for potential improvement of deep core quality and recovery is the use of a downhole motor to drive the core bit. For high temperature use this requires a turbodrill, a 40-m (120-ft)-long core barrel in front of the downhole motor, and special core bits. Downhole motor drives will improve bottom-hole bit dynamics compared to rotary drive coring systems. In addition, drill pipe stresses, fatigue, and wear will be significantly reduced, and up-hole casing wear minimized.

*Wireline Barrel, Downhole Motor Driven System.* A recommended system conceptual design is sketched in Figure 16. The system includes a 10-cm (4-in.) O.D. wireline barrel, a hollow shaft (high-temperature) turbodrill, a combined multi-cone and polycrystalline diamond compact (PDC) bit, and a PDC core trimmer included on the retrievable barrel. Significantly extended bit runs can possibly be achieved with such a system. This concept is based directly on the original Mohole and the DSDP coring system ideas, designs, and experience.

*Rotary Bit Improvements.* Improvement of hybrid multi-cone plus PDC core bits will improve rotary core recovery and extend bit life. Efforts to improve this existing type of bit for specific rock types should accelerate the development of improved core bits for hard rocks. Coring

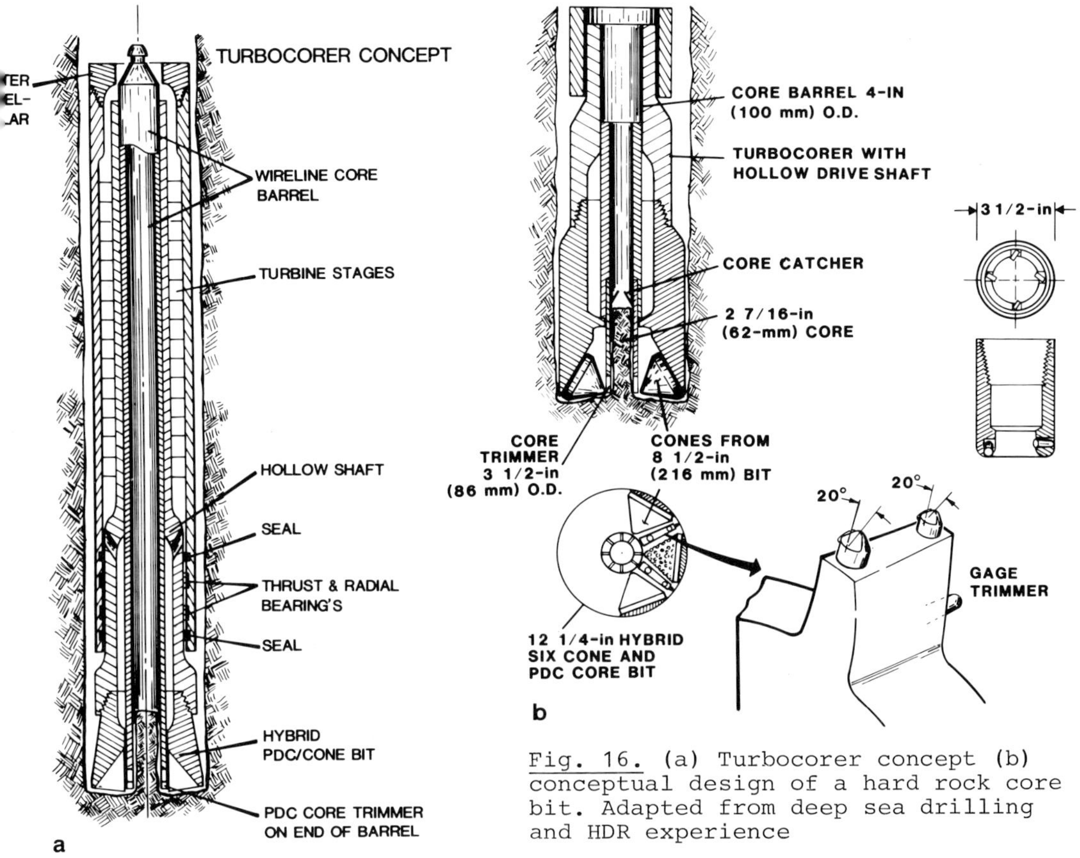


Fig. 16. (a) Turbocorer concept (b) conceptual design of a hard rock core bit. Adapted from deep sea drilling and HDR experience

effectiveness depends on long bit life, stable downhole bit dynamics, and effective core cutting and trimming; features that are possible with hybrid core bits.

## 3. Circulating Fluids

*Clear Water Systems.* Scientific coring and drilling usually requires that wellbore fluids and formation contamination be *minimized*. This often requires use of a clear water circulating fluid. Drilling problems associated with this approach are poor cuttings removal and ineffective bottom-hole cleaning. Solutions to these problems with clear water circulation include use of circulation of discrete high viscosity mud pills, and use of high fluid flow velocities. An additional problem with a clear water system, is the absence of filter cake (solids from the mud) deposited on the borehole wall. This condition leads to high rates of abrasive wear of downhole hardware and extremely high torque and pull friction drag forces on the drill string. A lubricity agent can be used to offset these latter conditions; however, an suitably inert and stable high-temperature additive does not yet exist.

*Cuttings Removal.* Removal of very fine cuttings from the drilling fluid is difficult, but important in scientific drilling or coring situations. The clear water systems are difficult to remove cuttings from the surface.

Use of high viscosity pills as mentioned above are an aid to solving this problem, but more effective filter and centrifugal mud solids removal systems are needed.

*Fluid Cooling.* High temperature downhole conditions will usually require use of a surface fluid cooling system. A heat exchanger system with low circulating fluid aeration or oxygen entrainment is most desirable in order to reduce corrosion control problems.

*Corrosion Control.* A carefully designed corrosion control system is important. Minimization of exposure to air and suppression of oxygen entrainment are necessary. Consideration of use of a nitrite additive as an oxygen scavanger is recommended. A complete and effective fluid chemistry and corrosion monitoring system is essential.

*Rig Pumps.* It is necessary to provide for sufficient excess rig pump flow and pressure capacity to handle high flow and pressure requirements needed for extended cooling and problem solving demands that will arise in deep, hot thermal regimes boreholes.

*Lightened Fluids.* Use of nitrogen gas, water mixture and use of nitrogen foams can be effective in fluid loss, corrosion and formation contamination problem situations. Lightened fluid columns with good cuttings removal can be achieved with foams. Development and fabrication of diesel exhaust nitrogen gas production systems (Osgerby and Durillo 1981) is strongly recommended. Coring operations with such lightened fluids and nitrogen gas must be developed.

*Higher Temperature Rated Muds.* It can be expected that deep formations that are unstable due to interactions with water or high pressure conditions will be encountered in some CSDP thermal regime targets. An improved high temperature mud system will be needed to cope with these problems. Capability beyond the present sepiolite clay-based systems (300°C) is required.

*Lost Circulation Control.* Improved high temperature "additives" and placement tools for lost circulation materials (LCM) should be developed. Lost circulation zones can jeopordize coring. Systems designed specially to control circulation lost to fractures need to be developed.

## 4. Fishing

*Fishing Tool Upgrade.* Downhole problems can be expected in any deep, hot hole, and therefore an essential task for drilling and coring procedures is to solve these problems effectively and rapidly. This requires a review, and then development effort, to upgrade the temperature rating of the usual downhole tools required to conduct such "fishing" operations. Typical equipment upgrading needs are

- Drilling and fishing jars, hydraulically operated units are most desirable.
- Mills, especially the cutting structure.
- Back-off high explosive shots, especially high temperature detonators and wirelines.
- Seals for grapples and overshots.
- Impression blocks.
- Cut-off tools.
- Free-point detection tools.

These improvements are important because the downhole problems often preclude the use of fluid circulation for cooling that can help marginal or underrated tools function.

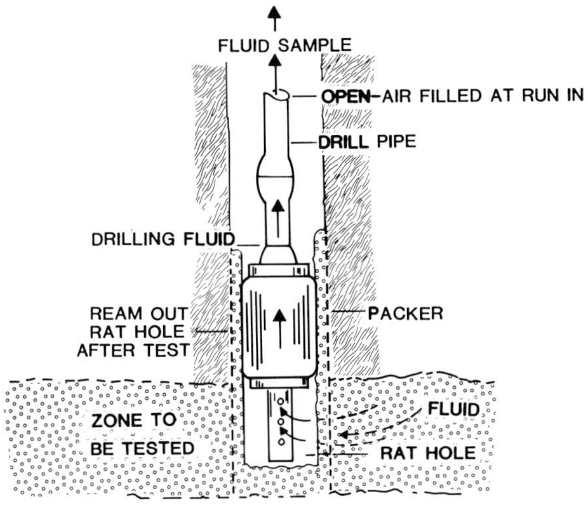


Fig. 17. Sketch of drill stem tester for deep crustal applications. Used to obtain formation fluid samples without contamination by circulating drilling fluid

*Experience.* An essential element in any successful fishing operation is the selection and availability of *experienced* personnel.

## 5. Special Tools and Instruments

*Sand Line.* The cables used to run various downhole tools in hot boreholes must have steel wire cores and be high temperature rated. Socket connections must be mechanical in design, because commonly used babbitt metals melt well below 300°C.

*Borehole Survey Instruments.* Drilling of straight or deliberately deviated holes demands good single- and multi-shot survey instruments. Both magnetic (compass) and gyro instruments for higher temperature service are needed. Steering tools for directional drilling monitoring and control of directional corrections must be available for the more extreme temperatures anticipated in CSDP thermal regimes projects. Crooked hole conditions are special hazards in the high-drag conditions encountered in boreholes in hard crystalline rock. Excessive crooked hole conditions ("doglegs") can seldom be eliminated or cured by even extended reaming with stiff reaming assemblies.

*Hole Caliper.* Knowledge of borehole wall rugicity, is essential to many drilling, coring, casing, and downhole tool-running operations. As an example, a top priority need exists for a caliper run before any drill stem test or packer setting operations are attempted.

*Drill Stem Tester.* It is often very desirable to obtain a fluid sample during drilling or coring operations. If a wireline core system is employed, then a sampler can be run within the drill pipe and, if contamination by drilling fluid is not considered to be a problem, a sample can be obtained by reversing flow into the sampler. However, if drilling fluid contamination is to be avoided, a drill stem tester (DST) as shown in Figure 17 can be used. The packer seal to the wellbore wall need not resist large pressure differences, because it functions mainly as an isolation seal from the drilling fluids in the annulus. The packer is only to be used once so high temperature non-elastomeric materials and compression-type packers, activated upon set-down, can be considered. The drill stem tester is run only partially filled with (clean) water so suction draws the sample from the zone being tested. A reliable high-temperature mechanically opened valve and a flow check valve are required. The scheme shown in Figure 17 includes a smaller diameter bottomhole drilled section, or rathole. Drilling this reduced section can enhance the wellbore surface finish for improved packer setting and sealing. Subsequent ream-out of the rat hole is possible, and spall, or hole sloughing, caused by the fluid sample withdrawal can be removed. Such a scheme also minimizes hole swabbing at retrieval of the DST system, and can alleviate packer washout during running into the borehole.

*Packer Systems.* Development of high temperature packer systems should be given a high priority. A variety of configurations, i.e., single set and straddle; both hydraulic and mechanical setting mechanisms, and inflatable and compression systems should be persued.

*Cementing.* Improved high temperature cements, cementing hardward and placement techniques are needed to assure adequate hole casing programs and to support many remedial operations; e.g., borehole sidetracking and hole stabilization. Reliable cement slurry formulations that are stable against adverse downhole chemistry, gases, and high temperatures are needed. The set-up and cured cement should be durable, of low permeability and retain strength. Improvements in placement hardware and procedures are also necessary.

*Downhole Coring Monitoring.* The requirements for high quality core and good recovery of scientific drilling projects dictate that the development of a downhole data-while-coring (DWC) system is warranted. Core quality and recovery are directly related to the control of the drilling parameters, and BHA and core bit dynamics. Therefore development of an instrumented, high temperature rated subassembly (collar) with appropriate drilling parameter and dynamic force and motion sensors is recommended. In addition, if a hollow shaft turbodrill, with wireline core barrel retrieval is developed, a sensor package can be included in the barrel, and the data telemetered on the cable. Because the drill string is not rotating in such a system the cable can be left latched onto the barrel and connected into the instrumentation during coring operations.

*Logging While Drilling or Coring.* A number of advantages would result from the development of a special drilling subassembly that contained an appropriate suite of geophysical measurements (logs) that could be monitored at the surface. Identification of lithologic changes could provide input for decisions on coring runs of scientifically significant intervals. The ability to review real time geophysical data and make spot coring decisions could ultimately result in more efficient coring procedures and operations than continuous coring or cuttings observations.

## Discussion and Conclusions

The Los Alamos Fenton Hill HDR drilling experience and several other deep, hot hole drilling projects have been used to provide this perspective on scientific drilling in the continental crust. The results from the two 2 km deep HDR wells in Cornwall, England; the three 3 km deep wells in British Columbia, Canada; the geothermal well at Puna in Hawaii; the Italian geothermal drilling experience, and the deep drilling in the Soviet Union have provided additional background information on drilling and coring in crystalline rock. Future CSDP thermal regimes projects can benefit from the drilling experience from geothermal production wells with bottom-hole static temperatures in the range of 300 to 400°C. These wells are now drilled rather routinely. However, coring and geophysical logging in that environment are not well-developed technologies.

The Soviet deep crystalline rock drilling experience, especially represented by the Kola peninsula borehole (abbreviated as UD-3) drilled successfully to a 10,780 m (35,367 ft) depth, supports several of the results and technology needs outlined above.

The Soviets report:

- Development of a high torque, low rotational speed turbodrill suitable for crystalline rock drilling.
- That the 21.4 cm (8-1/2-in.) diameter roller cone, TCI cutting structure bit is optimum relative to bit life and penetration rate in crystalline rock.
- Development of hollow shaft turbodrills.
- The desirability of continuous coring in scientific drilling projects.
- The general stability of long open hole sections in crystalline rock; and the advantages of open hole for both drilling and geophysical logging.

Experience has accumulated in drilling technology, indicating that for downhole tool and special surface equipment development, several general principles should be observed:

1. Always develop at least *two* prototypes of hardware.
2. Laboratory and field tests and trials are necessary prior to deployment of equipment to a CSDP drill site.
3. Avoid hardware and technology development efforts in scientific boreholes.
4. Develop and provide at least one (or more) backup procedure, approach or hardware solution to a drilling problem. Experience indicates that the best ideas and concepts may not apply downhole due to unanticipated borehole conditions. It is essential that more than one new hardware approach, source of a service, and method of solution be available. Competitive solutions are very important elements in obtaining satisfactory results downhole, and to devise effective solutions to drilling problems.
5. The CSDP thermal regimes environment is expected to be very challenging, and will require *all* the best drilling and coring development efforts that can be marshalled in order to achieve successful projects.

The worldwide experience of both drilling and coring of boreholes into hot crystalline rocks has demonstrated both the possibilities and problems of deep crustal drilling for scientific purposes. The drilling of 7 km of borehole in hot crystalline rock at Fenton Hill should serve as a guide for both conceptual and detailed planning of future CSDP projects. The major issues to be considered in such drilling project planning in the areas of casing design, bit selection, coring systems, downhole drilling assemblies, drilling procedures, drilling fluid systems, and on-line problem solving have been reviewed and recorded.

A list has been developed of the most significant problem areas and needed improvements to successfully approach deep drilling in crystalline rock and extreme temperature environments expected to be encountered in the deep CSDP thermal regimes investigations. These recommendations and judgements were derived from the experience gained at Fenton Hill and elsewhere, and include suggested approaches toward solutions.

*Note added in proof:* The excellent recent book by Koslovsky (1984) indicates that extensive development of a turbocoring system has been accomplished in the USSR in conjunction with the Kola superdeep borehole. A 10-m (30-ft) core barrel, or a reverse-flow-basket, is run below a turbodrill with a 10:1 speed reducer. A 21.4 cm (8-1/2 in.) diameter, four-roller-cone, core bit cuts a 7.5 cm (3 in.) core. Bit life is about 10 m (30 ft) requiring tripping of the BHA and drill string for each core run.

*Acknowledgments.* The Los Alamos Hot Dry Rock Geothermal project is conducted under the auspices of the Geothermal and Hydropower Technology Divison of the U.S. Department of Energy. The drilling experiences reported here are the results of the efforts of many of the staff of the HDR project at Los Alamos. The author wishes to thank those individuals who supported those drilling operations, and to acknowledge the contributions of the many firms who provided the equipment, instruments, services, advice, and supporting personnel. I wish to note with special appreciation and gratitude those companies that supplied innovative and developmental high-temperature-rated hardware and tools for evaluation and use.

## References

Bailey RA, Smith RL (1978) Volcanic Geology of the Jemez Mountains, NM. In: Hawley JW (ed) Guidebook to the Rio Grande Rift in New Mexico and Colorado. NM Bur Mines Mineral Res, Circ 163:186-196

Bailey RA, Smith RL, Ross CS (1969) Stratigraphic Nomenclature of the Volcanic Rocks in the Jemez Mountains, NM US Geol Surv Bull 1274-P:19

Campbell DA, Morris CW, Verity RV (1981) Geothermal well stimulation experiments and evaluation. Soc Pet Engrs Pap No 10216. 56th Annu Techn Conf Exhibit, San Antonio, TX, October

Carson C, Lin Y (1981) Geothermal well costs and their sensitivity changes to drilling and completion operations. Proc Int Conf Geothermal Drilling Completion Technol, 8-1 to 8-26, January 21-23

Continental Scientific Drilling Program (1979) NAS/NRC US Geodynamic Committee. Nat Acad Sci Wash DC

Dareing DW (1984) Drill collar length is a major factor in vibration control. J Petroleum Technol 36:637-644

Dareing DW, Kelsey JR (1981) Balanced pressure techniques applied to geothermal drilling. Transact Geothermal Resources Council 5:233-238

Dash ZV, Murphy HD, Aamodt RL, Aguilar RG, Brown DW, Counce DA, Fisher HM, Grigsby CO, Keppler H, Laughlin AW, Potter RM, Tester JW, Trujillo PE, Zyvoloski G (1983) Hot dry rock geothermal reservoir testing: 1978 to 1980. J of Volcanol Geotherm Res 15:59-99

DiPippo R (1983) Overview of worldwide geothermal power development. Geothermal Resources Council Bull (May) 12(5):3-9

Fridleiffsson IB, Gibson IL, Hall JM, Hall HP, Johnson NI, Christensen H-V, Schönharting G (1982) The iceland research drilling project. J Geophys Res 87(B8):6359-6361 (August 10)

Goff F, Waters AC (eds) (1980) Continental scientific drilling program thermal regimes: comparative site assessment geology of five magma-hydrothermal systems. Los Alamos Nat Lab report LA-8550-OBES

Heiken G, Goff F (eds) (1983) Geothermal energy of hot dry rock. J Volcanol Geotherm Res 15:1-3 (January)

Helmick C, Koczan S, Pettitt R (eds) (1982) Planning and drilling geothermal energy extraction hole EE-2. Los Alamos Nat Lab report LA-9302-HDR (April)

Herrick RG (1978) Roller cone/diamond drag core bit. Soc Pet Engrs Pap No 7517. 53rd Annu Techn Conf Exhibit, Houston, TX
Ioanesyan YR, Vasilenko AA, Matsiyevskiy VP, Simonyouts SL (1982) "Sovremenuoye Turbinnoye Burehiye", Tekhnitie, Kiev, 1977 (Translated as "Modern Turbine Drilling" by Moore DG and edited by Cooley WC. Terraspace Inc, Rockville, MD
Knight KE (1983) Technology transfer in the petroleum industry. Technol Transfer Soc, Int Symp Proc, 179-191, June 20-22
Kozlovsky YA (ed) (1984) Kol'skaia Sverkglobokaia (Kola Superdeep); Investigation of the deep structure of the Continental Crust with the help of drilling the Kola superdeep borehole. Nedra, Moscow Part III, pp 375-487
Larson VF, Robson VB, Foss GN (1981) Aids to conventional drilling result from deep sea project. World Oil, November 1981, pp 175-198; and deep ocean coring recent operational experience of the deep sea drilling project. Soc Pet Energy Pap No 9409. 55th Annu Fall Technic Conf SPE of AIME, September 1980
Laughlin AW (1981) The geothermal system of the Jemez Mountains, NM and its exploration. In: Rybach L, Muffler LFP (eds) Geothermal Systems: Principles and Case Histories. Wiley, New York, pp 295-320
Laughlin AW, Eddy AC, Laney R, Aldrich MJ (1983) Geology of the Fenton Hill, NM, Hot Dry Rock Site. J Volcanol Geotherm Res 15:21-41
Maurer WC (1980) Advanced drilling techniques. Petroleum Pub, Tulsa, OK, pp 541-591
Murphy HD, Tester JW, Grigsby CO, Potter RM (1981) Energy extraction from fractured geothermal reservoir in low-permeability crystalline rock. J Geophys Res 86(B8):7145-7158
Neudecker JW, Rowley JC (1982) High temperature directional drilling turbodrill. Los Alamos Nat Lab report LA-9231-MS (February)
Nuckols EB, Miles D, Laney R, Polk G, Friddle H, Simpson G (1981) Drilling fluids and lost circulation in hot dry rock geothermal wells at Fenton Hill. Proc Int Conf Geothermal Drilling Completion Technol 24-1 to 24-22 (January 21-23)
Oliver J (1982) Probing the structure of the deep continental crust. Science (Wash DC) 216:689-695
Osgerby IT, Durilla M (1981) Inert gas generation utilizing diesel exhaust. Proc Int Conf Geothermal Drilling Completion Technol. Albuquerque, NM, pp 29-1 to 29-20 (January 21-23)
Pettitt RA, Carden RS (1981) Sidetracking experiences in hot granite wellbores. Geotherm Res Council Transact 5:267-270 (October)
Pettitt RA, Laney R, George D, Clemens G (1980) Evolution of a hybrid roller cone/PDC core bit. Geotherm Res Council Transact 4:305-308
Potter RM (1973) Heat flow of the Jemez Plateau (Abstr). EOS Trans Am Geophys Union 54:1214
Reiter MC, Weidman C, Edwards CL, Hartman H (1976) Subsurface temperature data in Jemez Mountains, New Mexico. NM Bur Mines Miner Res Circ 151:1-15
Rezanov IA (1981) Ultra-deep drilling, Nauta (Science), Moscow, 160 pp (available translation DOE-tr-252)
Rowley JC (1974) Rock melting applied to excavation and tunelling. In: Advances in Rock Mechanics. Reports of Current Research, Vol 2, Part B. Proc 3rd Int Soc Rock Mechanics. Denver, CO, pp. 1447-1453
Rowley JC, Carden RS (1982) Drilling of hot dry rock geothermal energy extraction well EE-3. Los Alamos Nat Lab report LA-95-12-HRD (August)
Rowley JC, Watkins JW (1977) Tour of russian oil field technology. Soc Pet Engrs Pap No 6718. 52nd Annu Fall Technic Conf Exhibit, Denver, CO, October 9-12
Smith HTU (1938) Tertiary geology of the Abiquiu Quadrangle, NM. J Geol 46:933-965
Smith MC (1983) A history of hot dry rock geothermal energy systems. J Volcanol Geotherm Res 15:1-20

# OBSERVATIONS IN AN ACTIVE HYDROTHERMAL SYSTEM THROUGH DEEP DRILLING: VALLES CALDERA, NEW MEXICO

Dennis L. Nielson and Jeffrey B. Hulen[1]

## Introduction

Through a cooperative agreement between Union Oil Company (UOC), Public Service Company of New Mexico, and the U.S. Department of Energy (DOE), a large body of exploration data collected by UOC at the Baca project area in the Valles caldera between 1971 and early 1982 has been catalogued and open-filed. Using these data and cuttings from 19 deep UOC boreholes in the Redondo Creek portion of the caldera, we have completed several studies principally concerned with structural and stratigraphic permeability control in the high-temperature (up to 341°C) geothermal reservoir (Hulen and Nielson 1982, 1983; Nielson and Hulen 1983, 1984).

Reservoir data were collected at Baca in support of commercial geothermal exploration and development efforts, and so are generally not of the sample frequency required for purely scientific subsurface studies such as those proposed for the Continental Scientific Drilling Program (National Research Council 1979). Nonetheless, the Baca data are sufficiently complete that conclusions based on them will greatly facilitate planning for such studies, particularly those aimed at detailed characterization of thermal regimes and epithermal mineral resources.

Deep drilling and reservoir testing in the fractured, hot and corrosive Baca system has been complicated and risky. Serious drilling or completion problems plagued 74% of UOC's Redondo Creek boreholes; 29% were completely lost (Molloy and Laughlin 1982). Experience gained at Baca in coping with the hostile geothermal environment, therefore, will also help ensure maximum data recovery and reliability in any future scientific study of such systems.

Not only are subsurface reservoir measurements difficult, they are often strongly influenced by borehole drilling methods and completion procedures. In some cases, separation of induced effects from corresponding natural phenomena is nearly impossible. We have carefully evaluated the reservoir data used for this report, however, and believe that overall they reasonably approximate pre-drilling conditions. Departures from this conclusion are discussed individually in the text which follows.

Location of fluid entry zones in the Baca wells is based on information obtained during and after drilling by UOC and acquired through subsequent reservoir analysis by Grant and Garg (1981) and Riney and Garg

[1]Earth Science Laboratory, University of Utah Research Institute, Salt Lake City, UT 84108, USA

Observation of the Continental Crust through Drilling I (ed. by C.B. Raleigh)

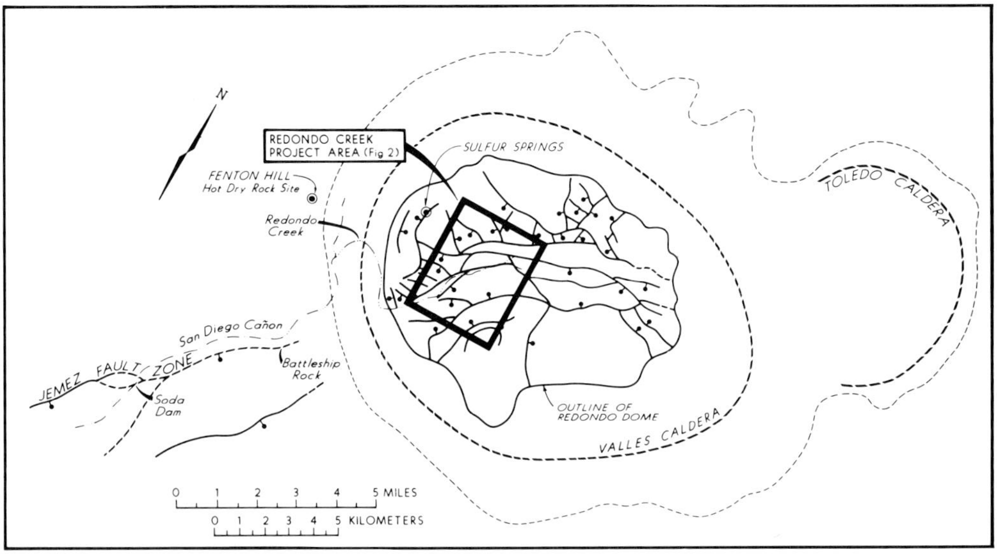


Fig. 1. Index map of the Valles caldera and vicinity, New Mexico

(1982). Since this paper is concerned with scientific aspects of the Baca system, fluid production zones will be discussed without regard to their commercial status.

Geologic and reservoir engineering data from the Baca geothermal system have been open-filed by the Earth Science Laboratory/University of Utah Research Institute (UURI) under funding from the DOE. These data are available for the cost of reproduction and handling.

## System Model

### 1. Introduction

The development of the Valles caldera is principally known from the work of Smith and Bailey (1966, 1968); Smith et al. (1970); Doell et al. (1968) and Bailey et al. (1969). Studies of subsurface samples (Hulen and Nielson 1982, 1983; Nielson and Hulen 1984) have added details to the stratigraphic framework and allowed inferences of the caldera's structural development. References to stratigraphy in this paper will utilize the nomenclature proposed by Nielson and Hulen (1984).

A generalized geologic map of the Valles caldera (Fig. 1) indicates several important features which will be discussed in this paper. The Baca project area is shown on Figure 1 and details of the well locations are shown in Figure 2. The Fenton Hill Hot Dry Rock site is also shown in Figure 1.

Samples from the deep Baca drilling (Fig. 2) are principally available in the central portion of the Redondo resurgent dome. Our studies have concentrated in this area and have been particularly concerned with the

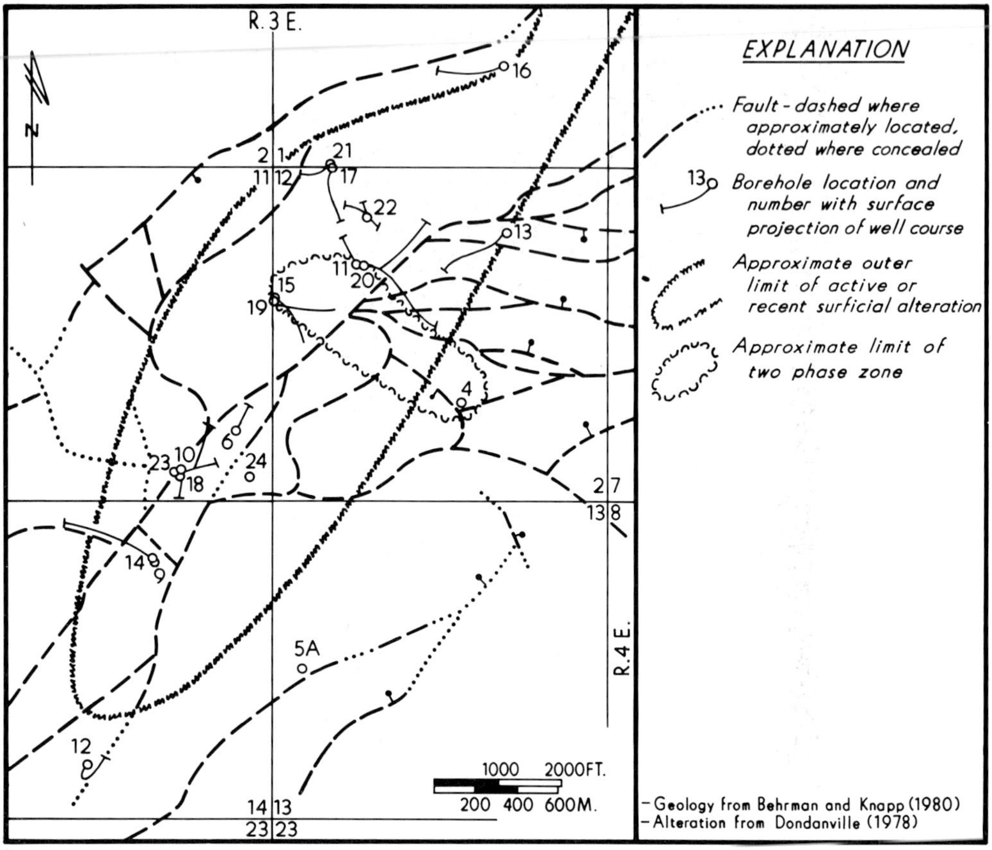


Fig. 2. Map of the Baca geothermal system, showing borehole locations, fault trends, and the approximate location of a probable two-phase zone

formation of the resurgent dome, the location of the magma beneath the dome, and the evolution of the hydrothermal system. Nielson and Hulen (1984) noted the fact that even the deepest hole (Baca 12, 3243 m) failed to intersect plutonic rocks which could be related to magma responsible for resurgent doming. They applied a model, developed by Johnson (1970) for the emplacement of laccoliths, to the formation of the Redondo dome to arrive at a predicted value for the thickness of roof rocks, given estimates of the diameter of the dome and the amplitude of domal uplift. This simplified model predicts a roof rock thickness of 4700 m, and thus is consistent with the absence of young intrusive rocks in the drill holes.

As a result of the doming process, the affected strata are subjected to differential stresses. In the upper portions of the dome, the rocks are under tension. This results in the development of faulting along the crest of the dome that forms most of the fracturing often identified as an apical (or keystone) graben. Approximately halfway down through the dome, a surface called the neutral plane separates the upper tensional environment from a lower region characterized by compression. In contrast to the normal faults above the neutral plane, the region below the neutral plane will be characterized by the development of conjugate shears.

Nielson and Hulen (1984) concluded that the idealized model of dome development was modified by the reactivation of older structures associated with the Jemez fault (Fig. 1). These structures were steep, throughgoing fault zones prior to the formation of the Redondo dome, and they have continued to maintain that character during and after the formation of the dome. These faults may extend to depths where they intersect crystallized magma which is the cause of the structural doming. It is our opinion that these reactivated faults are the principal conduits of upwelling hydrothermal fluids, and that successful wells in the field intersect either these faults or fractures or permeable stratigraphic horizons which communicate with them.

The timing of the initiation of the present hydrothermal activity remains undocumented. However, we can hypothesize that circulation of hydrothermal fluids has been an active phenomena since and perhaps before the initiation of explosive silicic eruptions. The present hydrothermal system thus has evolved with the magmatic and structural development of the caldera complex.

The chemistry of hydrothermal fluids sampled from the Baca reservoir has been described by Goff and Grigsby (1982) and White et al. (1984). The geothermal fluids from Baca are sodium chloride brines with concentrations of about 7000 ppm total dissolved solids. $^{18}O$ data indicate that the fluids are meteoric in origin with the probable zone of recharge being in the Jemez Mountains. The fluid chemistry has therefore evolved through the interaction of these fluids with the hot rocks of the hydrothermal reservoir.

The predominant zone of surface alteration within the caldera is located in the Sulfur Springs area (Fig. 1). Here acid sulfate alteration and the presence of mud pots, fumaroles, and acid springs indicate a boiling water table at depth. Holes up to 650 m deep have failed to intersect the water table in this area (Baca boreholes 1 and 3; Goff and Grigsby 1982).

Leakage of hot fluids from the Valles caldera into the Jemez fault and mixing of these fluids with fresh, cool waters have been responsible for hot spring development along San Diego Canon (Goff and Grigsby 1982).

In wells within the Baca field, a two phase-zone where liquid and steam co-exist has been reported in the upper portions of Baca-4, 15, and 11 (Atkinson 1980) and 20 (Christensen and Atkinson 1981). The approximate limit of this two-phase zone is illustrated in Figure 2.

Lambert and Epstein (1980) have performed a stable-isotope investigation of hydrothermally altered samples from Baca boreholes 7 and 4. They find that these samples have exchanged isotopically with meteorically derived fluids. They also conclude that the water-rock ratio has been very large – larger than at Wairakei or the Salton Sea fields. This is somewhat unusual in that the overall permeability of the Baca field has been found to be relatively low compared with other geothermal fields. Lambert and Epstein (1980) also find no evidence of cooling of the hydrothermal system through time. This is contrary to the evidence presented by Hulen and Nielson (1983) on the basis of alteration mineralogy.

## 2. Temperatures

Temperature logs from the Baca project area were acquired for exploration and production purposes and are not as detailed as those required

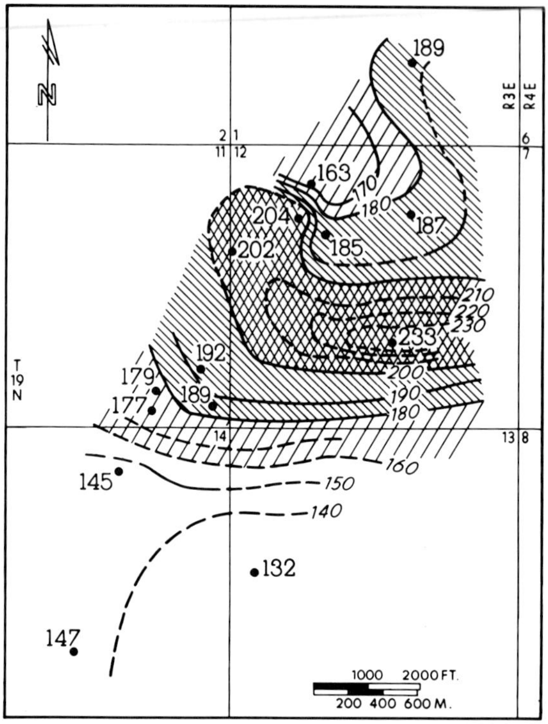


Fig. 3. Temperature distribution (°C) at 2200 m above sea level in the Baca geothermal reservoir

for scientific study. In general, the temperature coverage is spotty with data points often as widely separated as 152 m (500 ft). In addition, several holes do not have temperature logs, or have been logged only in their cooler upper portions. Our analysis of these data therefore will concentrate on overall temperature relationships and spatial variations.

Temperatures measured in the Baca boreholes have been contoured to show temperature distributions at three different elevations within the geothermal field. The temperatures, measured by UOC, are those that we consider static; that is, most likely to represent the thermal reservoir in its natural state prior to drilling. Data from Baca 5A has been omitted from Figures 4 and 5 because the temperature profiles show reversals which are not found in other Baca wells suggesting a possible influence from injection or the casing program (see also Grant and Garg 1981). Figures 3, 4, and 5 show contoured temperatures at 2200 m, 1700 m, and 1200 m above sea level. The temperature contours at 2200 meters are all within the Tshirege Member of the Bandelier Tuff. Comparison with an isopach map of the Tshirege (Fig. 11 of Nielson and Hulen 1984) shows that, in general, the temperatures are higher where the Tshirege is thinner. Comparison of Figure 3 with the structural setting shown on Figure 2 also suggests the influence of an east-west trending series of faults on the distribution of high temperatures.

The thermal regime at 1700 m is shown in Figure 4. The trend of higher temperatures is oriented northeast-southwest along the trend of the longitudinal graben which cuts the Redondo Dome (Fig. 1). Comparison with a structural contour map of the top of the Otowi Member of the Bandelier Tuff (Fig. 8; Nielson and Hulen 1984) suggests that this

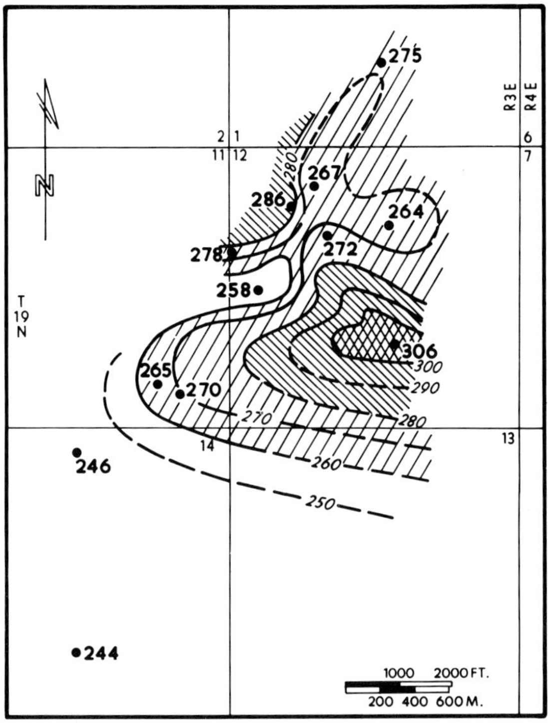


Fig. 4. Temperature distribution (°C) at 1700 m above sea level in the Baca geothermal reservoir

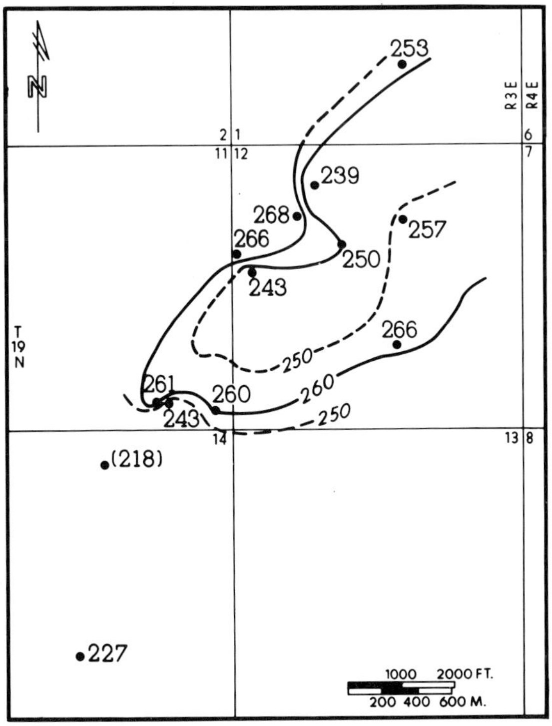


Fig. 5. Temperature distribution (°C) at 1200 m above sea level in the Baca geothermal reservoir

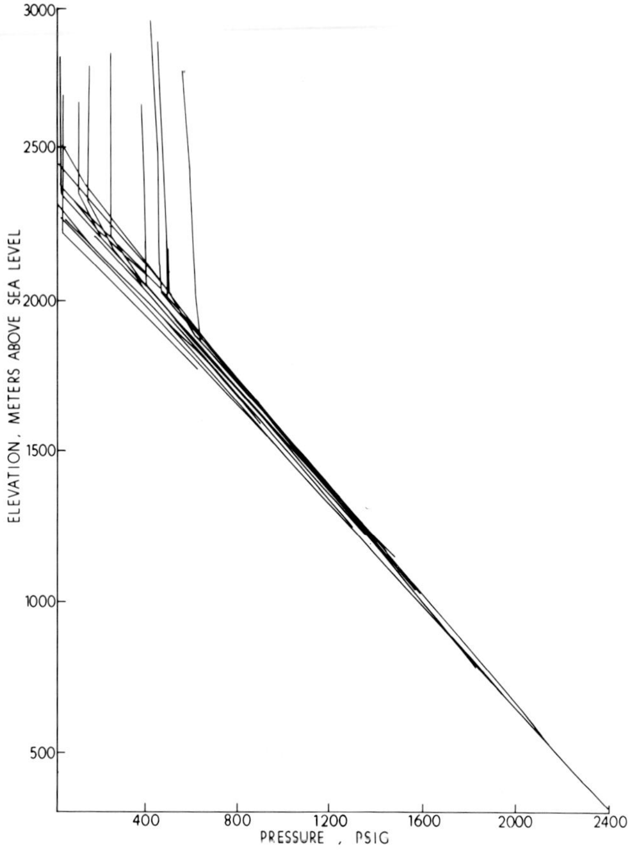


Fig. 6. Pressure survey plots for all wells drilled in the Baca field (Union Oil Co. 1982)

horizon may control temperature distribution at this level, presumably by channeling hydrothermal fluids along the contact zone.

The temperature distribution at 1200 meters elevation is shown on Figure 5. This map shows the high temperature for this elevation to be centered on Baca 4. Comparison with Figures 4, 5, and 6 of Nielson and Hulen (1984) shows that temperatures in this interval are largely controlled by the distribution of the Lower Tuffs, which separate the Paliza Canyon Formation from the Otowi Member of the Bandelier Tuff. The increase in temperature corresponds with an increase in the thickness of the Lower Tuffs. Figure 5 also suggests the possible importance of the east-west zone of faulting in controlling temperatures at this elevation.

It can be seen in Figures 3, 4, and 5 that temperatures are considerably lower in the southwestern part of the Baca field. We are led to the conclusion that fluid flow and resultant temperature distribution in the field are controlled by east-west and northeast-trending fault zones and also by the thickness of the Lower Tuffs. UOC recognized these relationships early in their exploration efforts, selecting as their target zone for commercial geothermal production the lower part of the Bandelier Tuff (the Lower Tuffs of Nielson and Hulen 1984) and the upper portion of the underlying Paliza Canyon Formation.

## 3. Pressures

Static pressure surveys for all wells in the Baca field are grouped in Figure 6, which is reproduced from UOC (1982). The approximate elevation of the water table, which can be interpreted as the point above which pressure profiles become essentially isobaric (for example, Celati et al. 1976, p. 1590), is shown on these surveys to be quite variable. Measured pressures above the water table also differ considerably. Reasons for this pressure variability are not not given by UOC (1982) but could include differences in water vapor temperature and partial pressure of other gases. The gradient below the water table is uniformly about 8.1 MP $km^{-1}$.

UOC (1982) has plotted pressures in the Baca boreholes below the water table at an elevation of 1676 m above sea level, documenting an increase in pressure from the southwestern to the northeastern portion of the geothermal field. This pressure increase corresponds to the northeasterly temperature increase previously documented not only at 1700 m elevation, but also at elevations of 1200 m and 2200 m. In other words boreholes which are hotter also show higher pressures. UOC (1982) concludes from this relationship that pressures are "controlled by a deep reservoir".

## 4. Hydrothermal Alteration

As part of our ongoing research on the Baca field, we have been investigating broad-scale and detailed hydrothermal alteration mineralogy and zoning as they relate to active and inactive fluid flow regimes. Initial efforts, presented in Hulen and Nielson (1983), were directed at characterization of high-level argillic alteration. This paper will briefly summarize deeper, higher-temperature assemblages. Relationships of these assemblages to specific active and sealed thermal fluid entry zones will be exemplified for selected wells in the following section on reservoir testing in the Baca field.

Alteration in the Baca system is of three distinctive types: argillic, propylitic and phyllic. Strong argillic alteration forms a blanket mostly above the water table (as deduced from pressure surveys) and is developed principally in formerly permeable non-welded tuffs. It is characterized by the assemblage smectite, mixed-layer illite-smectite, kaolinite, quartz and pyrite. Beneath the argillic blanket is a thick zone of pervasive propylitic alteration, within which original rock-forming minerals are partially converted to various combinations of chlorite, calcite, albite, epidote/allanite, illite, pyrite and hematite. Propylitization is very weakly developed in felsic host rocks such as the Bandelier Tuff, but commonly intense in deeply underlying intermediate-composition volcanic rocks. All known major thermal fluid entries at Baca occur within the propylitic zone. Locally interrupting the propylitic zone along favorable fault zones and stratigraphic aquifers are zones of moderate to intense phyllic alteration which are commonly but not invariably associated with active thermal fluid entries. Illite or sericite, pyrite and quartz, the principal minerals of the phyllic zones, typically replace the bulk of the host rock's primary constituents. Impermeable zones of phyllic alteration are interpreted as former fluid channels now sealed by hydrothermal alteration.

## Reservoir Testing

In spite of its usefulness in characterizing the Baca geothermal reservoir, hydrothermal alteration by itself cannot be used to differentiate between active and fossil thermal fluid channels. In fact, many of the most strongly altered zones in the Baca boreholes are those which had high initial permeability but have been hydrothermally sealed. Location of productive horizons must also rely on careful observation during drilling and especially on subsequent reservoir testing. In this section, we will briefly summarize various reservoir tests conducted at Baca, and discuss how such tests not only help pinpoint fluid entry zones but also measure, in situ, the full range of reservoir parameters likely to affect commercial production from the field. For more detailed analysis of Baca reservoir testing, the reader is referred to Union Oil Company (1982), Riney and Garg (1982) and Grant and Garg (1981).

Measurements in boreholes provide the only means of determining temperatures, pressures, fluid flow rates, and permeability in situ. But these measurements must be made with caution. Reservoir conditions are changed by the drilling process, and although temperatures and pressures will stabilize with time, they will not necessarily correspond to pre-drilling conditions (Grant et al. 1982). In addition, the physical character of formations adjacent to the wellbore can be modified during the drilling process, largely by infiltration of drilling fluid. Reaction of drilling mud with fluid and wall rock at geothermal temperatures can also extensively modify permeability.

Evidence from injection tests can be used to document permeability relationships at depth. Grant et al. (1982) cautioned that difference in pressure between a borehole and its wallrocks controls whether or not a particular zone will accept injected fluids. The distinction between zones which produce geothermal fluids and those which accept injected fluids has been noted in the Baca field by Riney and Garg (1982), and Nielson and Hulen (1983) have attempted to explain this relationship with structural models. We envision the following processes which differentiate the behavior of permeability zones at Baca. First, acceptor zones may be underpressured due to the lack of communication with a high-volume fracture. Second, injected fluids will tend to cool non-producers more than producers. Thermal contraction of fracture wallrock will increase the aperture of a fracture through time. Dash et al. (1983) have demonstrated a direct relationship between reservoir volume increase and extraction of thermal energy from the Fenton Hill Hot Dry Rock Site. Similarly, in hydrothermal fields the injectivity of a well generally improves during continued injection, partly due to solution or erosion of fractured wallrocks.

Following termination of injection, fluids will flow from fractures into the wellbore due to release of pressure, and measured temperatures will gradually rebound to pre-injection values. Thus, temperature recovery curves following the termination of injection are particularly valuable for understanding zones of fluid production or acceptance.

Pressure transient analysis has allowed definition of the permeability-thickness (kh) product of wells within the Baca field. This analysis makes assumptions which are violated in fractured reservoirs such as Baca, but it provides a useful measurement for comparison with other geothermal fluids. The method is described by Grant et al. (1982), and involves the following assumptions. First, the reservoir is horizontal and the permeability is homogeneous and isotropic. The geothermal fluid is uniform with constant compressibility, and flow into the well is horizontal and radial. Pressures at the midpoint of the reservoir are

Table 1. Permeability-thickness products for Baca wells based on Horner analysis of Pressure buildup tests (Union Oil Company 1982)

| Well | Depth (m) | k-h (md-m) |
|---|---|---|
| 4 | 1524 | 1581 |
| 6 | 1167 | 1954 |
| 10 | 1814 | 1570 |
| 11 | 2024 | 1054 |
| 13 | 1737 | 1939 |
| 15 | 1524 | 1189 |
| 19 | 1524 | 1076 |
| 20 | 1524 | 564 |
| 21 | 838 | 853 |
| 24 | 991 | 2911 |

assumed to be average for the reservoir as a whole. Pressures are then measured in the well following the termination of a flow test, and pressure buildup is plotted as a function of Horner time ratio $(tp + \Delta t)\Delta t$ on a semi-log plot, where tp is the length of the flow test, and $\Delta t$ is the shut in time. This data results in a straight line which is characterized by a slope m. Permeability-thickness is then found by

$$kh = \frac{2.303\ W\nu}{4\pi m}$$

where W is the mass flow and $\nu$ is the kinematic viscosity (Grant et al. 1982). Results of these pressure buildup tests for some of the Baca wells are shown in Table 1. In contrast with other geothermal production areas in the world, these values are low, indicating a relative lack of permeability in the Baca system.

One of the principal factors which violates the assumptions of a pressure transient test is the geologic complexity of thermal fluid entries into the well bore. As indicated by Hulen and Nielson (1982) these entries are both stratigraphically and structurally controlled. There are often multiple entries in a borehole, and as shown by Nielson and Hulen (1984) not all of the fractures penetrated carry hydrothermal solutions. Thus the values in Table 1 represent average determinations over the open portion of the well bore. A single fluid entry may be highly permeable, but may produce an insignificant kh when determined over the open length of the well bore.

Three Baca wells, B-4, B-18 and B-20 (Fig. 2) exemplify the advantages of combining detailed reservoir testing with careful analysis of geology and alteration in understanding permeability control in the active geothermal system.

*Baca 4*. Baca 4 is one of the best geothermal fluid producers in the field (Union Oil Company 1982). It is also unique in that, along with boreholes B-15, B-11, and B-20, it apparently produces from a two phase (liquid plus vapor) zone (Atkinson 1980), which Grant and Garg (1981) locate between depths of 1676 m and 1944 m (TD). Figure 7 shows tem-

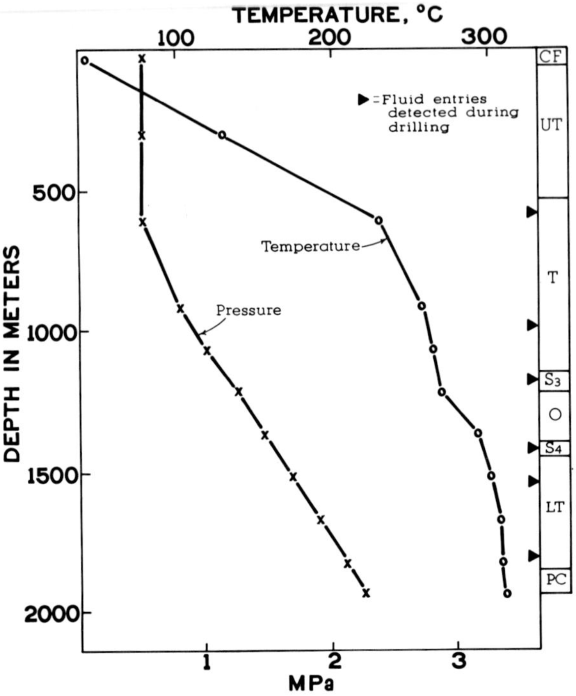


Fig. 7. Temperature and pressure surveys and generalized lithologic log for borehole Baca-4. *CF* Caldera-Fill sediments; *UT* Upper Tuffs; *T,O* Tshirege and Otowi Members of the Bandelier Tuff; *$S_3$* and *$S_4$* sandstones; *LT* Lower Tuffs; *PC* Paliza Canyon Formation. (Stratigraphic nomenclature after Nielson and Hulen 1984)

perature and pressure surveys and a generalized lithologic log of B-4, with steam and hot water entries reported during drilling.

Five flow tests on Baca-4 (UOC 1982) revealed that production was confined to the hot water portion of the hole. As summarized in Atkinson (1980) the well produced 75,750 kg $h^{-1}$ total mass flow at a well head pressure of 9.65 MPa and 27% of the flow flashed to steam. He also calculated a permeability-thickness product for the production zone of 1280 md-meters.

One injection test was run in Baca-4 (UOC 1982). Major fluid loss zones were identified between 1600.1 m and 1630.2 m (70%) and the remainder was lost between 1790.6 m and 1828.7 m. Thus all of the injected fluid was lost in the Lower Tuff interval.

Our sample distribution in Baca 4 is limited to 30.5 m composites, so exact correlation of documented fluid entries with specific alteration types is largely conjectural. However, entries do occur below the argillic zone and correspond with mostly propylitically zoned samples containing a few chips affected by intense phyllic alteration. Curiously, the Lower Tuffs, which both naturally produce fluid and accept it during injection in B-4, appear from the few samples available to be similarly altered to their counterparts in Well B-18 where the same zone is neither a fluid acceptor nor a producer.

*Baca-18*. Baca 18 is one of the few holes in the Baca field which intersects the Lower Tuffs and upper portion of the Paliza Canyon Formation, but is not a commercial geothermal fluid producer (UOC 1982). Figure 8 shows a series of temperature profiles for the hole and a simplified stratigraphic column. The temperature profile of 7/6/81 is considered

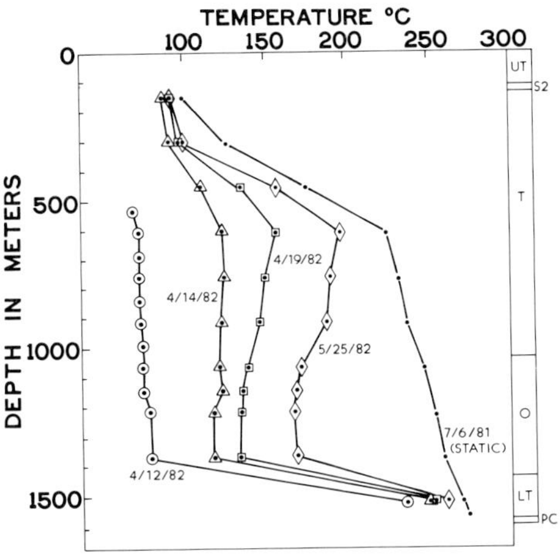


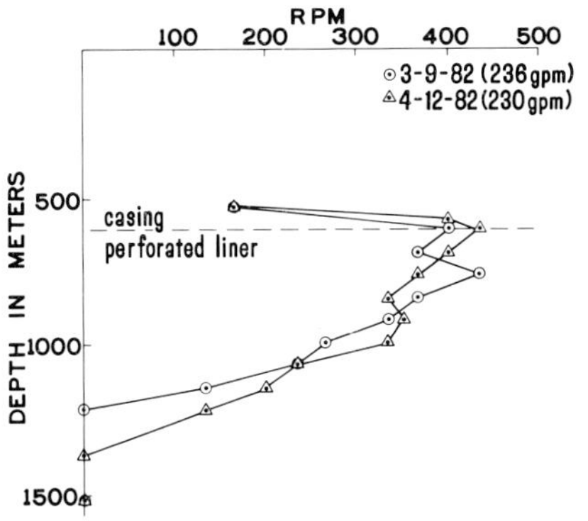


Fig. 9. Results of spinner surveys run in borehole Baca-18

Fig. 8. Temperature surveys and a generalized lithologic log for borehole Baca-18. *UT* Upper Tuffs; $S_2$ $S_2$ Sandstone; *T*,*O* Tshirege and Otowi Members of Bandelier Tuff; *LT* Lower Tuffs; *PC* Paliza Canyon Formation. (Stratigraphic nomenclature after Nielson and Hulen 1984)

to represent static conditions by UOC. A dramatic change in temperature gradient occurs approximately at the junction between 9 5/8" casing and 7" slotted liner (Fig. 9). This break also occurs within the Tshirege Member of the Bandelier Tuff, and there is no lithologic change to which we can attribute the change in temperature gradient. Thus, we suspect that it is an artifact of the casing program and has little geologic significance.

Since Baca 18 was not a commercial producer, UOC used it as an injection well to dispose of fluids generated during the flow testing of other wells. In conjunction with this injection, a spinner survey was run in the lower portion of the well, and its results are shown in Figure 9. It should be noted here that the presence of slotted liner in this interval makes the precise identification of fluid flow paths difficult to interpret, but the spinner survey can give a general idea of where the injected fluids are exiting the well. A plot of spinner survey results shows that nearly all of the fluid leaves the well between the top of the perforated liner and a depth of 1372 m with most of the loss occurring in the zone between 540 m and 990 m. Thus most of the fluid leaves the well in the lower portion of the Tshirege Member, and some in the upper portion of the Otowi Member. No fluid leaves the well in the Lower Tuffs or in the Paliza Canyon. Therefore, this normally productive zone is neither a fluid producer nor a fluid acceptor.

Following injection, temperature profiles were measured in B-18 to document thermal recovery of the well (Fig. 8). The temperature profiles confirm the results of the spinner survey which shows that the fluids have left the well between 610 and 1362 meters. The results also show that the zone which accepted the most fluid recovers at a more rapid rate, which we ascribe to a zone of weak hydrothermal fluid production.

Following the completion of the injection test in B-18, UOC (1982) determined a permeability-thickness product on the basis of Horner plots of pressure fall-off. They attributed a value of 2338 md-meters to

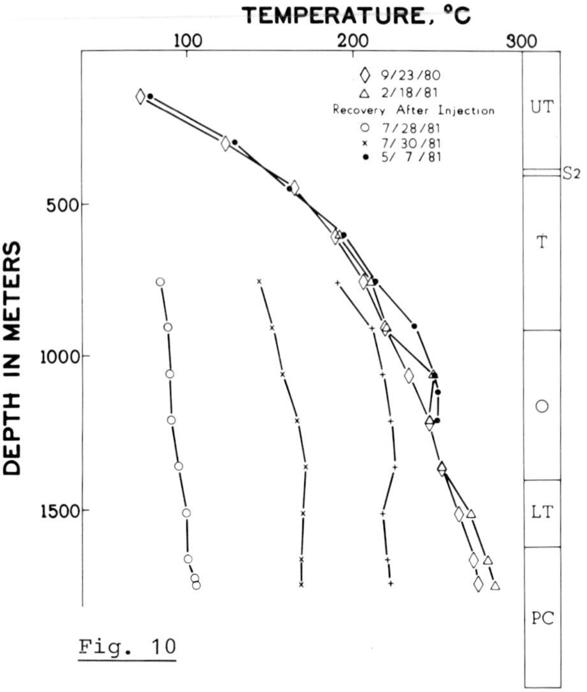


Fig. 10

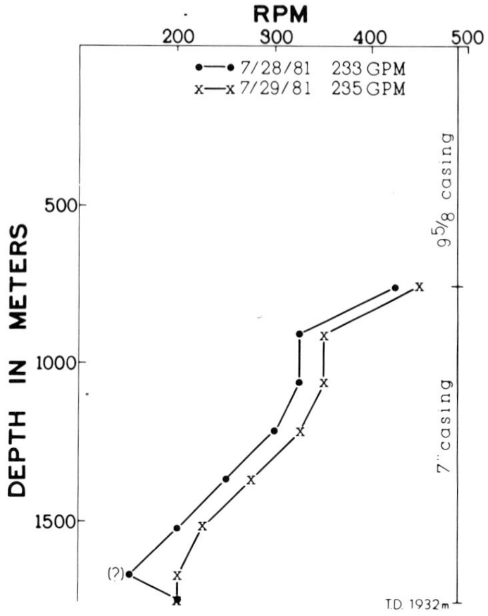


Fig. 11

Fig. 10. Temperature surveys and a generalized lithologic log for borehole Baca-20. *UT* Upper Tuffs; $S_2$ $S_2$ Sandstone; *T,O* Tshirege and Otowi Members of the Bandelier Tuff; *LT* Lower Tuffs; *PC* Paliza Canyon Formation. (Stratigraphic nomenclature after Nielson and Hulen 1984)

Fig. 11. Results of spinner surveys run in borehole Baca-20

Baca-18, which is relatively high for wells in the Baca field. In addition, there was also evidence of communication of this well with Baca-6 and Baca-24.

The Lower Tuffs in B-18 are moderately illitized and chloritized and can be interpreted as a zone of moderate propylitic alteration interrupted by narrow stringers of phyllic alteration, similar to that associated with numerous active fluid entries in other Baca wells. However, the results of the injection test have shown that any primary permeability in the Lower Tuffs of B-18 has been effectively sealed by this hydrothermal alteration.

*Baca 20*. Baca 20 is one of the more studied holes in the Baca field, from the standpoint of alteration (Hulen and Nielson 1983), reservoir engineering (Christensen and Atkinson 1981; Riney and Garg 1982), and fracture stimulation (Morris et al. 1982). Figure 10 shows a series of temperature measurements in Baca 20. The closest log to static conditions that includes the entire hole was that measured on 2/18/81. UOC (1982) considers the 5/7/81 log to be static, but this does not extend to the lower portions of the hole.

Christiansen and Atkinson (1981) report that fluids from B-20 had a surface flash of 55 - 65%. Since an isenthalpic flash of 35% is predicted for fluids at temperatures found in the well, they conclude that the well was producing from a two-phase reservoir. They also suggest that there may be a major fluid entry at about 1356 m where the well began to surge during drilling. This entry is in the densely welded, presumably weakly propylitized portion of the Otowi Member. Unfortunately, there were no cuttings returns from this interval. In addition, there is no notable increase in alteration adjacent to the zone. We have interpreted this zone as a minor interval of nonwelded tuff within the otherwise densely welded Otowi Member (Nielson and Hulen 1984).

Figure 11 shows the results of spinner surveys which were run during injection into Baca 20. Figure 11 shows the temperature recovery curves which followed the injection test. It can be seen that the area of injection cooled off rather uniformly, but that the lower portion of the Otowi Member heated up more rapidly with time, suggesting geothermal fluid flow in this portion of the stratigraphy.

## Summary and Conclusions

A wealth of reservoir engineering data and an extensive set of subsurface rock samples collected during geothermal exploration in the Valles caldera are being used to scientifically study structural/stratigraphic controls on thermal fluid migration and the relationship of hydrothermal alteration to active and fossil fluid conduits. Research has shown that fluid flow at Baca is controlled both by deeply-penetrating normal faults and by discrete stratigraphic aquifers in communication with these faults; many formerly permeable channels now are sealed by hydrothermal alteration. Although alteration is very useful in delineating broad-scale zoning characteristics of the system, it cannot be used by itself to distinguish active form inactive thermal fluid zones. To accurately locate these zones, alteration must be used in conjunction with stratigraphic and structural data and with careful reservoir engineering studies.

*Acknowledgements.* This work was funded by the U.S. Department of Energy, San Francisco Operations Office under contract DE-AC03-84SF12196. Cores and cuttings used in this study were provided by Union Geothermal Division of the Union Oil Company of California. Contribution number ESL-140 of the Earth Science Laboratory, University of Utah Research Institute.

## References

Atkinson PG (1980) Geothermal reservoir, initial state, Baca location No 1 NM, Redondo Creek field. Geotherm Resources Council Transact 4:435-438

Bailey RA, Smith RL, Ross CS (1969) Stratigraphic nomenclature in the Jemez Mountains, New Mexico. US Geol Survey Bull 1274-P:19

Celati R, Squarci P, Taffi L, Stefani GC (1976) Analysis of water levels and reservoir pressure measurements in geothermal wells. Proc 2nd UN Symp Devel, Use Geotherm Resources, 20-29 May 1975, pp 1583-1590

Christensen B, Atkinson P (1981) Baca-20, Redondo Creek field - summary and analysis of flow tests 1 through 4; 8/80 to 1/81. Union Oil Co int report. Open File report NM/BACA/110. Earth Sci Lab, Univ Utah Res Inst, Salt Lake City, Utah

Dash ZV, Murphy HD, Aamodt RL, Aguilar RG, Brown DW, Counce DA, Fisher HN, Grigsby CO, Keppler H, Laughlin AW, Potter RM, Tester JW, Trujillo PE, Zyvoloski G (1983) Hot dry rock geothermal reservoir testing: 1978 to 1980. J Volcanol Geotherm Res 15:59-99
Doell RR, Dalrymple GB, Smith RL, Bailey RA (1968) Paleomagnetism, potassium-argon ages, and geology of rhyolites and associated rocks of the Valles caldera, New Mexico. In: Coats RR, Hay RL, Anderson CA (eds) Studies in Volcanology. Geol Soc Am Memoir 116:211-248
Goff FE, Grigsby CO (1982) Valles caldera geothermal systems, New Mexico, USA. J Hydrol 56:119-136
Grant MA, Garg SK (1981) Baca geothermal demonstration project - interpretation of downhole data from the Baca geothermal field. Systems, Science and Software, Rept DOE/ET/271632, p 112
Grant MA, Donaldson JG, Bixley PF (1982) Geothermal Reservoir Engineering. Academic Press, New York, 369 pp
Hulen JB, Nielson DL (1982) Stratigraphic permeability in the Baca geothermal system, Redondo Creek area, Valles caldera, New Mexico. Geotherm Resources Council Transact 6:27-30
Hulen JB, Nielson DL (1983) Stratigraphy of the Bandelier Tuff and characterization of high-level clay alteration in borehole B-20, Redondo Creek area, Valles caldera, New Mexico. Geotherm. Resources Council Transact 7:163-168
Johnson AM (1970) Physical Processes in Geology. Freeman, Cooper, San Francisco, 577 pp
Lambert SJ, Epstein S (1980) Stable isotope investigation of an active geothermal system in Valles caldera, Jemez Mountains, New Mexico. J Volcanol Geotherm Res 8:111-129
Molloy MW, Laughlin AW (1982) Drilling problems. In: Goldstein NE, Holman WR, Molloy MW (eds) Final Report of the Department of Energy Reservoir Definition Review Team for the Baca Geothermal Demonstration Project. Lawrence Berkeley Lab Rept LBL-14132, pp 22-25
Morris CW, Harold RJ, Pearson CF (1982) Fracture stimulation experiments In: Goldstein NE, Holman WR, Molloy MW (eds) Final Report of the Department of Energy Reservoir Definition Review Team for the Baca Geothermal Demonstration Project. Lawrence Berkeley Lab Rept LBL-14132, pp 26-30
National Research Council (1979) Continental scientific drilling program. Natl Acad Sci (Wash DC), 192 pp
Nielson DL, Hulen JB (1983) Geologic model of the Baca geothermal reservoir, Valles caldera, New Mexico. 9th Workshop Geotherm Reservoir Engineering, Stanford Univ, pp 145-150
Nielson DL, Hulen JB (1984) Internal geology and evolution of the Redondo Dome, Valles caldera, New Mexico. J Geophys Res 89:8695-8711
Riney TD, Garg SK (1982) Evaluation of injection test data from a Baca well. Stanford Reservoir Engineering Workshop
Smith RL, Bailey RA (1966) The Bandelier Tuff: A study of ash-flow eruption cycles from zoned magma chambers. Bull Volcanol 29, Ser 2: 34-104
Smith RL, Bailey RA (1968) Resurgent cauldrons. In: Coats RR, Hay RL, Anderson CD (eds) Studies in volcanology. Geol Soc Am Memoir 116: 613-662
Smith RL, Bailey RA, Ross CA (1970) Geologic map of the Jemez Mountains, New Mexico. US Geol Surv Misc Inv Map I-571
Union Oil Company (1982) Final report Baca project, geothermal demonstration power plant. Report prepared for US Dept of Energy, San Francisco Operations office
White AF, Delany JM, Truesdell AH, Janik C, Goff FE, Crecraft H (1984) Fluid chemistry of the Baca geothermal field. 35th NM Geol Soc Field Guide, pp 257-263

# DEEP STRUCTURES OF CONTINENTS

# RATIONALE FOR A DEEP DRILL HOLE INTO THE SEISMICALLY ACTIVE AREA OF THE HOHENZOLLERNGRABEN, SOUTHWESTERN GERMANY*

Karl Fuchs[1]

## Introduction

Intraplate seismicity in continents is one of the important research topics of the International Lithosphere Program. It is expected that such research will improve our knowledge of deformation procedures and of the processes governing the release of seismic energy inside the continents, contribute to the mitigation of earthquake hazards, and provide a better understanding of intra-continental stress, and its spatial and temporal correlation with crustal heterogeneities. Deep continental drilling is a necessary and most important tool in this research, since in-situ physical, structural, compositional and rheological properties must be known at depth before an integrated model of intraplate seismicity with predictive capabilities can be established successfully. Drilling into seismic active areas in continents is therefore both a challenge and a necessity which requires the close cooperation of all geoscience disciplines.

Sites for deep continental drillholes in seismically active areas must meet several requirements. The seismicity zone must be well confined to ensure that the hole can be positioned relatively close to a future event of moderate size, and the rate of seismicity must be sufficiently high to provide a sufficient number of events for study during the lifetime of the drill hole. Finally, in view of the enormous costs of a deep drill hole, precautions must be taken to avoid the destruction of the hole by a major local earthquake. The biggest danger to the hole, of course, comes from the potential for fracture and off-set by actual fault rupture. This could lead to an immediate loss of the most interesting deep parts of the hole and must be avoided under all foreseeable circumstances. Seismic zones, therefore, with predominently vertical fault planes and strike slip motion are to be preferred, at least in the first stages of a deep continental drilling program in such areas. For magnitude 6 and greater earthquakes, nearfined accelerations may also endanger the stability of the hole. Initially therefore, confined areas of intermediate magnitudes earthquakes offer the best opportunity for positioning the deep hole sufficiently close to expected events to meet the research objectives.

In this paper the rationale for proposing a deep drill hole in a seismic active area in the Federale Republic of Germany is presented. The Hohenzollerngraben in Southwestern Germany will also serve as a case history for the defining process. This site is one of four possible

*In memoriam Henning Illies

[1]Geophysical Institute, Karlsruhe University, Hertzstrasse 16, 7500 Karlsruhe 21, FRG

Observation of the Continental Crust through Drilling I (ed. by C.B. Raleigh)

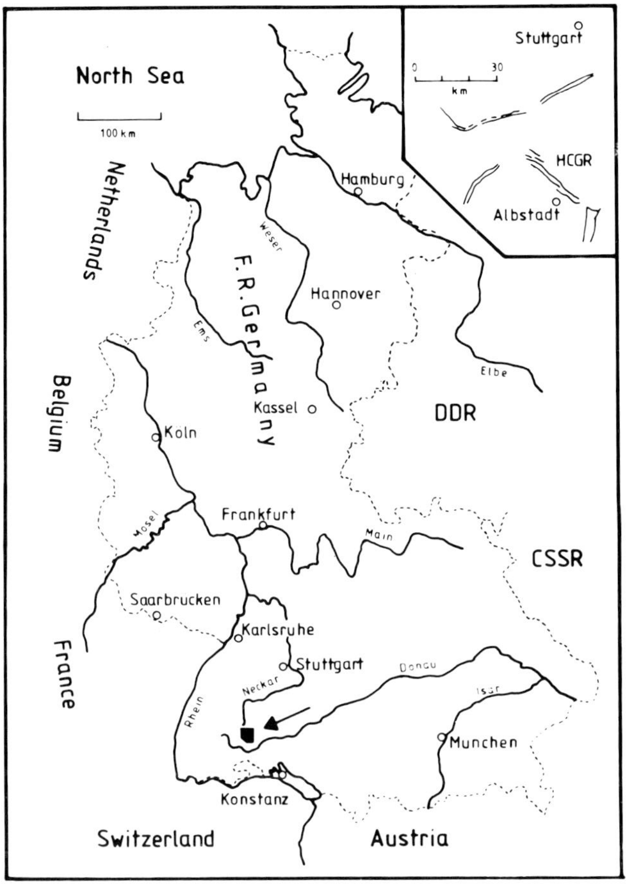


Fig. 1. Location map Hohenzollerngraben (HZGR) in southwestern Germany indicated by *black area with arrow*. *Upper right corner* displays this area enlarged. The HZGR is the NW-SE striking tectonic element north of Albstadt

locations currently identified in the German deep continental drilling program (Vidal et al. 1981; Alfred Wegener Stiftung 1984). Although in the selection process for the first deep continental drill hole in Germany priority has been given to drilling in deep crystalline basement, it is considered worthwhile to report the present stage of the presite survey in the seismic active area of the Hohenzollerngraben as a general contribution to international efforts in deep continental drilling.

## The Geological Setting

The Hohenzollerngraben (HZGR abbreviated in the following) with its distinctly localized intraplate seismicity is situated in the southwestern part of the South-German triangle about 100 km SSW of Stuttgart (Fig. 1) and is characterized by Herzynian basement rocks covered by gently SE-dipping Mesozoic strata. Its dimensions are 28 km in length and 1.5 km in width. The graben segment has subsided only 100-110 m. From the dip of the border faults the thickness of the sedimentary cover inside the graben is estimated to about 1200 m, with a horizontal extension of about 150 m. A distinct fault forms the northern rim of the graben, while its southern margin is developed as a series of faults. These findings suggest that the HZGR is a Y-shaped structure. There has been no tectonic movement in the graben area since the upper Miocene (Illies 1982).

The intraplate seismicity of the HZGR area is believed to be a consequence of the interaction of the regional stress-field, dominated by the Alpine orogeny in the south and the Rhinegraben taphrogenesis in the west, with local heterogeneities of the earth's crust. Illies (1982) proposed as the origin of the seismic activity the blocking of a sinistral shear motion on a bundle of N-S lineaments by the subsurface structure of the graben. The lineaments originate in the Alps and extend for several 100 km, well into Southern Germany. They can be observed on satellite images (Kronberg 1977; Bankwitz et al. 1979; Illies et al. 1981; Illies 1982) and are identified by numerous neotectonic investigations (Seibold and Stahlecker 1965; Stellrecht and Moelle 1980; Brux 1981).

According to Illies (1982) the shear lineaments were activated a few million years ago when the central European stress field rotated anticlockwise. This initiated strike slip motions along the master faults of the Upper Rhinegraben. However, soon the en-echelon structure of the Rhinegraben impeded this motion, and instead the continuing orogeny of the Alps formed of a new shear motion pattern in its foreland in Southern Germany to the east of the Rhinegraben.

It must be emphasized at this stage that the present seismic activity at the HZGR is certainly not to be attributed to a rifting process in the graben as this feature has been extinct since the upper Miocene. For the recent seismic activity, which starts below a depth of 2 km (Turnovsky 1981), the shallow wedge-shaped graben is most likely without significance. The zone of epicenters of the earthquakes has an extent of a few kilometers only that traversing the graben in a N-S direction near the town of Albstadt not too far from the famous Hohenzollern castle. Both, town and castle have suffered heavy damage from the seismic activity of the last seven decades.

## The Seismicity of the Hohenzollerngraben

Macroseismic investigations indicate that the seismic activity in the region of the HZGR can be traced back to about 1850 (Fiedler 1954). Before that a quiet period can be followed back to the 14th century (Schneider 1980). The actual scientific study of earthquakes in the western Swabian Alb started after the 1911-1913 earthquake series. The earthquake of 16 November 1911 with a surface wave magnitude ML = 5.6 is the largest seismic event in central Europe north of the south-alpine seismic belt. Gutenberg (1915), Schmidt and Mack (1912) and Sieberg and Lais (1925) covered the seismological and tectonic aspects of this series. In 1933 the first seismological station (Meßstetten, MSS) was installed close to the epicentral area (Hiller 1934). The observations from the period after the 1911-13 series have been compiled and interpreted by Hiller (1935, 1960).

A new climax in the seismic activity of the region occurred between 1934 and 1944, with the two strongest events in May 1943 causing considerable damage. After 1945 the network of seismic stations in the area was considerably improved (Berckhemer and Hiller 1960), and kinematic studies revealed the predominant sinistral strike-slip character of the events (Schneider et al. 1966; Schick 1968, 1970). A comparative

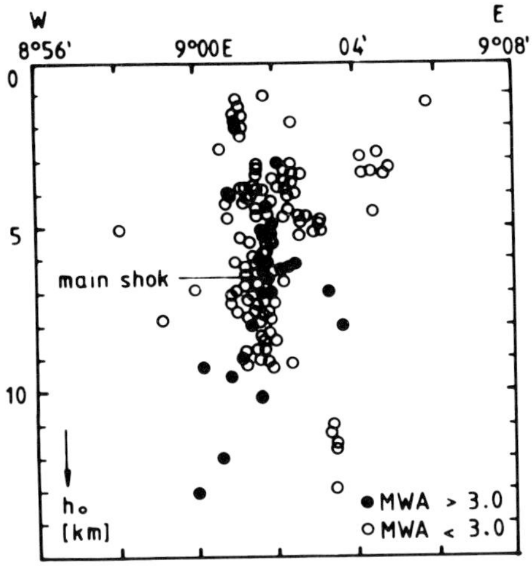


Fig. 2. E-W section through the seismic zone of the Hohenzollerngraben. Earthquake series of 1978 (Turnovsky 1981). The E-W extent measures only 3-4 km

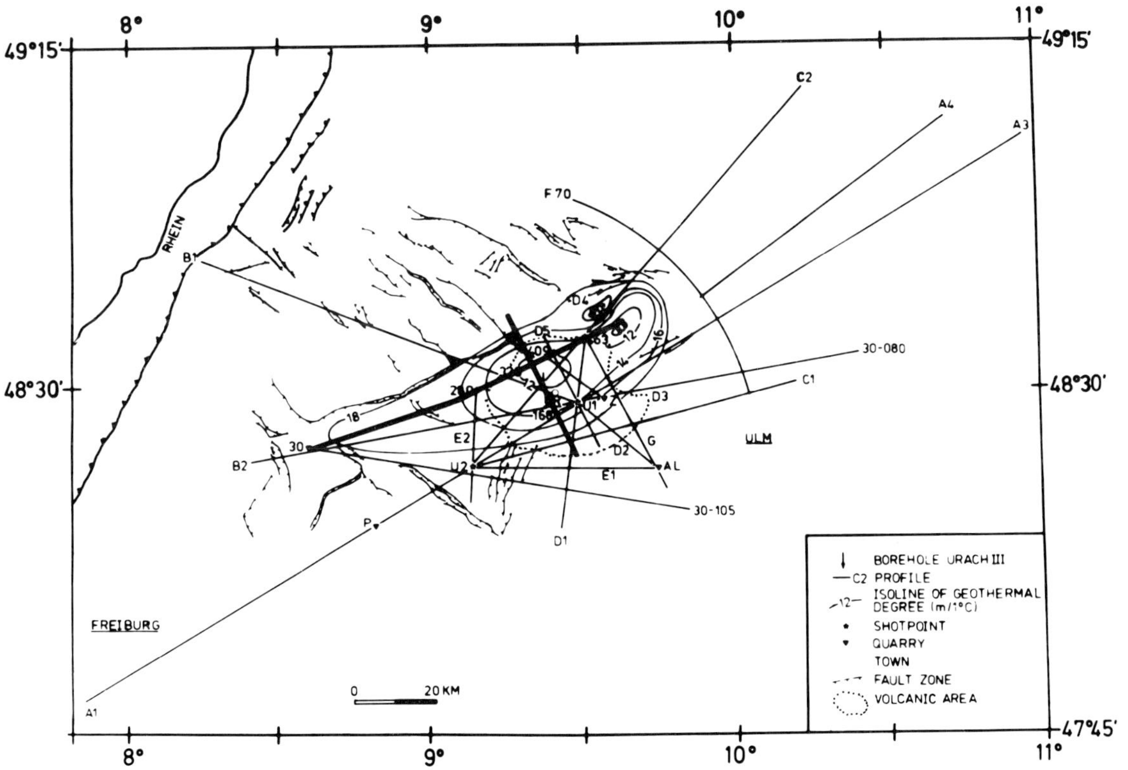


Fig. 3. Location map for deep seismic profiles in the neighbourhood of the geothermal research well Urach 3. *Thin straight lines* refraction seismic profiles; *thick crossing lines* reflection profiles

study of the activity since 1911 has provided evidence for its northward migration (Brenner 1961).

After a quiet period of about 10 years, a new series of seismic activity began in 1969 and culminated in the large earthquake of 3 September 1978 (MWA = 5.7) followed by long series of aftershocks (Haessler et al. 1980; Scherbaum 1980; Turnovsky 1981).

The hypocentral distribution on an E-W section through the HZGR during the 1978 aftershock series reveals that the recently active seismic zone is relatively narrow, with a thickness of only a few kilometers (Fig. 2).

## Crustal Structure from Deep Seismic Sounding

The site survey for the nearby geothermal research well, Urach 3 (Haenel 1982), about 50 km from the HZGR area, provides the first information on lateral variations of crustal structure. The survey comprised refraction and reflection experiments, with several refraction lines traversing the region of the HZGR (Fig. 3). The reinterpretation of the refraction lines in this region reveals the presence of strong lateral heterogeneities meeting in the vicinity of the HZGR: (1) an anomalous body of low P-velocity detected previously at the site of the Urach well in the upper crust both from refraction (Prodehl et al. 1982) and reflection measurements (Bartelsen et al. 1982; Trappe 1983) was shown to have its south-western margin extending into the region of the HZGR: low P-velocity combined with high density are special characteristics of the body required to match both kinematic and dynamic observations: (2) the lower crust to the SW of the HZGR is clearly slower and more homogeneous than the crust to the NE; (3) alternative models for the uprising of the lower crust in the vicinity of the HZGR have been developed to estimate the range of possible structures compatible with the seismic observations (Fig. 4; Gajewski and Prodehl 1984). The models show that the currently available data do not constrain the model sufficiently to define a site for a deep drilling location. More detailed surveys are required. However, the existing surveys do show that in the region of the HZGR, strong lateral heterogeneities exist. Their significance for the release of seismic energy will be discussed below.

Figure 5 shows the most important refraction record-sections which constrain the structure of the Urach body of low P-velocity to the NE of the HZGR (Gajewski and Prodehl 1984). Phase "b" is the wide-angle reflection from the bottom of the body. It is observed to the NE from both shot points U1 (Fig. 5b) and U2 (Fig. 5a); however, towards the SW this phase is only recorded from shot point U1 (Fig. 5c). The shot point U2 to the SW produces a record section (Fig. 5d) which characteristically is void of all phases between "$a_3$" (phase from the crystalline upper crust) and "c" (reflection from the crust-mantle boundary). A similar pattern of wide-angle reflections has been observed further to the SW of the HZGR (Deichmann and Ansorge 1983). The absence of phase "b" on profile U2-SW is the most important information on the southwestern margin of the Urach body. The model is also corroborated by the dynamical behaviour of phase "$a_3$" propagating through the upper crust. On top of the Urach body the phase strongly decays reaching a distance of only 50-60 km; it is screened off by the top of the Urach body, while in constrast to the SW from shotpoint U2 "$a_3$"-energy is propagated to distances greater than 100 km.

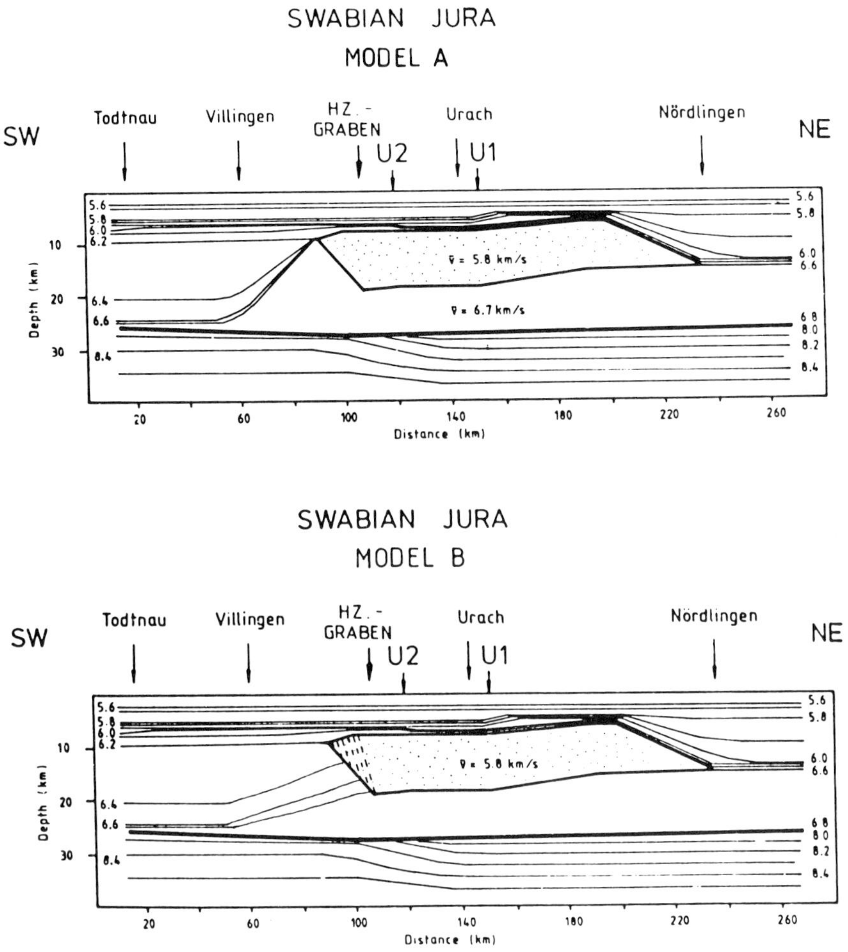


Fig. 4. Lateral crustal heterogeneities in the neighbourhood of the HZGR (HZ-Graben). *Top* Model A - 6.6 km $s^{-1}$ iso-lines rises to the top of the Urach body. *Bottom* Model B - 6.4 km $s^{-1}$ rises to the top of the Urach body, while the 6.6 km $s^{-1}$ iso-lines forms the base of the Urach body. *Dashed hatched area* of the SW margin of the Urach body signifies approximately the uncertainty of the margin from presently available data. (Gajewski and Prodehl 1984)

The two alternative models, A and B, both compatible with the observations, are shown in Figures 4a and 4b. In the original model A (Fig. 4a), the Urach body to the SW was bordered by a local rise of lower crustal material with P-wave velocity $v_p$ = 6.6 km $s^{-1}$. The synthetic seismogram section together with the associate ray diagram are shown in Figure 6. Note that phase "b" is absent in this section. In the alternative model B (Fig. 4b) the 6.4 km $s^{-1}$ iso-velocity line grazes the top of the Urach body while the 6.6 km $s^{-1}$ line merges with the bottom of the body dipping towards the SW. In this zone to the SW strong vertical velocity gradients must be absent to avoid the generation of phase "b" as a retrograde diving wave, as such a wave is not observed (see Fig. 5d).

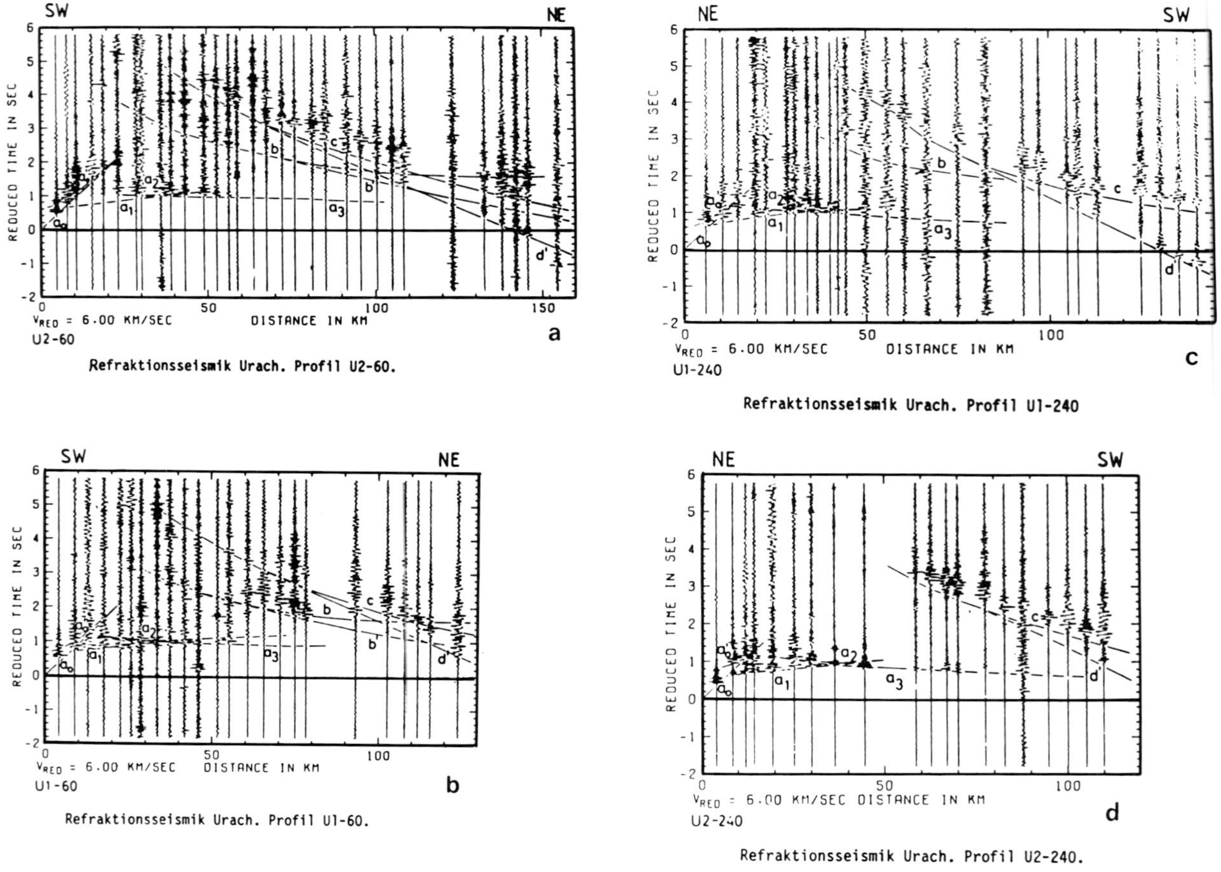


Fig. 5a-d. Record sections from refraction seismic survey of the Urach area of lines crossing the HZGR. a) U2 - 60: shot point U2 to NE; b) U1 - 60: shot point U1 to NE; c) U1 - 240: shot point U1 to SW; d) U2 - 240: shot point U2 to SW. For the positions cf. Figure 3. *Letters* signify the various seismic phases. Phase $a_3$ has travelled through the upper crust. Phase $b$ is the overcritical reflection from the crust-mantle boundary. Reduction velocity 6.0 km $s^{-1}$. (Gajewski and Prodehl 1984)

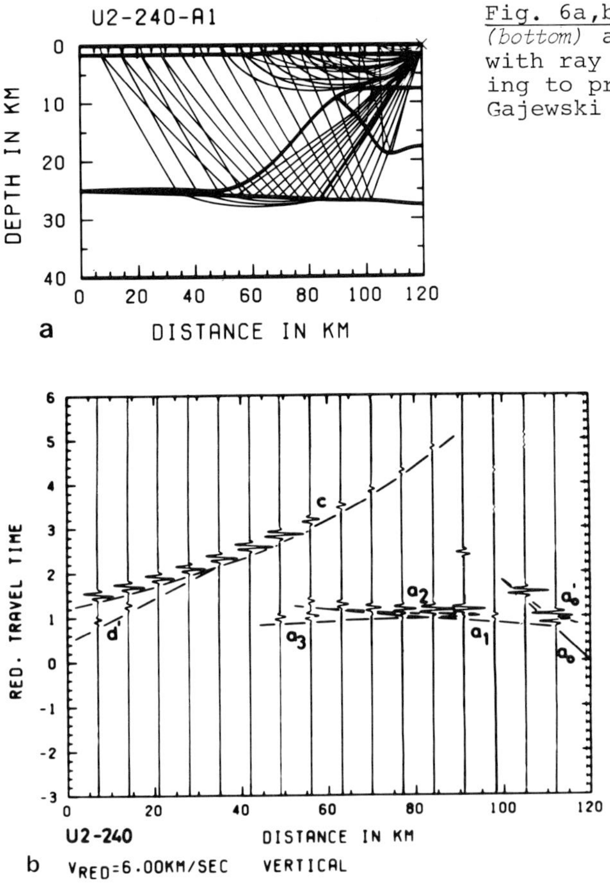


Fig. 6a,b. Synthetic record sections *(bottom)* and lateral heterogeneity with ray diagram *(bottom)* corresponding to profile U2-240 (cf. Fig. 5d; Gajewski and Prodehl 1984)

The hatched area at the southwestern margin of the Urach body (model B, Fig. 4b) indicates the uncertainty of the inversion of the currently available data with regard to the position of this margin.

The detailed structure of the anomalous Urach body in the central region of the geothermal anomaly of Urach has been established by a high resolution reflection survey with associated steep and wide-angle observations (Bartelsen et al. 1982; Meissner et al. 1983; Trappe 1983). Although the reflection profile (see Fig. 3) did not cross the HZGR, it corroborated the presence of the low velocity body and delineated its lateral extent to the west of Urach. Figure 7 is a reproduction of the 8-fold stacked record section from the SW-NE Urach relfection line (Bartelsen et al. 1982). The bottom of the anomalous body is more readily recognized than the top. From the paucity of reflections from the top of the Urach body and by matching acoustic impedances, Gajewski and Prodehl (1983), postulated a high density for the low velocity body. Although such a conclusion is rather tentative, the petrological significance of such an anomalous body poses a challenge to future field experiments, both at the surface and in a deep drill hole.

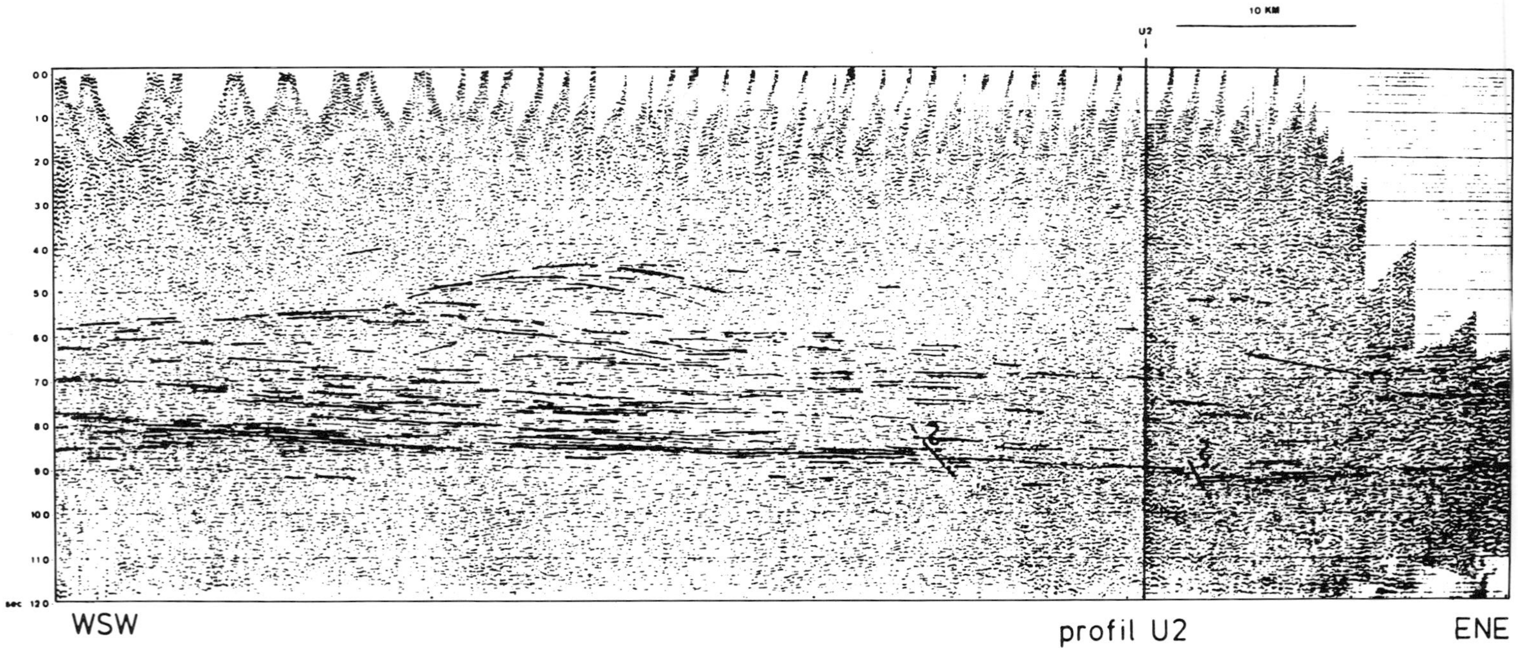


Fig. 7. Reflection record section (8-fold stack) from the approximately 80-km-long SM-NE line in the region of the Urach geothermal anomaly. (Bartelsen et al. 1982)

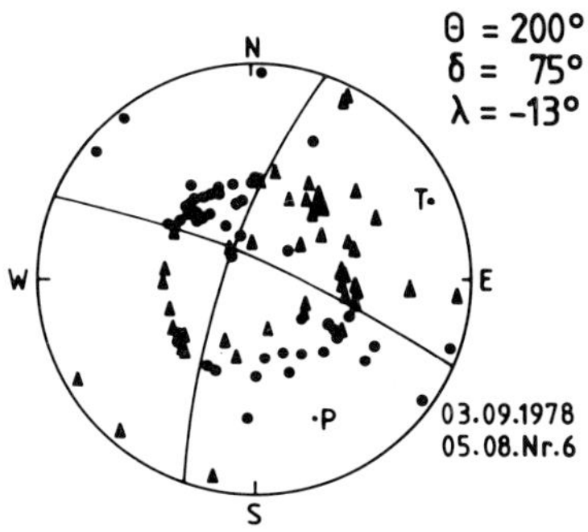


Fig. 8. Fault plane solution from short period observations of the HZGR-earthquake of 3, September 1978 (MWA = 5.7). (Turnovsky 1981)

## Stress Determinations in the Region of the Hohenzollerngraben

A knowledge of the crustal stress field is of special importance for the stability of a deep drill hole in the area of the HZGR and for understanding the nature of the seismicity there. A number of independent methods have been applied which provide insight into the stress regime of southwest Germany in general, and in the neighbourhood of the HZGR in particular, these include: fault plane solutions of local earthquakes, wall breakouts in the Urach well, and in situ- stress determinations by overcoring as well as by hydrofracturing.

Fault plane solutions of earthquakes in the Swabian Alb (Ahorner et al. 1972; Ahorner and Schneider 1974) reveal an extremely uniform stress field with a horizontal axis of compression in the direction N155°E. For the earthquake of 3.9.1978 (MWA = 5.7) near Albstadt, Turnovsky (1981) derives a solution with almost vertical fault planes (Fig. 8); the active sinistral fault plane striking N22.5°E. From spectral analysis, stress drops of about 1 MPa are found to be typical for nearly all events in the region of the HZGR.

In-situ stress determinations in the HZGR have been performed by a group from the Geological Institute of Karlsruhe University (Greiner 1974, 1978; Illies 1974, 1982; Baumann and Becker 1983), predominantly with the door-stopper method. These measurements reveal the largest horizontal main stress direction to be between 130° and 152°N with 1.7 MPa compressional stress. The smallest horizontal main stress amounts to -0.2 to -0.5 MPa dilatational stress (Greiner 1974). The most noteworthy result of the absolute determination of the stress is the large value of the horizontal shear stress (1.0 to 1.2 MPa). This value clearly exceeds that of the adjacent regions. Measurements with triaxial cells for a complete determination of the stress tensor are currently in progress (Baumann and Becker 1983; Bauman 1984). Furthermore two stations with continuous digital recording of the stress components (sampling rate 1 $h^{-1}$) have been installed in the region of the HZGR using the Gloetzl pressure cell and the Maihak sensor (vibrating wire) in 30-m-deep bore holes (for a description of the instruments, see Kovari 1979). Although some correlations with local seismic events are indicated it is still difficult to establish such a relation with certainty since the instruments are still settling and instrumental noise is still high (Baumann and Becker 1983; Bauman 1984).

The directions of the axes of the main stresses as obtained by the hydrofracturing method near the HZGR structure (Rummel and Jung 1975) are concordant with the directions derived by the doorstopper method in this region.

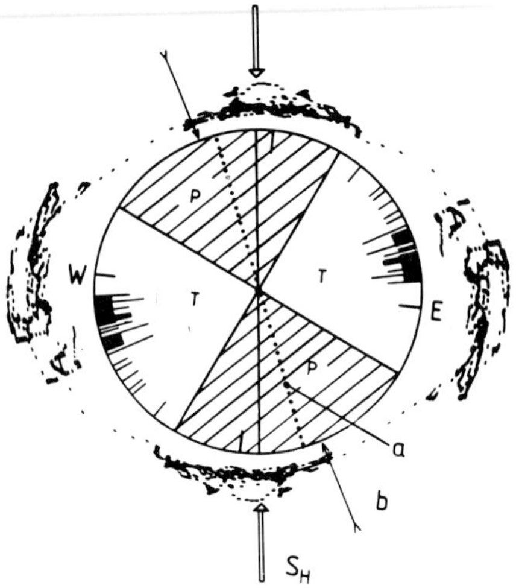


Fig. 9. Direction of horizontal component of main compressional stress axis ($S_H$) derived from wall breakouts of the deep geothermal research well Urach 3 and the corresponding direction from fault plane solutions in the neighbourhood of the HZGR. (Turnovsky and Schneider 1981). The form of the breakouts and the fault-plane solution are also given. (Bluemling et al. 1983)

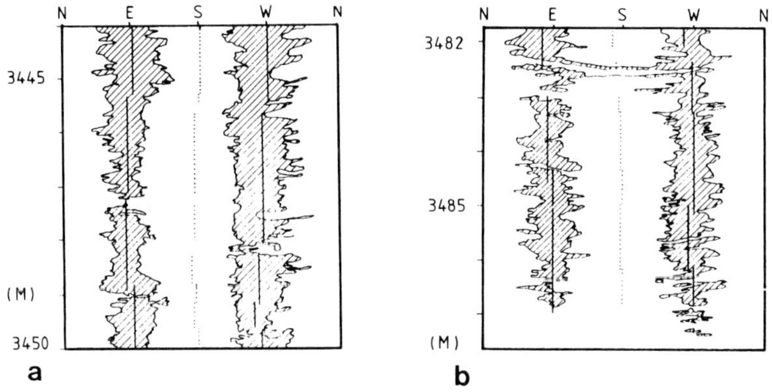


Fig. 10a,b. Display of televiewer data from various depth ranges of the deep geothermal well Urach 3. *Hatched areas* signify the azimuthal distribution of the breakouts. *Dotted lines* corresponds to the derived direction of maximal main stress. (Bluemeling, contribution to Alfred Wegener Stiftung 1984)

Wall breakouts in deep wells have recently been introduced as indicators of the orientation of crustal stress tensors (Babcok 1978; Bell and Gough 1979). The breakouts are oriented normal to the direction of the horizontal component of the compressional axis. This method has been applied to the crystalline part of the geothermal well (Bluemling et al. 1983) using a new technique in which the breakouts are not sampled discretely but continuously. In Figure 9 a wall-contour plott is shown together with a typical fault plane solution for the Swabian Alb. The direction of the compressional axis derived from the breakouts is N173°E ± 10°E, deviating about 15° from the direction bisecting the angle between the two fault planes. This is consistent with a coefficient of internal friction $\mu$ = 0.6. Breakouts have also been analyzed from televiewer images of the deformed well (Fig. 10). The results corroborate the direction derived from the 4-arm caliper data, however, with higher resolution.

Table 1. Temperature-depth estimates for the Hohenzollern-graben (Sattel, contribution to Alfred Wegener Stiftung 1984)

| Depth (m) | Temperature-range Without[a] (°C) | With[b] (°C) |
|---|---|---|
| 2000 | 98–108 | 94–104 |
| 4000 | 146–165 | 138–157 |
| 6000 | 196–234 | 178–206 |
| 8000 | 244–280 | 222–258 |
| 10000 | 290–334 | 262–306 |
| 12000 | 330–382 | 298–350 |
| 15000 | 388–460 | 352–418 |

[a]Radiogenic heat production; [b]2 μW $m^{-3}$ heat production in the crystalline part.

## Further Geophysical Investigations

Temperature-depth distributions in the HZGR can presently be estimated only within a broad range of uncertainty since no heat flow measurements have been obtained there so far. In its vicinity only a few BHT-determinations in early industrial wells of the BEB (Hannover, West-Germany) are available. Haenel (1980) estimated the heat flow value in the HZGR as 65 mW $m^{-2}$ from a regional interpolation of all available temperature data. This value is about 20 mW $m^{-2}$ less than in the distant center of the Urach geothermal anomaly 50 km away where abundant temperature and heat flow data are available both from a number of shallow research holes and from the geothermal well Urach 3 with its final depth of 3.3 km (Haenel 1982). Especially in view of the restricted lateral extent of the Urach body an extrapolation from Urach to the HZGR can only be regarded as a preliminary attempt to estimate the temperature depth distribution in this area. It certainly must be supported by local heat flow determinations.

The estimate of the temperature-depth distribution in the HZGR region is based upon the heat flow value extrapolated from Urach, upon the BHT-temperatures and upon the empirical linear relation between the depth-change of heat conductivity and of P-wave velocity (Sattel 1982). Using the velocity-depth distribution in the crust of the HZGR two estimates with upper and lower bounds have been derived: (1) for the case with radiogenic heat production of 2 μW $m^{-3}$, and; (2) for the case without (Table 1). These temperature estimates could change considerably if convective systems exist at depth in the HZGR area.

An improved determination of the temperature distribution in the HZGR region is important for an understanding of the fracture and creep behaviour of the material in and below the seismic zone (Meissner and Strehlau 1982). It is also required as basic technical information for the drilling of a deep well.

A detailed gravity survey of the Urach geothermal anomaly (Makris et al. 1982) also covered the HZGR region. Here the Bouguer gravity has small scale anomalies of about 5 mgal, and provides a constraint of

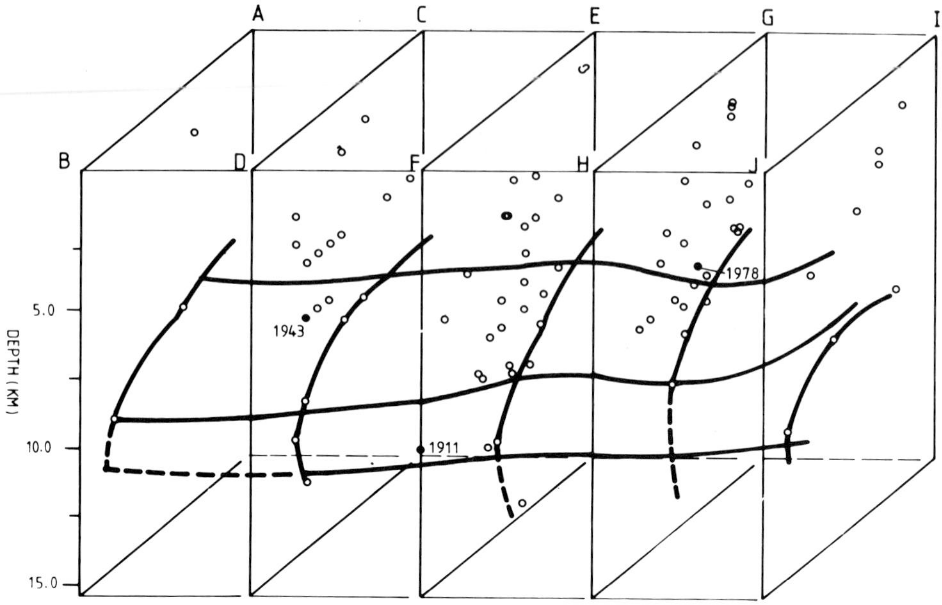


Fig. 11. Spatial distribution of hypocenters in the Hohenzollerngraben region for the period 1911-1978. The hypocenters have been projected onto the vertical N-S sections AB. CD, etc. see Fig. 12. The net of thick lines marks the envelope of the deepest hypocenters. Note its rise to the North

less than a few tens of kilometers on the lateral extent of the crustal heterogeneity at the HZGR, perpendicular to the SW-NE refraction line.

A high precision geodetic network for levelling and triangulation measurements has been established near Albstadt where the seismic zone crosses the HZGR. This will allow pre-, co- and aseismic movements at the earth's surface (Maelzer 1983) to be detected.

Magnetic anomalies of about 50 nT are present on the aeromagnetic map (Eberle 1973, BGR 1976) in the region HZGR-Urach which cannot be related to the crustal heterogeneities detected by DSS. The anomalies also appear in the map of isogones (Farkas 1973). For a better understanding of the earthquake generating processes the observation of temporal variations of the geomagnetic field is of considerable importance. Upto now only a secular change of -0.3° of the declination has been be determined for the epoch interval 1935.0-1981.5 (Farkas 1983, personal communication). The dimension and shape of the magnetic anomalies can only be determined by future more dense magnetic measurements.

A magnetotelluric depth sounding survey was carried out in the region around the geothermal research well Urach 3 (Berkthold et al. 1982; Berkthold and Kemmerle 1982). Shallow anomalies of high conductivity were related to water ducts near the surface, but in the deeper crust no electrical conductivities were encountered. In particular the Urach body of low seismic velocity and perhaps high density was not mapped as a region of anomalous conductivity.

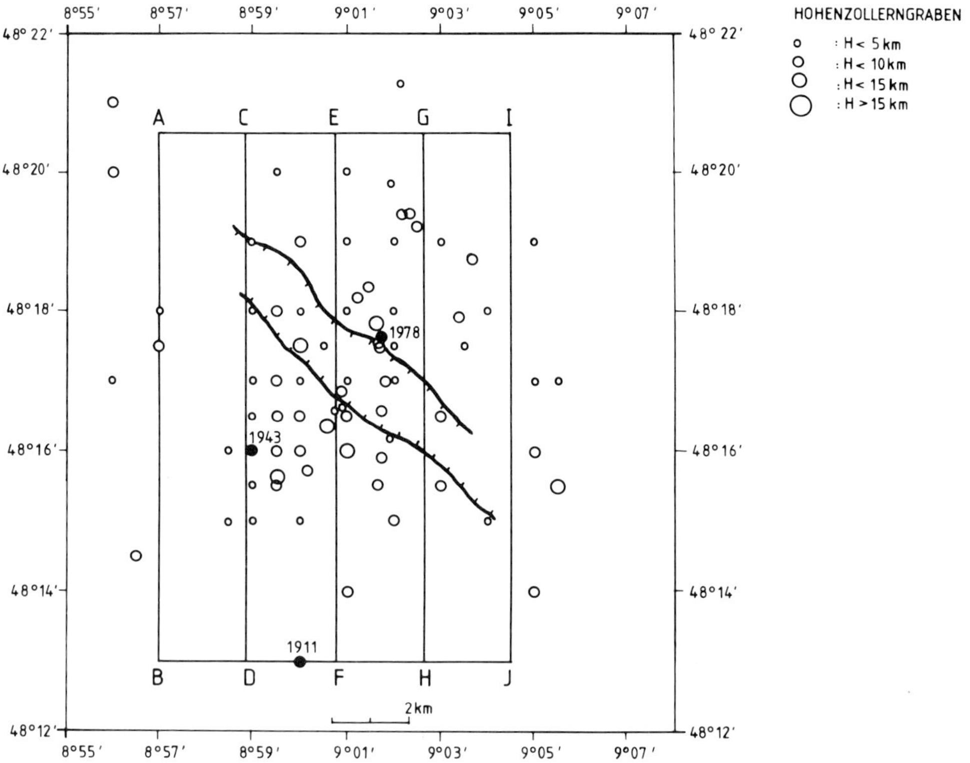


Fig. 12. Epicenters of the earthquakes in the Hohenzollern region for the period 1911-1978. The hypocenters have been projected into vertical sections through the NS lines AB, CD, etc. (see Fig. 11)

## Crustal Heterogeneities and Intraplate Seismicity

What is the mechanism that triggers the release of seismic energy at a particular place in a rather uniform crustal stress field and on a small element of a several hundreds of kilometers long bundle of shear lineaments? This is the key question for our understanding of the relatively strong seismicity occurring in the rather small, confined area at the HZGR. It is suggested that there is an excellent chance to answer this question by drilling a deep drill hole into the relatively well defined crustal structure of the HZGR region.

The strong lateral heterogeneities detected by deep seismic sounding demand a closer examination of the spatial distribution of hypocenters in the region of the HZGR. Figure 11 displays this distribution during the period 1911-1978. The position of the corresponding epicenters and the orientation of the N-S sections into which the hypocenters have been projected are shown in Figure 12. The hypocenters of the strong earthquakes of 1911, 1943 and 1978 are specially marked. In the northern part, the earthquakes are all less than 10 km deep. It can be seen that the envelope of the deepest events climbs steeply and with a slight curvature from south to north from focal depths of about 20 km to about 10 km. Although an apparent effect from the improvement of the network of seismic stations cannot be excluded we take this observation as support of our working hypothesis.

It is unlikely that this steep ascent is a purely geothermal effect. Although the increase of temperature towards the geothermal anomaly of Urach tends to move the brittle-ductile transition to shallower depth the isothermal surfaces at the border of the Urach anomaly will not rise as steeply as the envelope of the deepest hypocenters.

Therefore, it is suggested that in addition to the influence of temperature the position of the brittle-ductile transition is strongly influenced by the lateral changes in physical properties of the crust in the HZGR region. This can either be achieved by the ascent of the lower crust from SW to NE as in model A of Gajewski and Prodehl (1983) (see Fig. 4); and/or by the anomalously heated body of the Urach geothermal anomaly with a possibly strongly reduced viscosity due to thermal heating and the presence of fluid phases. As a consequence, the effective cross-section of the brittle part of the crust is thinned by a factor of about two approaching the HZGR from the south. This leads to a stress amplification (Kusznir and Bott 1977) which may elevate the stress above the fracture level, particularly in the region of the HZGR.

Intraplate seismicity linked to lateral crustal heterogeneities has been reported for other regions as well as the HZGR. A similar phenomenon can also be observed in the New Madrid rift zone (Zoback et al. 1980; Illies 1982; Mooney et al. 1983; Peters and Mooney 1983). In this region there appears to be a correlation between the trend of the epicenters and the shape of the 7.3 km $s^{-1}$ layer of the lower crust.

## Possible In-Hole Measurements

Although this paper is mainly concerned with developing the rationale for choosing the site of a possible deep drill hole in a seismically active area, it is also worthwhile mentioning briefly those in-hole measurements which would be of special importance for "seismic" holes and be the basis for their ultimate success. In-hole logging in "seismic" holes must include the depth distribution of physical, compositional and rheological properties that would be standard procedure for all deep drill holes regardless of their main target. However, special emphasis should be placed on the monitoring of temporal changes in physical fields, and physical and compositional properties. This requires continuous probing in the hole. The most important continuous in-situ experiments would include the temporal monitoring of changes with depth in: stress, hole deformation, seismic velocities, magnetic susceptibility, magnetization, gravity, electrical conductivity, and fluid and volatile phases. Of course, such measurements will also be performed in other bore holes with other targets, but for a "seismic" hole these observations must be continuous in order to detect changes with time. Continuous monitoring of the nearfield of the seismic sources by permanently installed broadband seismometers would also be a special feature of a "seismic" hole.

Continuous measurements in deep continental drill holes require new instrumental developments and also a careful setting of priorities in the selection of in-hole experiments since the number of continuous experiments that can be carried out simultaneously in one hole is certainly limited.

## Conclusions

During the reconnaissance stage of a pre-site survey for a possible location of a deep continental drill hole in a seismically active area, the currently available information on crustal structure, physical properties and hypocenter distribution has allowed the formulation of a working hypothesis on the triggering mechanism of earthquakes in a small zone at the HZGR. The presence of lateral crustal heterogeneities amplifies a regionally rather uniform stress field locally by a thinning of the brittle zone in the upper crust. The effect may be amplified by a local nearby thermal anomaly. This hypothesis can be tested in the future in several ways. A denser network of surface measurements will delineate the crustal structure with higher accuracy, and an improved more dense network of seismograph stations will improve the depth determination of local hypocenters. However, temperature extrapolations from heat flow measurements in shallow bore holes will always suffer from uncertainties increasing with depth. Thus it is impossible to distinguish between the influence of temperature and that of material properties on the location of the brittle-ductile transition. Even a high resolution of structure and of seismic velocities will remain ambiguous with regard to the corresponding petrological composition of the crustal heterogeneities. Numerical modelling without an adequate knowledge of physical, compositional and rheological parameters is not sufficiently constrained to distinguish between the two models of stress concentration in the HZGR.

The crux of the problem of the trigger mechanism of the HZGR intraplate seismicity is a better knowledge of physical, petrological and rheological properties "in vivo" at depth.

Therefore, the ultimate answer to the key question of the generation of intraplate seismicity – in the HZGR or any other zone of continental intraplate earthquakes – can only be obtained by in-situ information from a deep continental drill hole. It is hoped that the presentation of the rationale for a deep continental drill hole in the seismicity active area of the HZGR will serve as a useful case history for similar considerations in other parts of the world.

*Acknowledgements.* The presite survey of the Hohenzollerngraben region is a joint effort of geoscientists from all disciplines. A project study (Alfred Wegener Stiftung 1984, see vol. 2 for the location Hohenzollerngraben) was prepared by E. Althaus, F. Bauer, H. Baumann, H.-J. Bayer, P. Blümling, T. Farkas-Jandl, K. Fuchs (ed.), D. Gajewski, H. Langer, K. Lindner, H. Mälzer, J. Neuberg, C. Prodehl, C.-D. Reuther, G. Sattel, F. Scherbaum, R. Schick, G. Schneider, R. Stellrecht, F. Quiel, J. Walther, and K.W. Zippelt. This study received financial support from the German Research Society (Deutsche Forschungsgemeinschaft) within the KTB (Continental Deep Drilling Program) with funds made available by the BMFT (Federal Ministry for Research and Technology) and within the SFB 108 "Stress and Stress Release in the Lithosphere", a special research program at the University of Karlsruhe sponsored by the Deutsche Forschungsgemeinschaft as a German contribution to the International Lithosphere Program. M. Berry offered a number of critical comments in several aspects of this paper and kindly helped to improve the English. All this support is gratefully acknowledged. This paper is Contribution No. 76 of the SFB 108 of Karlsruhe University.

Henning Illies was the first to propose the Hohenzollerngraben as a possible site for a deep continental drill hole in a seismic active area (Illies 1982).

## References

Ahorner L, Murawski H, Schneider G (1972) Seismotektonische Traverse von der Nordsee bis zum Apennin. Geol Rundsch 61:915-942

Ahorner L, Schneider G (1974) Herdmechanismen von Erdbeben im Oberrhein-Graben und in seinen Randgebirgen. In: Illies H, Fuchs K (eds) Approaches to Taphrogenesis. Schweizerbart, Stuttgart, pp 104-117

Alfred-Wegener Stiftung (ed) (1984) Statusbericht des Kontinentalen Tiefbohrprogramms (KTB) der Bundesrepublik Deutschland, Bonn. 5 vol, ca 1400 p (January)

Babcock EA (1978) Measurement of subsurface fractures from dipmeter logs. Am Assoc Petrol Geol Bull 62:1111-1126

Bankwitz P, Bankwitz E, Frischbutter A (1979) Fototektonische Interpretation von Mitteleuropa nach Aufnahmen der sowjetischen Wettersatelliten Meteor 25 und 28. Akad Wissensch DDR. Zentralinst Phys Erde, Veröffentl 61

Bartelsen H, Lüschen E, Krey T, Meissner R, Schmoll H, Walter C (1982) The combined seismic reflection-refraction investigation of the Urach Geothermal Anomaly. In: Haenel R (ed) The Urach Geothermal Project. Schweizerbart, Stuttgart, pp 231-245

Baumann H (1984) Aufbau und Meßtechnik zweier Stationen zur Registrierung von Spannungsänderungen im Bereich des Hohenzollerngrabens - I. Resultate. Oberrhein Geol Abhandl 33 (in press)

Baumann H, Becker A (1983) Einrichtung einer Dauermeßstelle zur Registrierung von Spannungsänderungen im Hohenzollerngraben und Spannungsmessungen im südlichen Rheingraben. SFB-108 Berichtsband 1981-1983, Univ Karlsruhe, pp 23-40

Bell JS, Gough DI (1979) Northwest-southwest compressive stress in Alberta: Evidence from oil wells. Earth Planet Sci Lett 45:475-482

Berckhemer H, Hiller W (1960) Kurzperiodische Stationsseismographen mit Trägerfrequenzverstärker und mechanischer Registrierung. Z Geophys 26:1-8

Berkthold A, Haenel R, Wohlenberg J (1982) A model of the Urach geothermal anomaly as derived of geophysical investigations. In: Haenel R (ed) The Urach Geothermal Project. Schweizerbart, Stuttgart, pp 401-417

Berkthold A, Kemmerle K (1982) Distribution of electrical conductivity in the Urach geothermal area - a magnetotelluric and geomagnetic depth sounding investigation. In: Haenel R (ed) The Urach Geothermal Project. Schweizerbart, Stuttgart, pp 289-301

BRG (Bundesanstalt für Geowissenschaften und Rohstoffe) (1976) Hannover, Karte der Anomalien der Totalintensität des erdmagnetischen Feldes in der Bundesrepublik Deutschland 1:5000000

Bluemling P, Fuchs K, Schneider T (1983) Orientation of the stress field from breakouts in a crystalline well in a seismic active area. Phys Earth Planet Int 33:250-254

Brenner K (1961) Die Verlagerung der Bebenherde und des Herdmechanismus im Gebiet der Südwestalb seit dem Jahre 1911. Diplomarbeit, TH Stuttgart, 109 p

Brux C (1981) Vergleichende Untersuchung des Bruchtektonik im Bereich des Hohenzollern- und Lauchert-Graben (Schwäbische Alb) im Luftbild, Satellitenbild und Gelände. Dimplomarbeit, Geol Inst Univ Karlsruhe, 62 p

Deichmann N, Ansorge J (1983) Evidence for lamination in the lower continental crust beneath the Black Forest (Southwestern Germany) J Geophys 52:109-118

Eberle D (1973) Aeromagnetische Karte der Bundesrepublik Deutschland 1:10000000. Bundesanstalt für Bodenforschung Hannover

Farkas T (1973) Bestimmung der magnetischen Deklination und der Säkularänderung in Baden-Württemberg. Deutsche Geodätische Kommission, Reihe C, Nr 190

Fiedler G (1954) Die Erdbebentätigkeit in Südwestdeutschland in den Jahren 1800 bis 1960. Diss, TH Stuttgart, 152 p
Gajewski D, Prodehl C (1983) Die Struktur von Kruste und oberem Erdmantel im Bereich des Schäbischen Jura, abgeleitet aus seismischen Messungen. SFB-108 Berichtsband 1981-1983, Univ Karlsruhe, pp 177-208
Gajewski D, Prodehl C (1984) Crustal structure beneath the Swabian Jura, SW-Germany from seismic refraction investigation. (Submitted to J Geophys
Greiner G (1974) In-situ stress measurements - first results. In: Illies H, Fuchs K (eds) Approaches to Taphrogenesis. Schweizerbart, Stuttgart, pp 118-121
Greiner G (1978) Spannungen in der Erdkruste - Bestimmung und Interpretation am Beispiel von in-situ Messungen im süddeutschen Raum. Diss Univ Karlsruhe, 192 p
Gutenberg B (1915) Die mitteleuropäischen Beben vom 16. November 1911 und 20. Juli 1913. I. Bearb instrumentellen Aufzeichnungen. Veröffentl Zentralbl Int Seism Assoz, Strasbourg, 84 p
Haenel R (1980) Atlas of subsurface temperatures in the European community. Schäfer, Hannover, 79 p
Haenel R (1982) The Urach Geothermal Project. Schweizerbart, Zürich
Haessler H, Hoang-Trong R, Schick R, Schneider G, Strobach K (1980) The September 3, 1978, Swabian Jura earthquake. Tectonophysics 68: 1-14
Hiller W (1934) Eine Erdbebenwarte im Gebiet der Schwäbischen Alb. Z Geophys 9:229-234
Hiller W (1935) Erdbebenherde und Tektonik im Gebiet der Schwäbischen Alb. Z Geophys 11:15-19
Hiller W (1960) Die Erdbeben der Zollernalb. Amtliche Kreisbeschreibung des Landkreises Balingen, Bd 1. Staatl Landesamt Baden-Württemberg, Stuttgart, pp 1-40
Illies H (1974) Intra-Plattentektonik in Mitteleuropa und der Rheingraben. Oberrhein Geol Abh 23:1-24
Illies H (1982) Der Hohenzollerngraben und Intraplattenseismizität infolge Vergitterung lamellärer Scherung mit einer Riftstruktur. Oberrhein geol Abh 31:47-78
Illies H, Baumann H, Hoffers B (1981) Stress pattern and strain release in the Alpine foreland. In: Vykocil P, Green R, Mälzer H (eds) Recen Crustal Movements, 1979. Tectonophysics 71:157-172
Kovari K (ed) Field measurements in rock mechanics, vol 1 and 2. Balkema, Rotterdam, p 1026
Kronberg P (1977) Die regionale und überregionale Bruchtektonik Mitteleuropas. Bestandsaufnahme und Interpretation. Geotekt Forsch 53:7-41
Kusznir NJ, Bott MHP (1977) Stress concentration in the upper lithosphere caused by underlying visco-elastic creep. Tectonophysics 43: 247-256
Maelzer H (1983) Kriechende Spannungsumwandlungen: Rezente vertikale und horizontale Beobachtungen. SFB-108 Berichtsband 1981-1983, Univ Karlsruhe, pp 59-85
Makris J, Müller K, Tödt KH (1982) Gravity measurements at the geothermal anomaly Urach. In: Haenel R (ed) The Urach Geothermal Project. Schweizerbart, Stuttgart, pp 313-321
Meissner R, Strehlau J (1982) Limits of stresses in continental crusts and their relation to the depth-frequency distribution of shallow earthquakes. Tectonics 1:73-89
Meissner R, Lüschen E, Flueh ER (1983) Studies of the continental crust by near-vertical reflection methods; a review. Phys Earth Planet Int 31:363-376
Mooney WD, Andrews MC, Ginzburg A, Peters DA, Hamilton RM (1983) Crustal structure of the northern Mississippi Embayment and a comparison with other continental rift zones. Tectonophysics 94:327-348

Peters D, Mooney WD (1983) Tectonic evolution of the Mississippi Embayment. (in prep)
Prodehl C, Emter D, Jentsch M (1982) Seismic refraction studies of the geothermal area of Urach, Southwest Germany. In: Cermák V, Haenel R (eds) Geothermics and Geothermal Energy. Schweizerbart, Stuttgart, pp 277-283
Rummel F, Jung R (1975) Hydraulic fracturing stress measurements near the Hohenzollerngraben structure SW-Germany. PAGEOPH 113:321-330
Sattel G (1982) In-situ Bestimmung thermischer Gesteinsparameter aus ihrem Zusammenhang mit Kompressionswellengeschwindigkeit und Dichte. Diss, Univ Karlsruhe, 293 p
Scherbaum F (1980) Untersuchungen zur Struktur der P- und S-Phase im Epizentralgebiet. Diss, Univ Stuttgart, 76 p
Schick R (1968) Untersuchungen über die Bruchausdehnung und Bruchgeschwindigkeit bei Erdbeben mit kleinen Magnituden (M 4). Z Geophys 34:268-286
Schick R (1970) A method determining source parameters of small magnitude earthquakes. Z Geophys 36:205-224
Schmidt A v, Mack K (1912) Das süddeutsche Erdbeben vom 16. November 1911. Württemb. Jb Statist Landek, Jg 1912, 1. Heft, pp 96-139
Schneider G (1980) Das Beben vom 3. September 1978 auf der schwäbischen Alb als Ausdruck der seismotektonischen Beweglichkeit Südwestdeutschlands. Jahresber Mtt Oberrhein geol Ver NF 62:143-166
Schneider G, Schick R, Berckhemer H (1966) Fault-plane solutions of earthquakes in Baden-Württemberg. Z Geophys 32:383-393
Seibold E, Stahlecker R (1965) Kleintektonik und Albrand. Jahrb Geol Landesamt Baden-Württemberg, Freiburg 7:613-620
Sieberg A, Lais R (1925) Das mitteleuropäische Erdbeben vom 16. November 1911: Bearbeitung der makroseismischen Beobachtungen. Veröff Reichsanstalt Erdbebenforsch Jena, Heft 4, 106 p
Stellrecht R, Mölle KHR (1980) Bruchtektonisch-genetisches Modell des Hohenzollern- und Lauchert-Grabens, Südwestdeutschland. Jahresber Mitt Oberrhein Geol Ver NF 52:229-250. Schweizerbart, Stuttgart
Trappe H (1983) Eine Auswertung von Weitwinkelreflexionen auf dem Profil Urach und eine Korrelation zu den Ergebnissen der Steilwinkel-Reflexionsseismik. Diplomarbeit, Univ Kiel
Turnovksy J (1981) Herdmechanismus und Herdparameter der Erdbebenserie 1978 auf der Schwäbischen Alb. Diss, Univ Stuttgart, 109 p
Turnovsky J, Schneider G (1981) The seismotectonic character of the September 3, 1978. Swabian Jura earthquake series. Tectonophysics 83:151-162
Vidal H, Eder W, Fuchs K, Goerlic F, Illies H, Neumann J, Ziegler W (eds) (1981) Kontinentales Tiefbohrprogramm der Bundesrepublik Deutschland. Bold, Boppard, 70 p
Zoback MD, Hamilton RM, Grone AJ, Russ DP, McKeown FA, Brockman SR (1980) Recurrent intraplate tectonism in the New Madrid seismic zone. Science (Wash DC) 209(4460):971-976

# RATIONALE FOR SELECTING A SITE FOR AN ULTRA-DEEP DEDICATED SCIENTIFIC DRILL HOLE IN THE SOUTHERN APPALACHIANS

Robert D. Hatcher, Jr.[1], John K. Costain[2], Lynn Glover, III[2], Robert A. Phinney[3], Mark D. Zoback[4], Richard T. Williams[1], Pradeep Talwani[1], John B. Diebold[5], and Roger N. Anderson[5]

*False facts are highly injurious to the progress of science, for they often endure long; but false views, if supported by some evidence, do little harm, for everyone takes a salutary pleasure in proving their falseness.*

*Charles R. Darwin*
*The Descent of Man, 1859*

## Introduction

Most geologic studies of orogenic terranes are primarily two-dimensional, with some dependence upon down-plunge and down-dip projections of structures. Potential field and seismic reflection/refraction studies provide some information aobut the third dimension, but these data may be interpreted in a variety of ways without direct calibration using drilling. Drilling is therefore a necessary tool to provide the ultimate test of a scientific hypothesis regarding crustal evolution and structure. A hole may be drilled as an attempt to solve a problem or, more commonly, to prove existence of some potentially economic deposit. If we totally accepted the interpretations based upon structural cross-sections, seismic refraction and reflection studies, potential field studies, or studies which combine results from these and other techniques, it would never be necessary to drill except to recover economic minerals or hydrocarbons. However, the results of these types of analysis are generally not sufficiently precise to achieve to unique resolution of crustal structure and reduce the possibilities for alternative interpretations.

Drilling is a very expensive undertaking. Yet when weighed against the option of obtaining larger amounts of less precise data to test a hypothesis, or the existence of some economically important deposit, drilling must be considered the ultimate test of our knowledge. Borehole data are essential for extrapolations which are based upon all other geologic and geophysical data. Therefore, the opportunity for drilling a deep scientific hole to test a major scientific hypothesis and address major regional problems in geology and geophysics should

[1]Department of Geology, University of South Carolina, Columbia, SC 29208, USA
[2]Department of Geological Sciences, Virginia Polytechnic Institute and State University, Blacksburg, VA 24061, USA
[3]Department of Geological and Geophysical Sciences, Princeton University, Princeton, NJ 08540, USA
[4]Department of Geophysics, Stanford University, Stanford, CA 94305, USA
[5]Lamont-Doherty Geological Observatory, Palisades, NY 10964, USA

Observation of the Continental Crust through Drilling I (ed. by C.B. Raleigh)

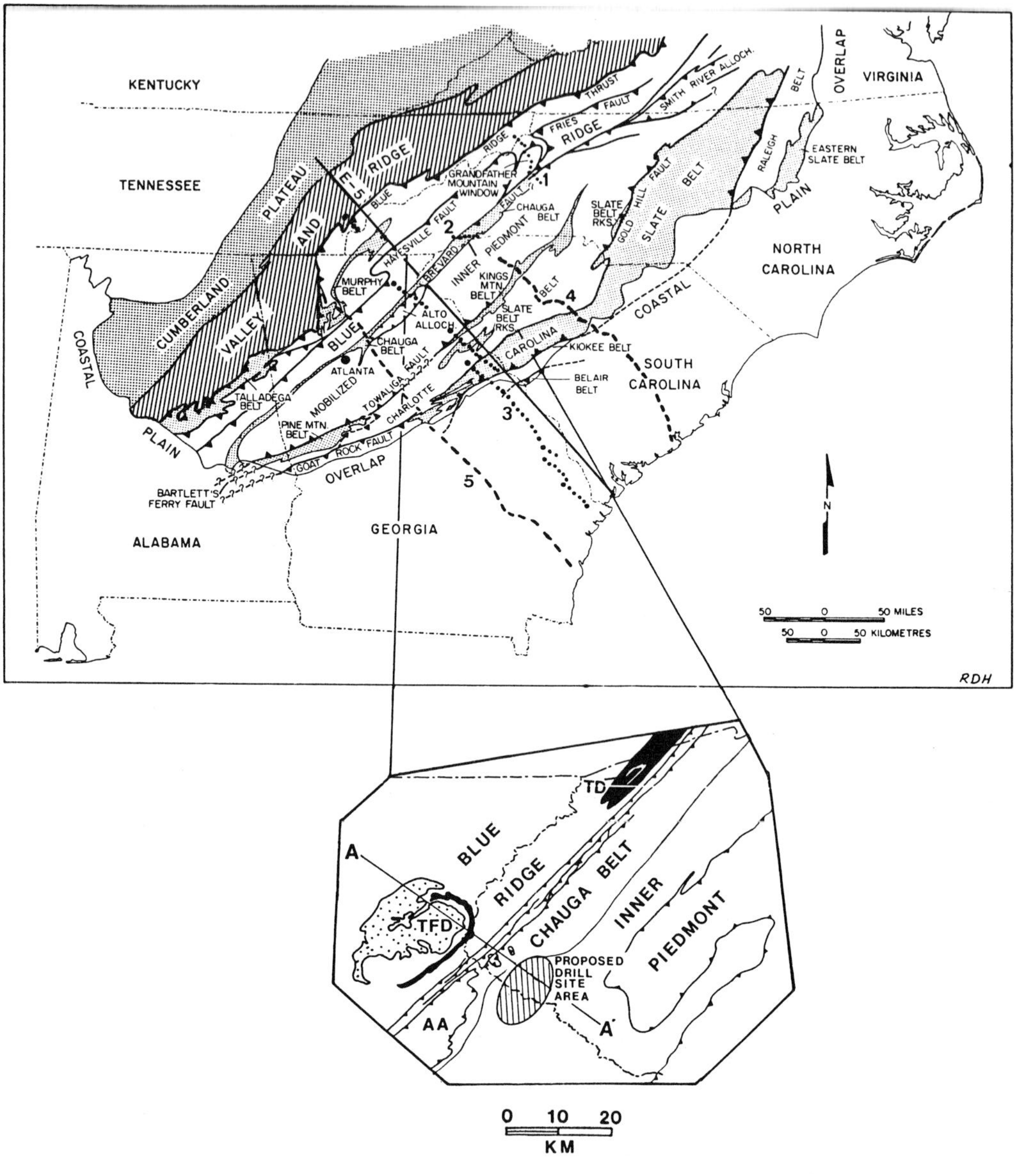


Fig. 1. Gereralized lithotectonic map of the southern Appalachians showing the major subdivisions and an inset with more details of the local geology in the vicinity of the proposed drill site area. *Solid line* labeled *E-5* on the smaller scale map is the location of the Geodynamic Transects Program E-5 section. *A-A'* in the *inset map* is an enlarged segment of the E-5 section in Figure 2. *Heavy dotted and dashed lines* are nearby seismic reflection lines in the crystalline Appalachians. Line *1* is the U.S. Geological Survey line through the Grandfather Mountain window. Line *2* is the Virginia Tech dynamite line. Line *3* is the COCORP VIBROSEIS line which is roughly coincident with a Seisdata, Inc. line. Lines *4* and *5* are the approximate locations of Seisdata, Inc. VIRBOSEIS lines. *TD* Toxaway dome; *TFD* Tallulah Falls dome. *Blacked-in areas* in the *inset* are Grenville basement rocks. *Stippled area* in the *inset* is the Tallulah Falls Quartzite

be regarded as an important and natural evolution in the study of the continental crust.

A major goal of the National Academy of Sciences Continental Scientific Drilling Committee (National Academy of Sciences 1979) has been for a number of years to initiate a program of deep drilling in the continental United States. The Continental Scientific Drilling Committee in November, 1983, endorsed the southern Appalachians as the prime target area for the first hole in this program (National Academy of Sciences 1984). Successful completion of this hole will prove an exciting undertaking as well as make significant contributions toward the advancement of science.

Our purpose here is to lay out the rationale for an interdisciplinary and collaborative geological/geophysical study of an area in the southern Appalachians which should lead to siting of a dedicated, approximately 10-km-deep scientific drill hole (Fig. 1). The ultimate objective would be to drill a hole which will penetrate several of the major thrust sheets which comprise the western part of the Appalachian orogen and pass into the North American basement beneath.

A deep scientific hole has been drilled to greater than 11.5 km to date in the Kola Peninsula in the Soviet Union. It was spudded in May, 1970, and is still being deepened (Rezanov 1981; Kozlovsky 1982). This hole is being drilled using turbine rather than rotary technology, and has helped to develop new drilling technology. Its scientific goal has been to provide information on the nature of mineralization and petrology of the rocks in this area. A similar deep hole for scientific purposes has never been drilled in the United States or in the western world, although the Federal Republic of Germany has initiated a major drilling program (Vidal and others 1981).

The proposed drill hole in the U.S. should provide additional information on the technology of drilling in crystalline rocks as well as a basis for the development of new drilling and logging techniques. Hopefully, innovations in drilling technology can be developed and implemented during the drilling of this hole.

It is appropriate that a project of this type be multidisciplinary in its goals and execution. The scientific and technological aspects of the project require involvement of investigators from several earth science and engineering disciplines. It is proposed herein that for the preliminary study several kinds of geologic and geophysical data be gathered on a collaborative basis to assist in the siting of the deep hole. These data will include surface geologic mapping and fracture studies, and subsurface data from a suite of shallow (300 m) boreholes, seismic reflection and refraction studies (using an industrial contract group for data acquisition), and potential field (gravity and magnetic) measurements. The main drilling phase of the project would involve an even broader spectrum of investigators in the sciences and engineering, from universities, government agencies and industry.

Several hypotheses have been advanced recently to focus the attention of the scientific community on the southern Appalachians. Principal among these is the hypothesis of large scale thin-skinned behavior of crystalline thrust sheets, a phenomenon actually observed over a century ago by Tornebohm in the Scandinavian Caledonides (Gee and Zacchrisson 1979) and in the Alps, where the Austroalpine thrust sheet gave rise to Oxburgh's (1972) concept of "flake tectonics". Confirmation of the nature of the Grandfather Mountain window Bryant and Reed (1970) provided a strong indication of the thin-skinned behavior of crystalline rocks in the southern Appalachians. Additional recognition of this phenomenon was made in interpretations by Hatcher (1971, 1972, 1978) and Hatcher and Zietz (1978, 1980). Seismic reflection data and interpretations by Clark and others (1978), Cook and others (1979, 1983), and Harris and Bayer (1979) have provided additional support for this interpretation of the crystalline southern Appalachians. Interestingly, the modern concept of thin-skinned tectonics was developed in the southern Appalachians by Rich (1934), but became embroiled in controversy (see for example, Rodgers 1949, 1964; Cooper 1961, 1964) until Bally and others (1966) settled the question for foreland fold and thrust belts using seismic reflection and drilling data for the Rocky Mountains of Alberta. Other structures in the southern Appalachians, such as the Brevard fault zone, have been the subject of lengthy controversies and have justly received the world-wide attention of the geologic community. Ironically, the Brevard fault zone ties into the thin-skinned crystalline thrust concept, because it appears to be stratigraphically controlled (Hatcher 1975, 1978).

The potential results of the main phase drilling project are severalfold: (1) They will provide a better understanding of the interiors of orogenic belts. (2) There should be a better understanding of the mechanisms of emplacement of large thrust sheets of crystalline rocks. (3) The crust beneath an orogenic terrane will be better understood. (4) The better understanding of a portion of the Appalachian orogen and confirmation or denial of the large scale transport of the Blue Ridge-Inner Piedmont thrust sheet above a less deformed and less metamorphosed basement will be addressed. (5) Solutions of more local problems will be addressed such as the angle of dip of the Brevard fault zone, whether or not thrusts and basement rocks of the Blue Ridge will be encountered in the hole, as well as the deformational processes attendent to these rocks and boundaries at depth will also be addressed. (6) A comprehensive program of downhole measurements will yield the first-ever mid-crustal measurements of the in situ stress field and the physical properties of mid-crustal rocks in place, the nature of deep-seated fractures, faults, and other discontinuities and the compositions of pore fluids.

A dedicated deep scientific drill hole must satisfy a number of scientific needs according to rigorously defendable criteria. In addition to being a multidisciplinary effort, a project of this magnitude must be carefully designed to maximize scientific gains, in the sense that interpretations based on all preceding data gathered by other techniques are thoroughly tested by the use of the drill. It is likely that regardless of how well the hole is designed and the program carried out, there will be surprises and unanticipated gains. These may account for the greatest scientific contributions of this and any subsequent drilling project.

## Criteria for Site Selection

We believe the process of site selection of a deep scientific drill hole should be rigorous and consider the maximum number of options before the site is selected for drilling, because of the large investment in both time and money that will be necessary to complete this project. Below is a list of equally ranked primary criteria for siting any deep drill hole.

1. The problem to be addressed can be solved most effectively by drilling.

2. The drill hole should address problems on several levels and from several viewpoints:

| | |
|---|---|
| Fundamental Principles<br>Regional Problems<br>Local Problems | Most or all principles and problems addressed should be multidisciplinary |

3. There should be intermediate level objectives in the hole so that, if the hole is lost, a large portion of the overall objectives of the study may be attained.
4. Despite the fact that additional data must be gathered before ultimate site selection can be made, those regions where an abundance of high quality geologic and geophysical data exist, but are insufficient to solve outstanding problems, should probably be considered first. Conversely, the best-studied areas may define the most fundamental outstanding problems.
5. The target bottom hole objective should have a suitable geometry so that the probability of hitting the target is maximized.

We believe the proposed deep hole in the southern Appalachians meets all of the above criteria. The application of these criteria is outlined in the following section.

## Potential Gains and Goals of the Study

We believe that the results of drilling at this site have the potential to shed light on major problems in geology and geophysics in the southern Appalachians that will have an impact on our understanding of the architecture of orogenic belts throughout the world. This proposed deep hole has goals at both intermediate and great depths. The deep drill hole, once completed, will encounter several levels of lithotectonic units and rock types. Opportunities will be present to investigate faulting processes, relationships of metamorphism to depth and overthrusting of large sheets, to more local questions related to the occurrence of transported Blue Ridge basement rocks at depth within different thrust sheets, as well as the nature and character of the sedimentary rocks encountered in the down-dip projection of the Brevard fault zone as well as the block above the basement. Seismic reflection character in crystalline rocks will be better calibrated, and the determination of regional seismic velocities will be improved. The nature of the crust in the area will be better understood from the direct study of wide angle reflections and refractions. A shallow corehole program, which is part of the preliminary study, will provide baseline data for in situ stress and heat production/heat flow, as well as a suite of near surface samples for rock physics measurements in the different lithologies in the area. These holes will permit detailed in situ investigation of the rocks to be encountered in the deep hole (Fig. 2). This will be important in predicting drilling rates, potential problems which might be encountered in the deep hole, as well as providing early insight into the physical properties of the rocks which will be drilled initially.

## Scientific Research Objectives

The main research objective of the preliminary site area investigations are: (1) to gather the necessary geophysical and geologic data

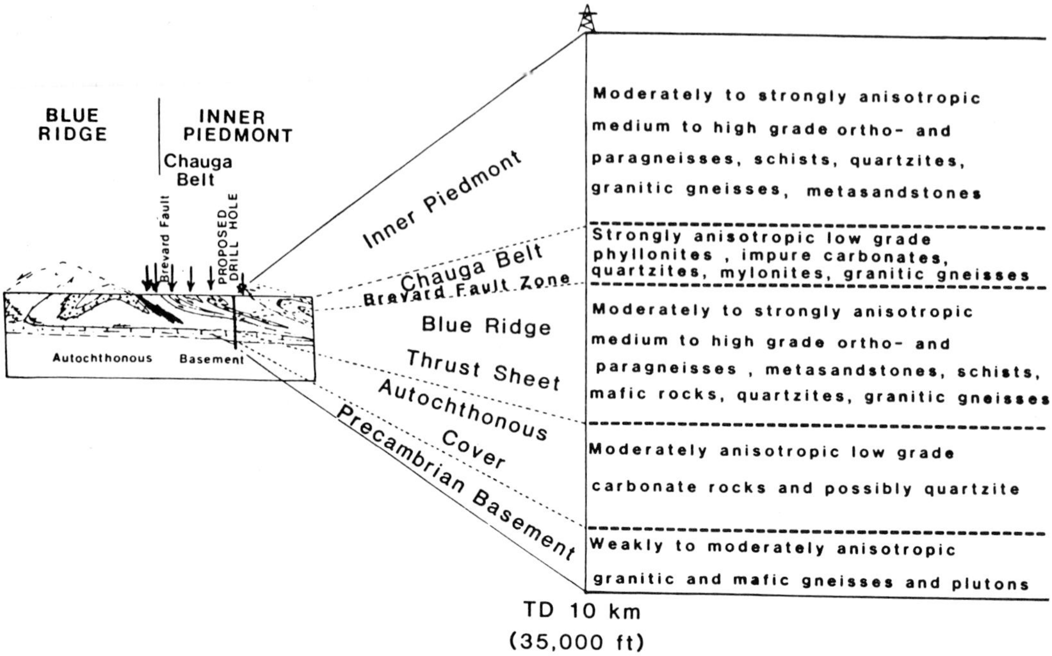


Fig. 2. Enlarged portion of Transect Section E-5 showing the possible tectonic units and generalized lithologies likely to be encountered if the deep hole were drilled directly on the line. *Striped area* is the Blue Ridge thrust sheet containing a highly deformed and transported mass of basement *(black)*. *Arrows* above surface indicate the approximate locations of six proposed shallow boreholes projected into the line of section. The entire section is 55 km in length. There is no vertical exaggeration. Location of this section is shown as A-A' in Figure 1

to site the Appalachian deep hole, (2) to anticipate physical conditions that may be encountered during the actual drilling (Fig. 2), and (3) to provide a geologic and geophysical framework for interpretation and extrapolation of data from the deep hole. Surface geologic studies will provide data on rock types and geometry of rock units that will probably be encountered in the deep hole. Results from fracture studies and studies of local geology in the vicinity of shallow boreholes will aid in interpretation of shallow borehole geophysical and petrologic data.

A detailed knowledge of seismic velocity structure is essential to spatially position seismic reflectors in the crust. The velocity structure will be obtained from the reflection stacking velocity, expanding spread profiles, in situ measurements in the shallow boreholes, and laboratory measurements on samples from the shallow boreholes. Combined with conventional and high resolution seismic profiles, this will permit detailed modelling of the subsurface geometry.

The surface geologic, seismic, and potential field data will be interpreted together in terms of the subsurface geologic structure and lithology. This local interpretation will then be examined in a regional context. All new data gathered, together with existing data, will be utilized by all investigators in the preliminary study to site the deep hole.

Our final objective is to provide data which will aid in anticipating problems which may arise during drilling of the deep borehole. These data will include subsurface lithology (from surface mapping and subsurface interpretation), in situ stress distribution, physical properties of core, and thermal measurements.

## Research Plan

Pertinent data necessary for siting the hole require the collection of a small amount of additional surface geologic data. Additionally, about 250 km of conventional seismic reflection data, high resolution data, and specially designed expanding velocity spreads will be gathered. The interpretation of seismic reflection data will be supplemented by detailed gravity measurements primarily along the route of the seismic reflection lines. A preliminary program of shallow drilling (4 holes to 300 m) will be carried out to determine terrestrial heat flow, in situ stress values and orientations. Seismic velocity will be measured in the holes and on core samples at different pressure conditions. All these data will have a direct bearing upon siting and potential drilling problems.

### 1. Surface Geology

Most of the proposed drill site area has been mapped in detail during the past 18 years. However, there are several areas which deserve additional attention. Additional mapping of about 350 $km^2$ (100 $mi^2$) in the Blue Ridge and Inner Piedmont in Georgia and South Carolina will provide accurate geometric control on configuration of surface units for down-dip projection to possible intersection by the proposed deep drill hole.

All the shallow holes are likely to be drilled in the area already mapped in detail. However, additional field studies of shallow borehole sites and the local site area for the deep hole will be made in greater detail than in the past to distances of up to 5 km from each site for accurate determination of surface geologic/geomorphic/engineering/environmental characteristics of all prospective borehole locations prior to selection of a final site and initiation of drilling activities.

### 2. Seismic Reflection Studies

While providing confirmation of the regional decollement, the COCORP line in this area does not resolve much detail in the upper plate. In consequence, the reflection lines proposed here are aimed at optimal resolution of the 10-11 km depth range appropriate to a deep hole (Fig. 3). A major objective during the site selection phase will be to optimize the site selection with respect to the number and quality of reflectors and correlations with surface geologic control. Selection of a deep drilling site that does not penetrate reflectors of regional significance restricts the potential contributions of a 10-11 km hole. The correlation of surface geologic control and subsurface reflector geometry is, therefore, a major condition for the selection of an acceptable deep drill site. Seismic reflection data acquired during the preliminary phase will be used to evaluate the potential for correlation with surface geology exposed in the Piedmont and Blue Ridge northwest of the drill site area with the tectonic stratigraphy south-

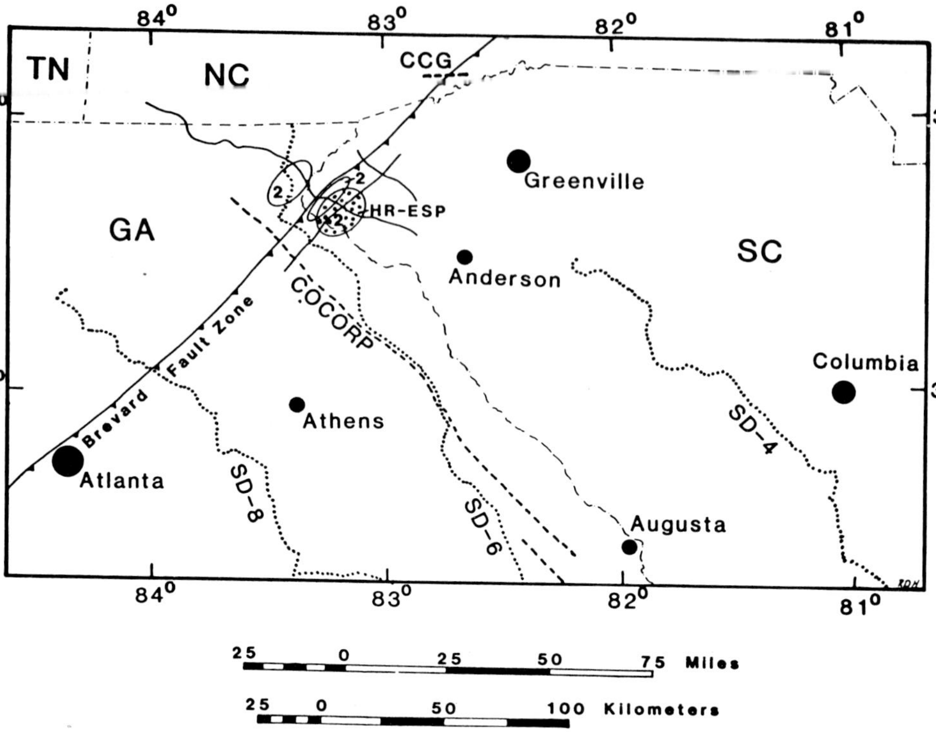


Fig. 3. Map showing locations of proposed seismic reflection lines *(solid)* relative to existing data *(dashed* and *dotted lines). Dotted lines* labeled 5D-4, SD-6, SD-8 are the approximate locations of Seisdata, Inc. lines. The COCORP line *(dashed)* is almost coincident with SD-6. The area labeled HR-ESP *(stippled)* is the site study area where high resolution an expanding spread seismic profiling will be carried out. *Dashed line labeled CCG* is the Virginia Tech dynamite line. The ellipses are areas where the preliminary shallow boreholes for in situ stress and heat flow measurement will be drilled. *Numbers inside the ellipses* indicate the number of holes to be drilled in each area

east of the proposed drill site. An equally necessary condition is that the site be selected with the expectation of solving major problem(s) in geology and geophysics in the southern Appalachians that will have an impact on our understanding of the architecture of orogenic belts elsewhere. We propose a siting strategy using data from surface geology, reflection seismology, in-situ stress, and potential field data to ensure optimum placement of the deep drill site in a regional context.

The existing seismic reflection data base is inadequate to ensure continuity between available seismic lines. Seismic reflection data in the site area of the deep hole will delineate reflector geometry to: (1) anticipate geologic conditions that might affect drilling progress; (2) obtain velocity information from the surface to 15 km to complement correlations based upon geologic projections from the topographic surface; (3) aid in the determination of depths to major changes in rock properties; and (4) refine the selection of the deep drill site by locating it over critical reflectors.

In addition to conventional seismic reflection profiling, we propose two additional experiments. To obtain a detailed velocity structure, it would be desirable to conduct a program of ESP's (Expanding Spread Profiles). A short high resolution reflection profile is planned for the vicinity of the planned drill site to develop detailed knowledge of the structure there. The results from these studies are particularly needed to understand and interpret data obtained from the deep hole – particularly for rock physics and compositional data.

## 3. Shallow Borehole Program

Six shallow coreholes will be drilled in an area of about 250 $km^2$ centered about the probable location of the proposed deep hole in South Carolina and Georgia. The following measurements should be made in boreholes: in situ seismic velocity, in situ stress, the nature and distribution of fractures, hydraulic conductivity, heat flow, and heat production. Additionally, a variety of petrophysical, paleomagnetic, and petrologic measurements will be made on core samples. These holes will be kept open and available for subsequent geophysical and seismological investigations. As shown in Figure 2, one hole will be drilled to the northwest of the deep drill site in basement orthogneiss and metasedimentary rocks northwest of the Brevard fault zone. Southeast of the Brevard fault zone, two holes will be drilled in the Chauga belt, one in granitic rocks and one in metasedimentary rocks. The four tectonic units being sampled by the shallow holes will all possibly be encountered in the deep drill hole.

The data from the shallow holes will directly benefit the deep drill hole in several ways: they will aid calibration of geophysical logs in the deep hole, this will provide a more regional context for measurements made in the deep hole, the seismic velocity data will help calibrate seismic reflections and refracted data, and it will aid interpretations of physical property measurements made in the deep hole since similar measurements will be made in the same rocks only at lower temperatures and stresses.

## 4. Heat Flow Measurement

Heat flow measurements will be made to determine if there will be difficulties in drilling produced as a function of thermal properties and will provide for definition of the heat flux near the deep hole site. The sites will be planned so that they also satisfy the requirements for defining both the heat flow and state of crustal stress in the region of the deep hole. If the shallow and deep hole site are properly chosen, the heat flow data may help resolve conflicting hypotheses put forth to explain the linear relation between heat flow and heat generation that has been observed in crystalline rocks.

## 5. In Situ Stress

Although the proposed deep drill hole will provide the first opportunity ever to make mid-crustal stress measurements, there are few reliable in situ stress data of any type in this region, even at shallow depths (Fig. 4). The magnitude and orientation of stresses will be obtained as a function of depth in the six shallow holes. In addition to helping provide data on stress magnitude and orientation in the site area, when combined with stresses observed in the deep hole, they will provide the most detailed three dimensional knowledge of the

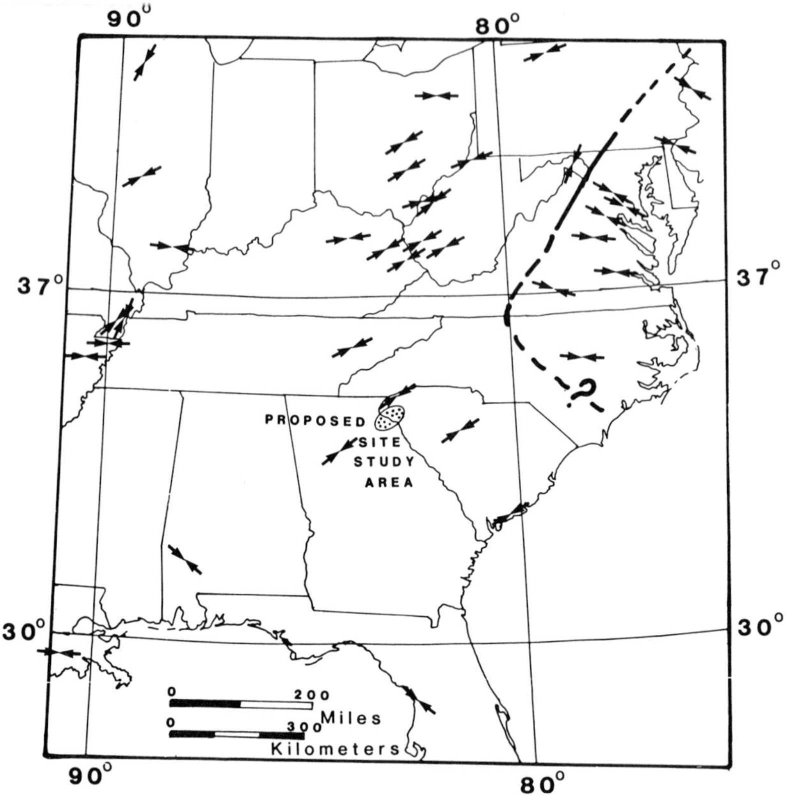


Fig. 4. Available data on the directions of the maximum horizontal compression in the southeastern United States. Data came from earthquake focal mechanisms, hydraulic fracturing, stress-induced wellbore breakouts, and geologic fault offsets

stress field anywhere. These shallow data will also help make judgments about stresses to be anticipated during drilling of the deep hole.

The stress field will be studied in these holes using a combination of borehole televiewer logging and hydraulic fracturing. The borehole televiewer is a well-logging tool principally used for studying the location and orientation of subsurface fractures. Recent work has shown that it can also be used to detect stress-induced wellbore breakouts which indicate the orientation of the least horizontal principal stress.

Hydraulic fracturing stress measurements will be made in the wells to determine both the magnitude and orientation of in situ stresses.

The opportunity to study fractures and joints at depth in the shallow wells will greatly help to improve our understanding of fracture and joint development in this region and anticipate conditions in the deep hole. Independent in situ stress data will be available in each well, and the variety and nature of fracture filling(s) or lack of them, along with the variety of types of rocks being drilled will permit detailed study of the chronology of fracturing.

Borehole televiewer logging will be used to study the attitudes and distribution of fractures. This will be supplemented by detailed local surface joint mapping and studies on core samples.

Very few seismic velocity and other geophysical measurements have been made in wells drilled in igneous and metamorphic rocks. A suite of P- and S-wave seismic velocity measurements will be made in the shallow wells to study the effect of depth on seismic velocity of various rock types. Laboratory velocity measurements on core samples will also supplement these studies and indicate how to extapolate the shallow-hole measurements to greater depth. These data will be used to calibrate the seismic profiles and tie together the deep drill hole with the regional seismic coverage. The site of geophysical measurments in the shallow holes (an on core samples from the holes) will be extremely important for interpretations of geophysical logs in the deep hole where core recovery will be limited.

### 6. Gravity Measurements

To provide additional constraints to the interpretation of seismic reflection data, up to 1500 new gravity stations will be obtained along seismic survey lines, at a station spacing of about 0.5 km, and elsewhere in the study area as needed to supplement extisting regional data, so that detailed coverage of the study area is obtained. Two- and three-dimensional interpretation of these data will aid in the definition of the sub-crustal structure and the selection of the drill site.

## Communication and Involvement of the Scientific Community

We regard the prospect of drilling a deep scientific hole in the southern Appalachians as a very exciting undertaking. It is to be a multidisciplinary project that addresses fundamental as well as more regional problems. We hope that there will be extensive involvement of others in the scientific community and industry in the preliminary, main and post-drilling phases of the study.

## References

Bally AW, Gordy PL, Stewart GA (1966) Structure, seismic data, and orogenic evolution of southern Canadian Rocky Mountains. Bull Can Petrol Geol 14:337-381

Bryant B, Reed JC Jr (1970) Geology of the Grandfather Mountain window and vicinity, North Carolina and Tennessee. US Geol Surv Prof Pap 614, 190 p

Clark HB, Constain JK, Glover L III (1978) Structural and seismic reflection studies of the Brevard ductile deformation zone near Rosman, North Carolina. Am J Sci 278:419-441

Cook FA, Albaugh DS, Brown LD, Kaufman S, Oliver JE, Hatcher RD Jr (1979) Thin-skinned tectonics in the crustalline southern Appalachians; COCORP seismic reflection traverse across the southern Appalachians. Am Assoc Petrol Geolog Stud Geol 14, 61 p

Cook FA, Brown LD, Kautmann S, Oliver JE (1983) The COOCORP seismic reflection traverse across the southern Appalachians. Am Assoc Petrol Geolog Stud Geol 14, 61 p

Cooper BN (1961) Grand Appalachian field excursion. Va Engr Exper Sta Exten Ser Geol Guidebook 1, 187 p
Cooper BN (1964) Relations of stratigraphy to structure in the southern Appalachians. In: Lowry WD (ed) Tectonics of the southern Appalachians. Dept Geol Sci, Virginia Tech Memoir 1:81-114
Gee DG, Zachrisson E (1979) The Caledonides in Sweden. Sveriges Geol Undersokning, Ser CNR 769, 49 p
Harris LD, Bayer KC (1979) Sequential development of the Appalachian orogen above a master decollement: a hypothesis. Geology (Boulder) 7:568-572
Hatcher RD Jr (1971) Structural, petrologic and stratigraphic evidence favoring a thrust solution to the Brevard problem. Am J Sci 270: 177-202
Hatcher RD Jr (1972) Developmental model for the southern Appalachians. Geol Soc Am Bull 83:2735-2760
Hatcher RD Jr (1975) The second Penrose field conference: the Brevard zone. Geology (Boulder) 3:149-152
Hatcher RD Jr (1978) Tectonics of the western Piedmont and Blue Ridge, southern Appalachians: review and speculation. Am J Sci 278:276-304
Hatcher RD Jr, Zietz I (1978) Thin crystalline thrust sheets in the southern Appalachian Inner Piedmont and Blue Ridge: interpretation based upon regional aeromagnetic data. Geol Soc Am abstracts with programs 10:417
Hatcher RD Jr, Zietz I (1980) Tectonic implications of regional aeromagnetic and gravity data from the southern Appalachians. In: Wones DR (ed) The Caledonides in the USA. Virgina Polytech Inst Geol Sci Memoir 2:235-244
Kozlovsky YA (1982) Kola super-deep: interim results and prospects. Episodes 1982:9-11
National Academy of Sciences (1979) Continental Scientific Drilling Program, US Geodynamics Committee, Geophysics Research Borad, National Research Council, 192 p
Oxburgh ER (1972) Flake tectonics and continental collision. Nature (London) 239:202-209
Rezanov IA (1981) Ultra-deep drilling. Nauka (Science) Moscow, US Dept Energy Trans 252, 169 p
Rich JL (1934) Mechanics of low-angle overthrust faulting as illustrated by Cumerland thrust block, Virginia, Kentucky and Tennessee. Am Assoc Petrol Geol Bull 18:1584-1596
Rodgers J (1949) Evolution of thought on structure of middle and southern Appalachians. Am Assoc Petrol Geol 33:1643-1654
Rodgers J (1964) Basement and no-basement hypotheses in the Jura and the Appalachian Valley and Ridge. In: Lowry WD (ed) Tectonics of the southern Appalachians. Virginia Polytech Inst Dept Geol Sci Memoir 1:71-80
Vidal H, Eder W, Fuchs K, Goerlich F, Illies H, Neumann J, Ziegler W (1981) Continental deep drilling program of the Federal Republic of Germany. Comiss Geosci Coop Res Information XI, US Dept Energy Trans 251, 69 p

# DISCOVERY OF TWO PROTEROZOIC GRANITE-RHYOLITE TERRANES IN THE BURIED MIDCONTINENT BASEMENT: THE CASE FOR SHALLOW DRILL HOLES

M.E. Bickford and W.R. Van Schmus[1]

## Introduction

In this paper we will use a case history, the discovery of two distinctive Middle Proterozoic granite-rhyolite terranes in the buried midcontinent basement, to make a strong case for the value of shallow drill holes. As we shall show, the terranes described record an important event in the evolution of the North American continent but would not have been recognized without the study of material from drill holes. There are still important unanswered questions about the extent, nature, and tectonic setting of these terranes that could be answered with additional samples from carefully selected holes drilled for basic scientific purposes.

## Granite Rhyolite Terranes in the Buried Midcontinent

The presence of high-silica rhyolite, dacite, and related epizonal granitic plutons in the buried crystalline crust of the southern and eastern midcontinent region was first recognized by Muehlberger and others (1966) and by Lidiak and others (1966), who studied cuttings and cores from wells to basement that were available then. These siliceous, commonly pyroclastic, volcanic rocks, and their associated shallow granitic plutons are known from only a few exposures. The largest of these is the St. Francois Mountains region of southeastern Missouri, but other exposures occur in northeastern Oklahoma and in the eastern Arbuckle Mountains of southeastern Oklahoma.

Muehlberger, Lidiak, and their co-workers correctly identified the extent of these rocks (Fig. 1), showing that they stretched from northern Ohio at least to the Texas Panhandle. Their geochronological studies, however, mostly yielded ages that were too low because they were done by Rb-Sr and K-Ar methods on single minerals and single whole-rock samples, because the rocks studied mostly have very low total Sr abundances, and because most have been affected by pre-Paleozoic weathering and alteration.

We began studying rocks of this region in 1968 with geochronological and petrological investigations in the St. Francois Mountains region of southeastern Missouri, an area in which about 900 $km^2$ is underlain by exposed rhyolite, rhyolitic ash-flow tuff, dacite, and epizonal granite rocks. Subsequent studies have shown that the St. Francois Mountains exposes a typical volcano-plutonic terrane in which primary

[1]Department of Geology, University of Kansas, Lawrence, KS 66045, USA

Observation of the Continental Crust through Drilling I (ed. by C.B. Raleigh)

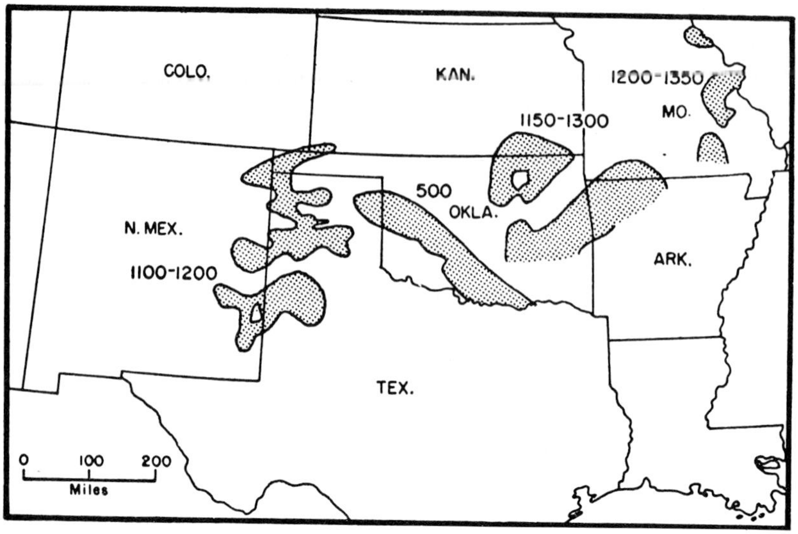


Fig. 1. Map showing extent of rhyolite in outcrop and subsurface in a portion of the southern midcontinent. (After Muehlberger et al. 1966)

magmas have evolved to yield derivatives with increasingly greater silica and incompatible element contents (Bickford et al. 1981a; Cullers et al. 1981). These magmas erupted from numerous cauldrons to form extensive pyroclastic sheets and caldera-filling units and minor rhyolitic to dacitic flows (R.E. Anderson 1962, 1970; Berry and Bickford 1972; Shuster 1978; Nusbaum 1980; Sides et al. 1981).

Extensive Rb-Sr age determinations on these rocks (Odom 1969; J.E. Anderson et al. 1969; Mose 1971) yielded ages of 1320 to 1408 m.y. but the isochrons showed considerable scatter and uncertainties in the ages were large. Subsequent U-Pb age determinations done on zircons separated from samples of four of the granitic plutons and one ash-flow tuff yielded ages of 1480 m.y., but a fifth pluton, the Munger Granite Porphyry (Tolman and Robertson 1969) yielded a zircon age of 1380 m.y. (Bickford and Mose 1975). At the time, 1380 m.y. seemed an anomalously low age for rocks from this region; Bickford and Mose (1975) concluded that most of the rocks were formed about 1480 m.y. ago and that the Rb-Sr system was disturbed.

As our understanding of the St. Francois Mountains grew it became increasingly clear that the terrane could not be understood in isolation. We needed to know the extent, age, and chemical characteristics of rocks of this type in the subsurface. Thus, we began collecting samples of buried basement rocks from the midcontinent region, obtaining cuttings and, occasionally, cores from the libraries of state geological surveys, oil and gas companies, and mineral exploration companies. In many cases we obtained enough material to separate zircons for age determination, and in most cases we were able to get useful petrographic data. In some instances, when core or very clean cuttings were obtained, we analyzed the samples for major elements. The results of this work have been published in papers beginning in 1979 (Bickford and Lewis 1979) and continuing to the present (Bickford et al. 1981b; Van Schmus and Bickford 1981; Hoppe et al. 1983).

Fig. 2. Photomicrograph of basement sample from Greenwood County, Kansas showing granophyric texture that is typical of many epizonal granite plutons in the subsurface of the southern midcontinent

It is now clear that large volumes of rhyolite, dacite, and related shallow granitic plutons underlie much of the midcontinent area; the terrane extends northeast to southwest for about 2000 km and is about 200 km wide. If these rocks formed a layer only one km thick, they would still constitute $4 \times 10^5$ $km^3$ of quite silicic material, a major addition to, or reworking of, the continental crust. Igneous rocks other than granite and rhyolite are quite rare. Moreover, sedimentary or metasedimentary rocks are rare and there is no evidence of compressional deformation. Granitic plutons are typically epizonal and many show granophyric or micrographic texture (Fig. 2). In many samples of rhyolite the groundmass is recrystallized to fine granular quartz and feldspar but in some the rocks are remarkably well preserved and the structure of shards is clearly visible in thin section (Fig. 3).

When it first became clear that granite and rhyolite-dacite made up the basement of such a large region, we envisioned a great belt of 1480 m.y. old rocks extending across the continent (Van Schmus and Bickford 1981). Further, plutons of this age are known to intrude older rocks to the north; a major example is the Wolf River batholith of central Wisconsin (Van Schmus et al. 1975; Anderson and Cullers 1978). However, as additional samples became available from Missouri, southern Kansas, and Oklahoma, that simple model began to change. The Spavinaw granite, which crops out in northeastern Oklahoma, yielded a zircon age of 1370 m.y., and rocks from exposures in the eastern Arbuckle Mountains of southeastern Oklahoma yielded ages of 1370 to 1400 m.y. (Bickford and Lewis 1979).

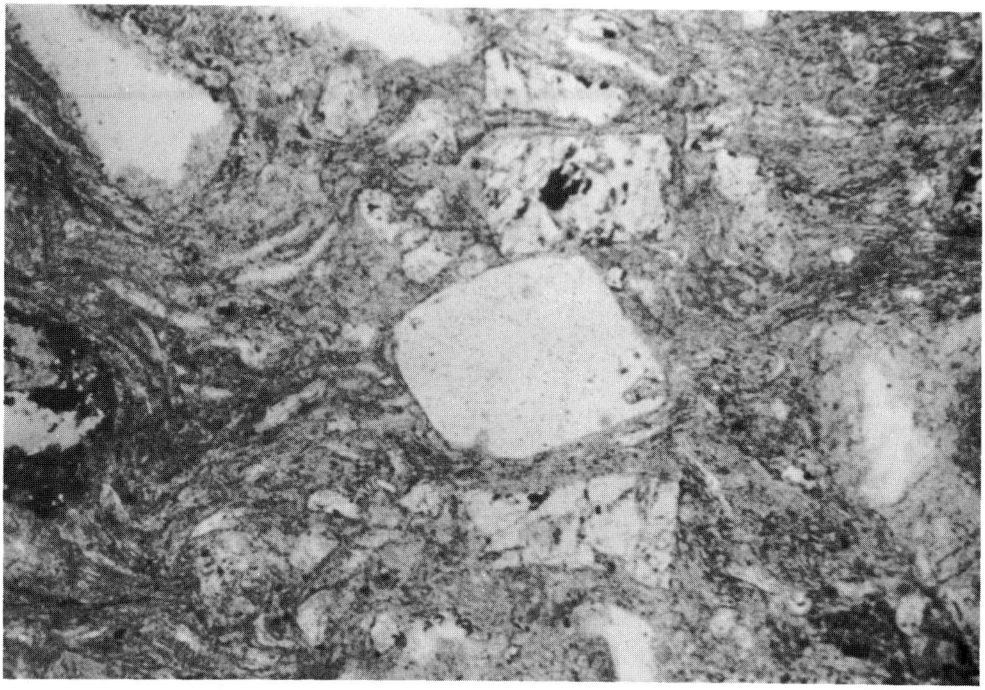

Fig. 3. Photomicrograph of rhyolitic ash-flow tuff sample from subsurface in Stafford County, Kansas showing excellent preservation of primary pyroclastic textures and structures

The magnetic anomaly map of Kansas (Fig. 4) reveals a number of small, sharp, circular positive anomalies. With D.W. Steeples, a colleague at the Kansas Geological Survey, we obtained funding from the Department of Energy and Los Alamos National Laboratory to do "add-on" drilling to deepen wells over two of these anomalies in eastern Kansas (sites 1 and 2, Fig. 4) and to core them. The rocks encountered are granite with about 2 percent magnetite. Zircons from these samples yielded ages of 1339 and 1361 m.y. (Figs. 5 and 6); a similar age was obtained from core from western Missouri, also obtained from drilling over a sharp positive magnetic anomaly (Steeples and Bickford 1981; Bickford et al. 1981b).

We obtained still more basement material from the Oklahoma Geological Survey for much of eastern Oklahoma and from the Kansas Geological Survey for southern Kansas. In addition, Prof. W.M. Muehlberger of the University of Texas, Austin, sent us core from basement wells in the Texas Panhandle. Analysis of these samples showed that rocks 1340-1400 m.y. old make up the basement of southwestern Missouri, most of Oklahoma, southern Kansas, and Panhandle Texas. The Munger Granite Porphyry of the St. Francois Mountains of Missouri, which intrudes volcanic rocks evidently belonging an older, 1470 m.y. old, terrane is thus currently the most easterly example of the 1340-1400 m.y. old rocks. Still more recently, work on the San Isabel batholith of the southern Wet Mountains of Colorado (with R.L. Cullers of Kansas State University) showed that this large, mesozonal granitic intrusive crystallized 1360 m.y. ago. It intruded migmatitic metasedimentary rocks

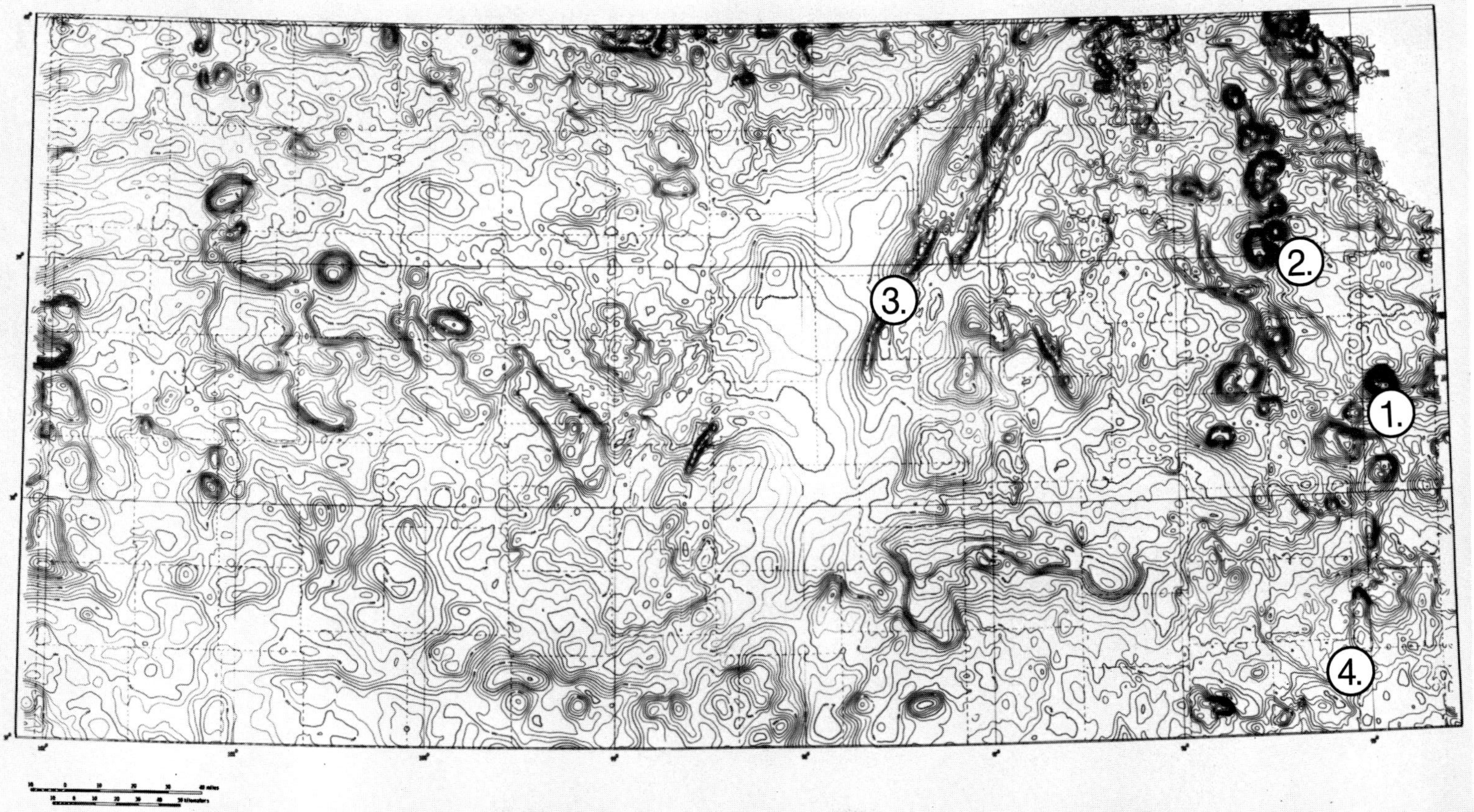


Fig. 4. Aeromagnetic map of Kansas with 50 $\gamma$ Contour interval. Map compiled by H. Yarger et al., Kansas Geological Survey. This figure, after Steeples and Bickford (1981) shows the *locations of four proposed "add-on" drilling sites*

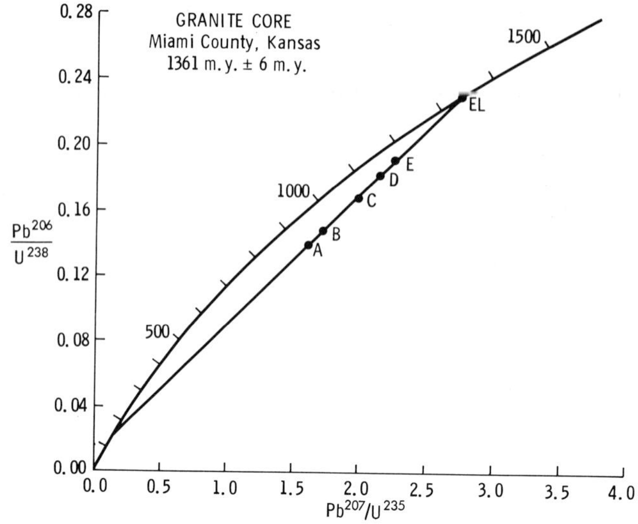


Fig. 5. Concordia plot for zircons from subsurface granite, Miami County, Kansas (site 1, Fig. 4)

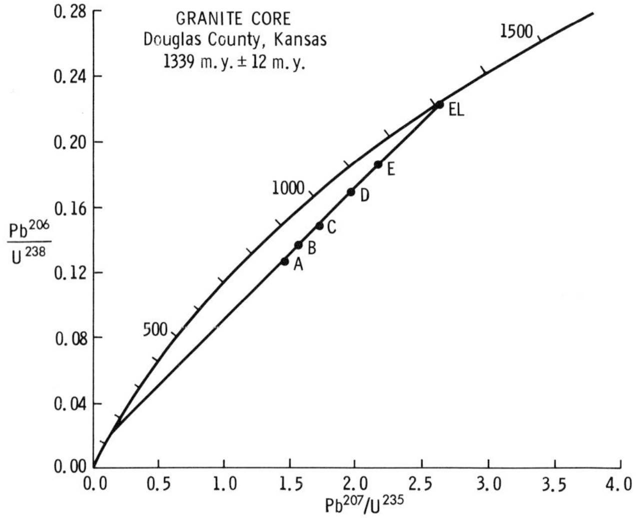


Fig. 6. Concordia plot for zircons from subsurface granite, Douglas County, Kansas (site 2, Fig. 4)

Fig. 7. Map showing inferred distribution of 1450-1490 m.y. old and 1340-1400 m.y. old granite-rhyolite terranes in the southern and eastern midcontinent basement. Symbols represent locations of basement samples dated by U-Pb on zircons, mostly from drill core or cuttings, in the midcontinent. Symbols in the Rocky Mountains of Colorado and New Mexico show approximate locations of plutons of corresponding age, dated by Rb-Sr or U-Pb methods. Geology of northern midcontinent, etc. after Van Schmus and Bickford (1981, 1983). *Numbers in parentheses* refer to approximate bedrock ages of regions

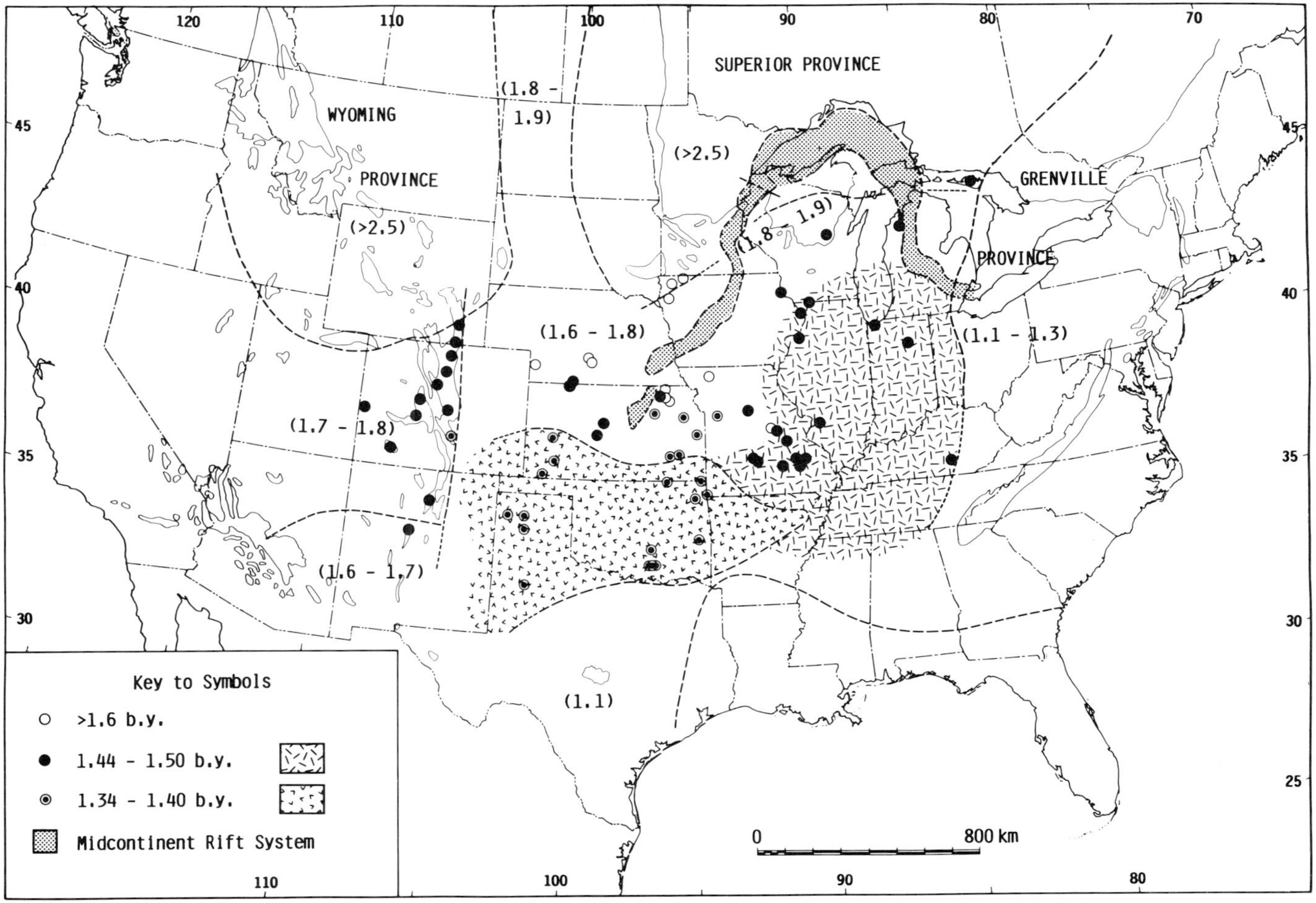
SUPERIOR PROVINCE
WYOMING
PROVINCE
GRENVILLE
PROVINCE
(1.8 - 1.9)
(>2.5)
(>2.5)
(1.8 - 1.9)
(1.6 - 1.8)
(1.1 - 1.3)
(1.7 - 1.8)
(1.6 - 1.7)
(1.1)
0
800 km
Key to Symbols
>1.6 b.y.
1.44 - 1.50 b.y.
1.34 - 1.40 b.y.
Midcontinent Rift System

Fig. 7.

that are 1700-1800 m.y. old and is the most westerly known example of the 1350-1400 m.y. old suite (Bickford et al. 1984).

The distribution of these granitic or rhyolitic rocks of the southern midcontinent is shown in Figure 7. Rocks formed 1450-1490 m.y. ago underlie northern Missouri, Illinois, Indiana, and parts of Ohio, and plutons of this age intrude older crust from Labrador to the southwestern United States (the so-called suite of "anorogenic granites"; Silver et al. 1977; Anderson 1983). The second terrane of granite and silicic volcanic rocks was formed 1340-1400 m.y. ago and underlies southwestern Missouri, southern Kansas, most of Oklahoma, and the Texas Panhandle (Thomas et al. 1984). There are also plutons of this age that lie to the north of the volcanic rocks (e.g. the plutons drilled in eastern Kansas), but at present it cannot be demonstrated that they intrude older crust. These terranes, which represent a major and perhaps unusual event in the Proterozoic evolution of the North American continent, would not be known were it not for recovery of material from drilling. Their relationship to rocks of the northern midcontinent is still obscure because material and data are scarce in Nebraska, Iowa, and South Dakota. Furthermore, since the drilling of exploratory wells for oil, gas, and minerals is dictated mostly by structures and lithologies in the Paleozoic rocks overlying the basement, it may be a very long time before samples that will elucidate many of these questions are available. Ten or twenty carefully located, dedicated shallow drill holes could produce major scientific results.

Finally, what is the origin of these terranes? They clearly do not represent a great volcanic arc stretching from Ohio to the Texas Panhandle as we previously thought. Rather, volcanism and shallow plutonism in a quite restricted high-silica compositional range occurred in a large region of the eastern midcontinent about 1480 m.y. ago and then an entirely similar igneous event occurred in the southern and western midcontinent about 100 m.y. later. It seems likely that the rocks of both terranes are a veneer, perhaps only a few kilometers thick, that lie upon, and were derived from, older crust. The Sm-Nd analyses of samples of these rocks by Nelson and DePaolo (1985) indicate that their mantle separation ages are 1600-1800 m.y. Thus, they are probably underlain by crust of that age and were probably derived from it by crustal anatexis. The presence of the 1360 m.y. old San Isabel batholith emplaced at mesozonal depths into crust 1700-1800 m.y. old suggests that in the Wet Mountains we may see a deeper section of crust, similar to that beneath the midcontinent, that was exposed by Laramide and Tertiary uplift in the southern Rocky Mountains. A drill hole 4 to 5 km deep in northeastern Oklahoma might easily pass through the veneer of volcanic rocks and penetrate older crust and could yield a major contribution to understanding of crustal evolution.

*Acknowledgments*. The research cited in this paper has been supported mostly by grants from the National Science Foundation, the most recent being EAR 821937. Additional support has come from the Nuclear Regulatory Commission through contract NRC-04-77-017 to the Kansas Geological Survey; by grants from Los Alamos Scientific Laboratory (4-LP9-771G-1) and the Department of Energy (DE-AS07-79ET27204); and by grants from the Kansas Geological Survey and the General Research Fund of the University of Kansas.

## References

Anderson JL (1983) Proterozoic anorogenic granite plutonism of North America. In: Medaris LG Jr, Byers CW, Mickelson DM, Shanks WC (eds) Proterozoic Geology: Selected papers from an international symposium. Geol Soc Am Memoir 161:133-153

Anderson JL, Cullers RL (1978) Geochemistry and evolution of the Wolf River batholith, a Late Precambrian rapakivi massif in north Wisconsin, USA. Precambrian Res 7:287-324

Anderson JE, Bickford ME, Odom AL, Berry AW Jr (1969) Some age relations and structural features of the Precambrian volcanic terrane, St. Francois Mountains, southeastern Missouri. Geol Soc Am Bull 80: 1815-1818

Anderson RE (1962) Igenous petrology of the Taum Sauk area, Missouri (unpublished) PhD Diss, Wash Univ, St. Louis, Missouri

Anderson RE (1970) Ash-flow tuffs of Precambrian age in southeast Missouri. Rep Invest Mo Geol Surv 46, 50 p

Berry AW, Bickford ME (1972) Precambrian volcanics associated with the Taum Sauk Caldera, St. Francois Mountains, Missouri, USA. Bull Volcanol 36:308-318

Bickford ME, Mose DG (1975) Geochronology of Precambrian rocks in the St. Francois Mountains, southeast Missouri. Geol Soc Am Spec Pap 165, 48 p

Bickford ME, Lewis RD (1979) U-Pb geochronology of exposed basement rocks in Oklahoma. Geol Soc Am Bull 90:540-544

Bickford ME, Sides JR, Cullers RL (1981a) Chemical evolution of magmas in the Proterozoic terrane of the St. Francois Mountains, southeastern Missouri I. Field, petrographic, and major element data. J Geophys Res (Granite-Rhyolite issue) 86:10365-10386

Bickford ME, Harrower KL, Hoppe WJ, Nelson BK, Nusbaum RL, Thomas JJ (1981b) Rb-Sr and U-Pb geochronology and distribution of rock types in the Precambrian basement of Missouri and Kansas. Geol Soc Am Bull 92:323-341

Bickford ME, Cullers RL, Van Schmus WR (1984) U-Pb zircon chronology of Early and Middle Proterozoic igneous events in the Gunnison, Salida, and Wet Mountains areas, Colorado. Geol Soc Am abstracts with programs 16:215

Cullers RL, Koch RJ, Bickford ME (1981) Chemical evolution of magmas in the Proterozoic terrane of the St. Francois Mountains, southeastern Missouri II. Trace element data. J Geophys Res (Granite-Rhyolite issue) 86:10388-10401

Hoppe WJ, Montgomery CW, Van Schmus WR (1983) Age and significance of Precambrian basement samples from northern Illinois and adjacent states. J Geophys Res 88:7276-7286

Lidiak EG, Marvin RF, Thomas HH, Bass MN (1966) Geochronology of the midcontinent region, US, IV. Eastern area. J Geophys Res 72:5427-5438

Mose DG (1971) Chronology of Precambrian volcanic rock units in the center of the St. Francois Mountains, Missouri (unpublished) PhD Diss, Univ Kansas, Lawrence, 259 p

Muehlberger WR, Hedge CE, Denison RE, Marvin RF (1966) Geochronology of the midcontinent region, US. III. Southern area. J Geophys Res 72:5409-5426

Nelson BK, DePaolo DJ (1985) Rapid production of continental crust 1.7 to 1.9 b.y. ago: Nd isotopic evidence from the basement of the North American mid-continental. Geol Soc Am Bull 96: 746-754

Nusbaum RL (1980) A Precambrian collapse caldera boundary in the St. Francois Mountains, southeast Missouri (unpublished) MS Thesis, Univ Kansas, Lawrence, 77 p

Odom AL (1969) Rb-Sr geochronology of Precambrian ignimbrites in the St. Francois Mountains, southeastern Missouri (unpublished) MS Thesis, Univ Kansas, Lawrence, 48 p

Shuster RD (1978) Chemistry and petrography of caldera-related ash-flows, St. Francois Mountains, Missouri (unpublished) MS Thesis, Univ Kansas, Lawrence, 45 p

Sides RJ, Bickford ME, Shuster RD, Nusbaum RL (1981) Calderas in the Precambrian St. Francois Mountains terrane, southeastern Missouri. J Geophys Res 86:10349-10364

Silver LT, Bickford ME, Van Schmus WR, Anderson JL, Anderson TH, Medaris LG (1977) The 1.4-1.5 b.y. transcontinental anorogenic plutonic perforation of North America. Geol Soc Am, abstracts with programs 9: 1176-1177

Steeples DW, Bickford ME (1981) Piggyback drilling in Kansas: an example for the continental scientific drilling program. EOS 62:473-476

Thomas JJ, Shuster RD, Bickford ME (1984) A terrane of 1350-1400 m.y. old silicic volcanic and plutonic rocks in the buried Proterozoic of the midcontinent and in the Wet Mountains, Colorado. Geol Soc Am Bull 95:1150-1157

Tolman C, Robertson F (1969) Exposed Precambrian rocks in southeast Missouri. Missouri Division of Geological Survey and Water Resources Report of Investigations 44, 68 p

Van Schmus WR, Bickford ME (1981) Proterozoic chronology and evolution of the midcontinent region, North America. In: Kroner A (ed) Precambrian Plate Tectonics. Elsevier, Amsterdam, pp 261-296

Van Schmus WR, Bickford ME (1983) Accretionary provinces of the Proterozoic basement in the continental interior of the US. Geol Soc Am, abstracts with programs 15:711

Van Schmus WR, Medaris LG, Banks PO (1975) Geology and age of the Wolf River batholith, Wisconsin. Geol Soc Am Bull 86:907-914